ALSO BY DAVID POGUE

Mac Unlocked

iPhone Unlocked

macOS: The Missing Manual

Windows 10: The Missing Manual

iPhone: The Missing Manual

Pogue's Basics: Tech

Pogue's Basics: Money

Pogue's Basics: Life

Abby Carnelia's One and Only Magical Power

The World According to Twitter

Opera for Dummies

Classical Music for Dummies

Magic for Dummies

How to Prepare for Climate Change

A Practical Guide to Surviving the Chaos

David Pogue

Simon & Schuster

NEW YORK LONDON TORONTO SYDNEY NEW DELHI

Simon & Schuster
1230 Avenue of the Americas
New York, NY 10020

First Simon & Schuster trade paperback edition January 2021

Interior design by Paul Dippolito

Manufactured in the United States of America

10 9 8 7 6 5 4 3 2 1

Library of Congress Cataloging-in-Publication Data

Names: Pogue, David, author.
Title: How to prepare for climate change : a practical guide to surviving the chaos / David Pogue.
Description: First Simon & Schuster trade paperback edition. | New York : Simon & Schuster, 2020. | Includes bibliographical references and index. | Summary: "A practical and comprehensive guide to surviving the greatest disaster of our time, from New York Times bestselling self-help author and beloved CBS Sunday Morning science and technology correspondent David Pogue"—Provided by publisher
Identifiers: LCCN 2020020295 | ISBN 9781982134518 (paperback) | ISBN 9781982134587 (ebook)
Subjects: LCSH: Climatic changes. | Preparedness. | Emergency management. | Environmental disasters.
Classification: LCC QC903 .P64 2020 | DDC 363.738/747—dc23
LC record available at https://lccn.loc.gov/2020020295

ISBN 978-1-9821-3451-8
ISBN 978-1-9821-3458-7 (ebook)

For Kell, Tia, Jeffrey, and Nicki . . .
my reasons to prepare

Contents

Introduction

MAYBE YOU'RE LIBERAL, MAYBE YOU'RE CONSERVATIVE. Maybe you think the climate crisis is man-made, maybe you think it's just natural cycles. Maybe you think the whole thing is a Chinese hoax.

Guess what? It doesn't matter. The world is getting hotter, natural systems are going haywire, and you should begin to prepare.

Even if we stopped burning fossil fuels and chopping down forests tomorrow, we wouldn't stop climate change. We wouldn't stop land ice from melting, millions of people from enduring forced migration, thousands of animal species from going extinct, and thousands of people from dying every year from insane-weather events that are hitting steadily more frequently.

That's because 93% of our new, improved heat has gone into the *oceans*, which take decades or centuries to heat up or cool down. As a result, the planet's climate would take a lifetime to reset.

In short, the time for bickering about who or what is at fault is long gone.

The Intergovernmental Panel on Climate Change (IPCC) is the world's most authoritative climate assembly: 1,300 scientists from all over the world. Every few years, they release a sort of "How Screwed Are We?" report card. In the latest report, the IPCC projects that the earth's average temperature will rise between 2.5°F and 10°F in the next 100 years.

And that's a very conservative estimate. After all, it had to earn a consensus of government bureaucrats.

It's time to accept the new realities—of extreme weather, sure, but also what that weather will mean for our everyday lives: a lot of conflict, cost, and chaos.

In 2007, the *New York Times* interviewed John Holdren, Barack

Obama's senior science advisor. What he said has become famous in climate circles:

"We basically have three choices: mitigation, adaptation, and suffering. We're going to do some of each. The question is what the mix is going to be."

Once you start reading about climate science, you encounter these terms a lot.

- **Mitigation** means trying to *stop* climate change: Replace fossil fuels with clean power. Eat less beef. Fly less. Grow trees instead of clear-cutting them. Adopt smarter farming and industrial techniques. Drive less. Have fewer children.
- **Adaptation** means *coping* with climate change: Build seawalls. Raise houses. Move farmland to cooler regions. Plant heat-tolerant trees. Buy out homeowners in flood-prone areas.
- **Suffering.** Well, you'll be reading plenty about that.

Here's an analogy: Suppose you're all dressed up, and suddenly it starts raining hard. *Mitigation* would be aiming a blow dryer at the sky in hopes of drying up the rain. *Adaptation* would be opening an umbrella.

There's no longer any intra-expert arguing over mitigation versus adaptation; it's too late for that. We have to do *both*, as hard and as fast as possible.

Reams have been written about what you, as an individual, can do to pursue mitigation—and you should. You should mitigate the hell out of your home, your family, your town, your employer, your voting record.

But so far, the only people doing much *adaptation* are governments, corporations, and institutions.

Where does that leave you, the individual? *You* can't build a seawall this weekend. *You* can't persuade farmers to move north. *You* can't develop drought-proof seeds.

That's where this book comes in. It's a practical guide to adaptation steps that you can take, as one person—for your own benefit, your family's, and your community's. It's about where to live, how to invest, what to eat, how to build, what insurance you need, how to talk to your kids.

COVID-19 and Climate Change

The COVID-19 virus has killed people by the hundreds of thousands. It has flattened the economy. It has brought our way of life to a screaming halt.

The pandemic and the climate—two crises that might seem unrelated at first—actually have quite a bit in common. Both are unanticipated consequences of human encroachment into undisturbed natural areas (the 2020 coronavirus apparently crossed to humans from bats). The air pollution that's changing the climate also drives up the COVID-19 death rate. And both the virus and the climate crisis cause disproportionate suffering among people of color and low incomes.

The pandemic did, however, introduce one tiny scrap of good news: a global pause in planet-heating human activity.

At the depths of the pandemic, air travel was down 95%. Road traffic in locked-down countries dropped 75%. Factories shut down or scaled back operations. Millions of office buildings, restaurants, stores, and malls sat empty, their lights and air-conditioning shut off.

The skies over Los Angeles, usually the smoggiest in the country, were shockingly clear. In many cities, thanks to a 90% drop in noise levels, you could hear birdsong for the first time. And without boats churning up sediment, the canals in Venice became so clear that you could see fish at the bottom.

With so many planes and cars sitting idle, petroleum consumption cratered; soon, there was nowhere to store the oil that producers continued to pump. The bizarre and historic result: In April 2020, the contract price for oil sank to a *negative* $37.63 a barrel. If you had a place to store it, oil traders would have *paid* you to haul away their crude.

That spring, all of this together produced a stunning 17% global drop in CO_2 emissions. It was the biggest pollution pullback in human history.

And yet, as you may recall, there wasn't much in the way of cheering, fireworks, and beer bashes.

That's because the coronavirus didn't produce a Great Stopping—only a Great Pausing. It was a temporary fluke.

Once locked-down countries began lifting restrictions, greenhouse-gas emissions climbed right back up. Two months after that April 2020 low point, emissions had climbed back to within 5% of their 2019 levels.

Not depressed yet? Then consider this: Once the virus is under control, our CO_2 emissions are likely to climb even *higher* than their original levels. That's what we do. During 2009's Great Recession, for example, global emissions dropped by 1.4%, but they more than rebounded the following year (up 5.1%). In other words, an economic crisis may cause emissions to drop—but we more than make up for it once the crisis passes.

The bottom line: When the history of our planet is written, the coronavirus emissions dip won't even merit a mention. It was only a hiccup in the larger trend line.

Indeed, by the middle of 2020, scientists quietly noted the breaking of an upsetting new human record. The density of the carbon-dioxide blanket surrounding the earth had reached 417 parts per million—its highest level in 10 million years.

It's also about how to prepare for the extremes that are coming soon to weather near you: floods, fires, heat waves, droughts, superstorms, water shortages, food shortages, power failures, and social disruption.

The Adaptation Era

All over the world, adaptation is underway. Starbucks is developing new, climate-resilient coffee beans. Miami developers are siting their projects farther from the shore. Monsanto is investing millions to create genetically modified, climate-resistant crops. Every major insurance company has hired climatologists and statisticians to incorporate extreme weather into their mathematical models.

Governments around the world are taking adaptation steps, too, by building some of the biggest public-works projects ever conceived, in hopes of protecting the land from the rising seas and flooding. New Orleans will spend the next 50 years constructing the most ambitious (and expensive) coastal-protection system in history. Similar projects are underway in New York City (a 10-mile system of movable walls and raised parks), London (a 16-mile-long "Super Sewer," 24 feet across), Tokyo (250-foot-deep underground cisterns), Jakarta, Indonesia (an 80-foot-tall seawall, 25 miles long—the biggest ever made), and China (permeable pavement, artificial ponds and wetlands, rain gardens, and underground storage tanks in over 600 cities).

Venice, the city whose "streets" are canals, now floods routinely; it's building a series of 78 massive, bright yellow, mobile sea walls, ten tons per slab, that can be raised during storms to protect the city. The government of the Netherlands, a low-lying country, spends $1.35 billion a year on flood protection. And in the United States, FEMA (the Federal Emergency Management Agency) has bought 43,000 homes in neighborhoods that repeatedly flood—then demolished them for good.

Figure I-1. "The Big U" is New York's plan to protect the low-lying southern tip of Manhattan from rising sea levels.

Figure I-2. The Netherlands has built massive floodgates that, in times of North Sea storms, swing closed like eyebrows knitting to keep out the floodwaters.

Plenty of institutions intend to get *ahead* through adaptation, too. Oil companies are preparing to exploit melting arctic sea ice to drill for more petroleum. The melting ice sheets in Greenland are unlocking vast deposits of uranium, gold, and rare-earth metals; over 100 new mines are in development. And as usable farmland becomes increasingly valuable, "the Chinese are running around right now buying up every shred of decent arable land in the world," says Nick Nuttall, former communications director for the UN Framework Convention on Climate Change.

Now, you don't hear much about *individuals* preparing for a new climate era, but that's only because they don't make headlines.

Some of their adaptations are small. In Phoenix, Arizona, where it can get over 120°F, people have learned to keep oven mitts in the car. According to resident Sadie Lankford, "The steering wheel and any metal you have to touch on your car will burn you."

Other adaptation steps are somewhat bigger—such as moving to safer cities. Sometimes, people are pushed out by extreme-weather disasters, like the wildfire that drove Bob and Linda Oslin out of Paradise, California. They've decided not to return.

Others are moving preemptively, like 55-year-old Karen Colton and her wife, who are moving from Florida to Asheville, North Carolina. They're tired of stressing about the next big hurricane.

And some people, contemplating a future of weather extremes, shortages of food and water, and more disease, are choosing not to have children. "It would be bringing a life into a future that does seem ever more desolate," 23-year-old Hannah Scott told the *Guardian*.

The planet we occupy is changing fast, and that's new information. It's affecting how people are making decisions about their lives: sometimes because they have a choice, and sometimes because they don't.

Why the Planet Warms

For 200 years, we've powered our society by burning coal, gas, and oil. The by-products of that combustion, carbon dioxide and other gases, pour into the air—152 million tons of it a day. We're using the sky, as climate activist and former vice president Al Gore puts it, "as if it were an open sewer."

The carbon dioxide collects in the atmosphere, trapping heat that would otherwise have escaped into space. The planet warms, and *boom*: That's your greenhouse effect.

Parked-car-in-summer effect would be a more helpful term, actually. You know what happens to a dog locked in a car all day in the hot sun, right? Right. We're the dog.

Now, greenhouse gases have always been around. They occur naturally. For all of human history, they've served as a thin blanket around the planet, keeping it from becoming an ice ball. And over the millennia, greenhouse gas levels have ebbed and flowed on their own, roughly every 100,000 years (see Figure I-3).

But see the nearly vertical spike at the right edge of that graph? *That* has never before happened. If you zoom in on it, you discover that this sudden spike began about 150 years ago, when we started burning oil and coal during the industrial revolution. As a result, our planetary blanket is getting a lot thicker.

That's no 100,000-year cycle. That's *150 years*. This time, the intensity of that warming blanket of gases has shot up *suddenly*—many times faster than at any other time in human history—and to levels much *higher* than

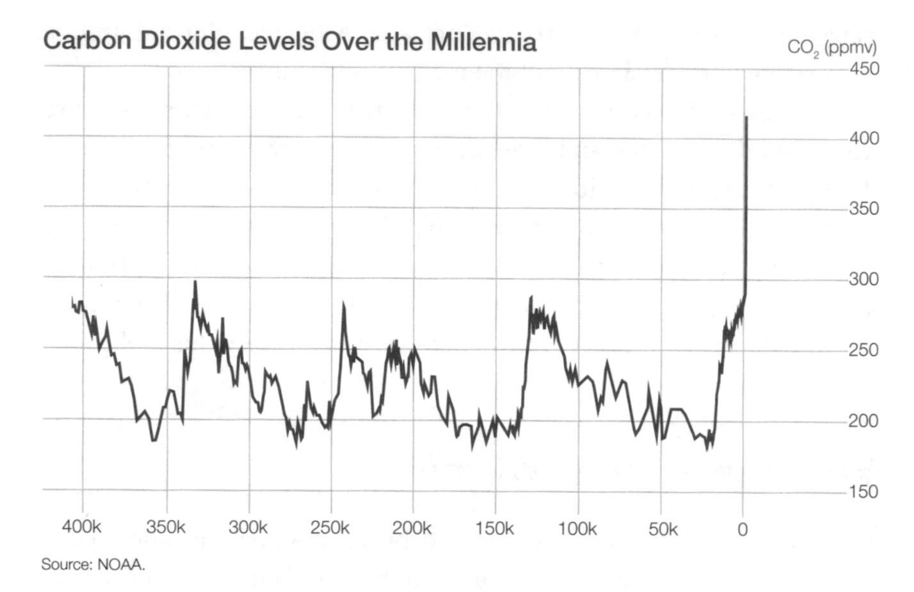

Figure I-3. Carbon dioxide levels have risen and fallen for at least the last 400,000 years (horizontal axis). Just never this high, this fast.

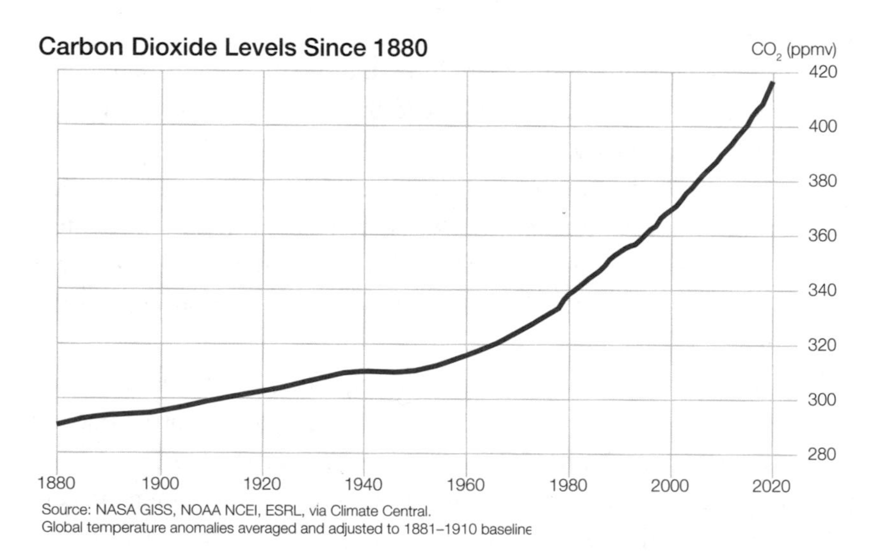

Figure I-4. The carbon dioxide concentration of the atmosphere since 1880, represented in parts per million (molecules per million).

at any other time in human history. And we've been setting new records for atmospheric CO_2 every single year.

Scientists measure CO_2 in the atmosphere in units called parts per million (meaning molecules per million molecules; see the sidebar on page 10).

In 2020, the carbon dioxide concentration in the atmosphere hit 417 parts per million. The last time the world experienced this level of carbon dioxide in the air was 10 million years ago.

If you look again at that graph of carbon dioxide levels since 1880 (Figure I-4), but this time superimpose the temperature readings over the same period (Figure I-5), you get a pretty good idea why scientists think that global warming is somehow related to global CO_2 buildup.

Now you can see why a lot of smart people are in favor of *stopping* burning fossil fuels, to *stop* carbon from continuing to rise, to *stop* the temperature from shooting up even more.

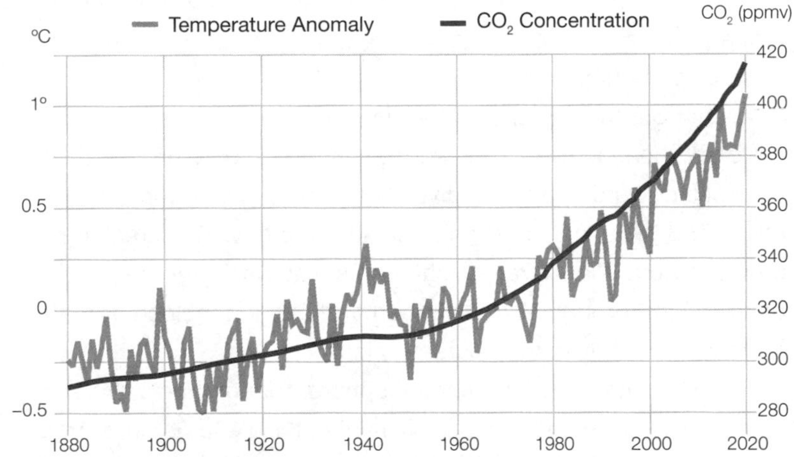

Figure I-5. The rising planetary temperature plots eye-poppingly well with the rising CO_2 levels. The dark blue line shows the increasing CO_2 levels since 1880. The light blue line shows the temperature anomaly (variation from normal) over the same period.

Why 417 PPM Matters

In the graph in Figure I-5, and in many a scientific discussion of CO_2, you run into the opaque term *parts per million*. You might read, "For the last 10,000 years, the average CO_2 concentration was 260 parts per million; today, it's 417 ppm."

Welcome to another episode of *Scientists Muffling Their Own Message with Jargon.*

Parts per million means "*molecules* per million *molecules*." So 417 ppm CO_2 means that out of every million molecules in the atmosphere, 417 of them are carbon dioxide. The other 999,583 molecules are mostly nitrogen and oxygen. (We don't count water vapor in these tallies.)

That sounds like an unbelievably small proportion. It's four-hundredths of one percent of the atmosphere! Who cares?

Ah, but that's like learning that only 417 ppm of your drinking water is concentrated poison. You would care *a lot*.

To understand why those trace amounts of carbon dioxide make such an outsize difference, consider the way the sun's rays interact with our planet.

Light from the sun shoots through the atmosphere and hits the earth, heating it up. That warmth, in the form of infrared light, bounces upward again. (Hold your hand over parking-lot pavement after the sun has set on a hot summer day. You'll feel that radiating heat.) The infrared light, on its way back out to space, passes right through all the nitrogen and oxygen molecules in the atmosphere. They're transparent to infrared.

The CO_2 molecules in the atmosphere, though, are another story. They *scatter* infrared rays. The effect is like a partial mirror, bouncing some of that heat back down to the earth, trapping it.

And why does carbon dioxide reflect infrared light, where nitrogen and oxygen do not? The chemical and electrical explanation would make your eyes glaze over—but the short version is that the more atoms a molecule has, the more likely it is to scatter infrared energy—and the more damage it causes as a greenhouse gas. Nitrogen and oxygen molecules, N_2 and O_2, have only two atoms each; CO_2 has three.

In theory, you now understand why methane (CH_4), with five atoms, is an even worse greenhouse gas than CO_2. In methane's first 20 years of release into the atmosphere, it traps heat 80 times more effectively than carbon dioxide. It's nasty stuff.

All right, then, what about water? After all, an H_2O molecule has the same number of atoms as CO_2, and traps heat just as well. In fact, it's responsible for at least half of the greenhouse effect. Why don't we hear about that?

Mainly because *we're* not pumping it directly into the atmosphere, as we do carbon dioxide, methane, and the other villainous gases. "Human activities are responsible for increasing amounts of these gases in the atmosphere, but human activities are not directly relevant to the amount of water vapor in the air," says chemist and science educator Jerry Bell. "The amount of water vapor in the air is determined by the temperature of air and is entirely 'natural'—that is, its contribution to the greenhouse effect has almost nothing to do with human activities."

Still, as the earth warms, there *is* more atmospheric water vapor that traps heat, and it *does* add to the climate-change problem.

The Effects of Warming

Global warming isn't a fantastic term. It makes people think that all we're worried about is *warmer weather.* (People like Senator James Inhofe, who, in February 2015, carried a snowball into Congress as evidence that global warming is a hoax.)

But warmer water and air are only the triggers. Those factors release an enormous, complicated chain of chaos.

For the record, *climate* is not *weather.* Climate is measured in decades; weather is an hourly or daily measurement. Climate is regional or global; weather is local. Climate is the *average* of weather.

That's why the effect of climate change is not just "hotter summers, milder winters." It's warmer weather *and* colder weather, record heat waves *and* record cold snaps. It's more flooding *and* more droughts. It's wet areas getting wetter, hot areas getting hotter, and dry areas getting drier. It's much heavier rainfall and much nastier superstorms. It's tornadoes getting stronger and spreading to new states. It's more wildfires that last longer and do more damage.

But believe it or not, the bigger problem is not in the air; it's in the water. Water absorbs much more heat than air—four times as much, pound for pound. The seas have been warming steadily since we started burning fossil fuels.

As a result of that warmer water, the sea levels are rising. Superstorms and hurricanes are growing more violent and more intense. Fish, birds,

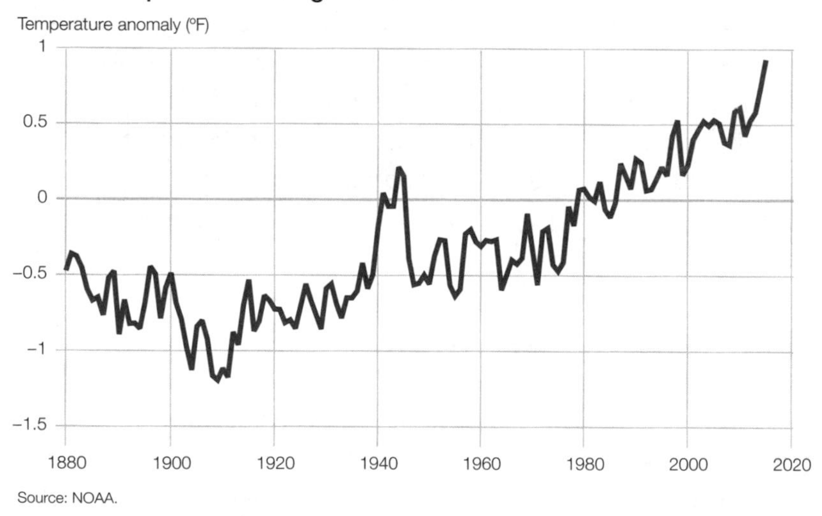

Figure I-6. The average surface temperature of the world's oceans, shown here in degrees F deviation from the 1971–2000 average, has climbed steadily since 1880.

turtles, and other marine life are dying off because warmer water holds less oxygen and becomes more acidic. About half of all coral reefs have died since 1980, and the rest may be gone by 2100.

Climate Chaos

The problem with *both* the terms *global warming* and *climate change* is that they imply knowability. They suggest that we know what the change *is*.

But we can't know what the change is when everything is changing, and all at once—oceans, atmosphere, plants, animals, permafrost, weather. Every corner of the globe is affected, every growing season, every living thing. *Climate chaos* is the better term. Or *global weirding*.

Some of the effects have already made headlines: crop disruption, animal extinction, the melting permafrost, an explosion of the tick and mosquito populations, and infestations of beetles in the Pacific Northwest, which are killing as many as 100,000 trees a *day*. So far, they've destroyed 85 million acres of American Western forest.

We're an affected species, too; we're losing *our* habitat. As water encroaches on the coasts, as food and water supplies become unreliable, and as farms grow less productive, people are being displaced. By 2050, the number of "climate refugees" is expected to grow as high as 1 billion people. By 2070, 19% of the planet's land surface will be uninhabitably hot (up from 1% today). As usual, the poorest countries—the ones that produce the fewest greenhouse gases—suffer the most. Even within wealthier countries, though, climate migrations are becoming standard; Hurricane Katrina alone displaced over a million Americans.

But global weirding is also producing freakish changes that you'd never predict. Studies show that higher temperatures lead to lower productivity; lower PSAT test scores; more bar fights, shootings, rapes, car thefts, and murders; more suicides; more power outages; and smaller beaches.

More lightning strikes, more volcanic eruptions, more kidney stones, smaller goats, less Belgian beer, more expensive chocolate—scratch the surface of any human or natural activity, and you'll find more chaos.

The Numbers

Now, as you'll find out in Chapter 17, there are hundreds of hopeful signs—indications that *somebody* is starting to take climate change seriously.

Unfortunately, no amount of turning off lights and driving electric cars will rewind the air and oceans to 1880 levels within our lifetime, or our children's, or our grandchildren's. The greenhouse effect has already pumped enough new heat into the oceans to keep warming the planet for decades.

The world has changed, and it will keep changing for the rest of your life. It's time to prepare.

Chapter 1

Acclimating to Climate Change

MOST PEOPLE, MOST OF THE TIME, CONCEIVE OF THE CLImate crisis in terms of its effects on the physical world: weather, buildings, agriculture, land, animals. But if you're a human being, there's a less publicized challenge: preparing *mentally* for the new era.

The psychological costs of climate change include spikes in grief, anger, helplessness, shame, fear, disgust, cynicism, and fatalism—feelings that lead to real-world consequences like stress, drug abuse, strain on relationships, and increases in aggression, violence, and crime. "Children and communities with few resources to deal with the impacts of climate change are those most impacted," notes the American Psychological Association.

Depression is paralyzing, and that's why this is the first chapter in this book: You can't take *any* action, in any realm of your existence, if you're mired in hopelessness. Preparing to handle your own feelings about our planet's destruction, therefore, must be your first step before you can take any others.

In fact, the changing climate can take two kinds of emotional toll: post-traumatic and pretraumatic.

- **Post-traumatic impacts** are the ones you feel when you've lived through an extreme-weather event. Every time a hurricane, flood, wildfire, heat wave, drought, or another eco-disaster hits a region, the affected population suffers a spike in anxiety, alcoholism, drug use, depression, suicides, and psychiatric hospitalization.

It's not hard to imagine why: These events usually mean experiencing damage or loss to your family, your home, your stuff, a pet, or your livelihood. Meanwhile, the stress of the situation is often magnified by disruptions in society's infrastructure: cell service, access to food and water, medical facilities, and so on.

People who live through one of these events may also suffer from *solastalgia*, a cool word for a terrible feeling. It's homesickness while you're still at home, because you no longer recognize the place. In the climate-crisis era, more and more people experience solastalgia—when their homes are destroyed by extreme weather, or when they're forced to move because their community is flooded or has dried up. It's possible to feel solastalgia even when part of your town is destroyed but *not* your own neighborhood.

But even after the nuts and bolts of your life have been restored, the mental damage may not recede. Post-traumatic stress syndrome (PTSD), a chronic disorder that persists long after the original disaster, is common among extreme-weather victims. After Hurricane Katrina, for example, half of the affected population developed an anxiety or mood disorder such as depression, and one in six people suffered PTSD. And after the 2003 California wildfires, one-third of all surviving adults fell into depression; one-quarter suffered PTSD.

- **Pretraumatic stress** stems from despair for our world before it has even finished becoming uninhabitable. It's anxiety, panic, despair, and mourning over the planet's ruined future. It's a depression that's made worse by the sense that nobody seems to be *doing* anything about it.

  It's a big problem. In the United States, anxiety diagnoses are up 40% since 2016. Depression cases are up 33% since 2013. The suicide rate has ballooned 50% since 2003. Climate news isn't solely responsible, but it's not helping; 40% of Americans, for example, feel "helpless" about the deteriorating climate.

As though those psychological depressors aren't enough, our moods in the new climate era are also affected more directly—by the heat. Dozens of studies have established a link between hot weather and impatience, irritability, and violence. For example, for each degree increase in average temperature, 2% more Americans report mental-health issues.

Validation

How's this for good news? There's a whole new mental-health field called ecotherapy. It's staffed by psychologists and psychiatrists who specialize in treating environmental grief, depression, or panic.

These therapists have various approaches and backgrounds, but they all seem to agree on one thing: *You're normal.*

"There's a normal range of anxiety, depression, and grief that's associated with these issues. Any sane, feeling person would have them," says climate psychologist Thomas Doherty.

But that doesn't mean that you shouldn't try to understand, manage, and accept these feelings. "Too much anxiety is inhibiting. It robs people of their creativity, of their joy, their quality of life," Doherty says.

He notes that some people are inherently more vulnerable to anxiety or depression right out of the gate.

If you were a regular patient of one of these ecotherapists, the first project they'd work on would probably be *validating* your struggles—reassuring you that they're real, reasonable, and sane.

Four Fueling Factors

Four compounding factors swirl around that process, which you can overcome only by recognizing them and picking them apart.

- **Social pressure.** Validation of your feelings can be helpful no matter what problems you're facing. But the polarizing and controversial nature of climate change makes acknowledging your dread more complicated. You may live in a society, a workplace, or even a family that's populated by deniers, doubters, and defeatists.

  "Interacting with people who aren't feeling the same things you are can feel very confusing, very alienating," says environmental psychologist Renee Lertzman. "You then tend to minimize or question your own experience, like, 'I'm just overreacting. I'm being too sensitive. I should be handling this better. I should do this. I shouldn't be feeling that.' But it's very hard, generally speaking, to move forward in any constructive way while we're attacking ourselves."

- **Negative bias.** The world hasn't *yet* heated up 10 degrees. The sea levels aren't *yet* eight feet higher. We still have a chance to change our fate.

  But it's still easy to freak out now, before the worst has even come to pass—because we're wired to fear the worst.

  "We're built with what's called a negative bias: We scan the environment for dangers, rather than scanning for beauty," says Leslie Davenport, a therapist specializing in climate psychology and the author of *Emotional Resiliency in the Era of Climate Change: A Clinician's Guide.* Our excellence at spotting and fleeing potential threats has long been an evolutionary advantage for our species. For thousands of generations, it's helped us survive.

  But as the climate changes, negative bias is choking us with stress and anxiety. "On one level, it has served us," says Davenport. "But right now, it's working against us."

- **Avoidance.** Climate change is not just any old bad news. It's not headlines saying, "Last Blockbuster Video Store Closes Its Doors." Instead, it's deeply unsettling news that affects us to the core.

  "The climate crisis is nothing if not a metaphor for our gradual decline and death," says psychiatrist and environmental activist Lise Van Susteren.

  Because the topic is so dark and so threatening, we don't like to talk about it. Many of us keep our terror bottled up, which makes climate anxiety even more crippling. "People do their very best to avoid these topics," says Van Susteren. "It takes me only seconds to shut a dinner party down by bringing up certain issues on climate."

- **Threats to identity.** In wealthy countries like the United States, the thought of human-caused climate change can be hard to acknowledge for yet another reason: It contradicts so much of what we believe about ourselves.

  For generations, the American Way has meant that we measure success by how much we own and consume. That we're rewarded by hard work with a first-world lifestyle. That each generation lives more comfortably than its parents.

  "All these core values that, for the most part, we've innocently invested ourselves in, we're now being told, 'That's all wrong. This doesn't work,'" says Davenport.

The way that message hits many people, according to Renee Lertzman, is, "You have to change everything. You cannot exist the way you have been existing, and therefore you have to change your entire sense of who you are, and yourself."

That's why what Lertzman calls climate melancholia can be such a devastating affliction. We're concerned about the fate of the earth and the species, yes. But because our lifestyle is part of our identity, and that lifestyle is responsible for the problem, our sense of self is threatened.

All right, you get it: Climate despair is complex, fraught, and wired into some of our deepest psychological scaffolding.

If you do manage to complete Step 1—establishing that your feelings are legitimate—then you're ready for Step 2. That, as it turns out, is taking action.

The Action Antidote

Depression, the clinical condition, isn't just feeling down. It's the feeling that your situation is terrible *and you can't do anything to change it.*

"Anytime there's a sense of helplessness, hopelessness, a sense of 'I'm a victim, there's nothing I can do,' doing *something* has always been a therapeutic intervention," says Leslie Davenport.

No amount of reading or thinking is guaranteed to make you feel better about the climate problem. Taking some kind of action is the *only* therapy that always works.

"That's the most effective way of dealing with climate despair: to do something," says Richard Heinberg, senior fellow at the Post Carbon Institute. "Otherwise, it just sits inside you and churns, and you end up spending hours every day looking at the computer for more evidence of crisis and breakdown and collapse. That's not a good basis for psychological health."

Davenport is quick to add that action doesn't have to mean "carrying a sign and screaming until someone handcuffs you." Action can take hundreds of forms, in broad categories like lifestyle, advocacy, self-help, group conversation, and so on. And it can begin with very small steps.

Mitigation

One satisfying way to begin—one that doesn't require having enough courage for public presentations or confrontations—is simply to lower your own carbon footprint. Minimize the degree to which *you* and your family are contributing to the problem.

Start, for example, by calculating how much CO_2 your current lifestyle is pumping into the atmosphere each year. Free online calculators like carbonfootprint.com make this job easy and pleasant (well, as pleasant as such a horrifying exercise can be).

The world brims with lists of lifestyle changes that you can make to reduce your carbon emissions. The Big Three: your transportation, your diet, and your home. So, you know, fly less, eat less red meat, drive an electric car (or ride an electric bike), take public transportation, set the thermostat at 68°F (winter) and 76°F (summer), install LED bulbs instead of incandescents.

You should feel great about joining the fight against plastic, too, because producing plastic requires massive amounts of heat, which comes from burning fossil fuels—and because seven of the ten largest plastic producers are oil and gas companies. The more plastic we use, the more petroleum they'll extract (and the more plastic will wind up in the oceans—currently 8 million metric tons a year).

And finally, the general rule is, consume less, because the manufacture and shipping of everything produces emissions.

For a more complete list of practical ways to reduce your carbon footprint, see the free downloadable appendix to this book, "Your Carbon Footprint." You can download it from www.simonandschuster.com/p/how-to-prepare-for-climate-change-bonus-files.

Now, you may wonder what difference *your* actions could possibly make. You, as an American, contribute 20 tons of CO_2 a year? Well, the airlines pump out 168 *million* tons a year. Cows and sheep belch and fart out 442 *million* tons a year. Burning coal produces 1.3 *billion* tons. And those are just the U.S. numbers. What possible effect could *your* actions have, puny mortal? Making changes to *your* lifestyle would be like rearranging the crackers on a plate on a deck chair on the *Titanic.*

It is true that if you can make only one gesture toward solving the carbon problem, you're most effective as part of a *group.* "I'd rather go encourage my mayor to develop a plan for rising seas for the community than to build a barricade around my house," says Ben Strauss, CEO of Climate Central.

But there are three huge reasons why you, as a tiny, noncorporate entity, should attempt to minimize the greenhouse gases you produce:

- **There's a cumulative effect.** You're joining millions or even billions of *other* people making similar small shifts, and they do add up.
- **There's a social-norms effect**, also called behavioral contagion. It means that what you see other people doing affects what *you* do.

  "Think about it: Four people go to a restaurant. Three order the salad and fruit cup. Are you really going to order the steak with cheese fries?" says Lise Van Susteren.

  The same thing goes when it comes to climate action. For example, for every home in a neighborhood that gets solar panels, the number of *other* people installing them goes up. In other words, whatever you do gets magnified by your effect on other people.

  "If I'm careful about my carbon footprint, and I try to eat very little carbon-intensive food, or I say, 'I'm taking the train, I'm not flying,' or I'm busy turning off the lights or putting up solar panels—that influences the people around me," Van Susteren says. "You don't even have to mention climate."
- **You'll feel better.** As you now know, cultivating a sense that you're *doing* something about your problems—taking some action, no matter how small—gives you a sense of control. And you feel better.

Political Pressure

You might feel as though a single voter couldn't possibly have any effect on the voting patterns of your elected officials. Incredibly, though, that's not entirely true. Each member of Congress, for example, employs a bank of young phone answerers and email readers whose job is to record the calls, letters, and emails from voters. Those are tallied in a software app and handed, as a report, to the congressperson.

True, some officials simply don't care what their constituents think. As Senator Ted Cruz's press secretary told the *New Yorker*, "The senator was elected based on certain values and ideals, and he's going to keep fighting for those, even though some of his constituents might disagree."

But other times, history has shown that a public outcry can succeed. Floods of calls and emails from the public helped to shut down Congress's attempt to give itself a 50% raise in 1989, for example, as well as the SOPA bill (an overreaching digital-privacy bill) in 2012, and the Trump administration's 2017 attempts to repeal the Affordable Care Act. Then, of course, the 2020 protests following the death of George Floyd led directly to a rapid cascade of changes intended to address systemic racism in America: the removal of Confederate statues, building names, and flags; changes in casting and availability of TV episodes and movies; and a flurry of new laws, policies, and proposals for reforming police departments. Public pressure can work.

So how, exactly, should you reach out?

- **Visit the office.** In theory, you, or a group that you represent, are actually allowed to *talk to* your congresspeople. You can visit them in Washington or at their state or district offices. In a poll of members of Congress, 97% said that these meetings are the best way to sway an undecided official.

  Well, great—except that *everybody* wants face time with these people. Sure, you can call the member's office and ask for an appointment, but you may be in for some *dis*appointment.

  Some members of Congress are more accessible than others. How to help your odds: Mention a relationship ("My dad went to high school with the senator"). Be a donor (donor requests for meetings are honored three times as often as nondonors'). If you've got the bucks, hire a company like SoapboxConsulting.com, which does nothing but try to get constituent appointments with Congress members.

  If you do get in the door, you'll have about fifteen minutes, and it'll probably be with a junior staff member—not the actual senator or representative. Plan what you're going to say, backed up with data.

Stress how the issue affects *you* and your state or district. Know what you're asking for (voting a certain way on a bill? Contacting a certain federal agency?). Follow up by email.

- **Editorials in local newspapers** are an excellent way to get your Congress member's attention, according to their staffers.

- **Showing up at a town meeting or public form**, which all Congress members routinely hold, is effective. These meetings are public, which magnifies your message, and they're face-to-face meetings (or video chats), which beats being reduced to a tally mark in a staffer's software program. Practice what you're going to say—make it good—and, as always, tie the message into how the issue affects *you.*

  This master calendar lists upcoming town meetings: https://townhallproject.com.

- **Phone calls** are effective because they draw the assistant's attention and time. Be prepared to sit on hold, or to call back. Keep your call short. Refer to a particular bill number, if that's what you're calling about.

  Here are some ways to make your call *less* effective: Be rude, be a crackpot, read from a script some group has written for you, or call someone who doesn't represent *your* state or region.

- **Emails and faxes** all get read and entered into the tally database and passed along to the official.

- **Paper mail** works, but it's slow. Every envelope Congress receives makes a first stop at a security bureau, where it's opened, tested, decontaminated, and finally scanned and sent to the recipient as an electronic image.

- **Facebook and Twitter, online petitions, comments sent from apps such as Countable, and boilerplate emails that come from advocacy-group websites.** Don't waste your time. Congresspeople don't trust these channels; they're too easy to hack, game, or blast out en masse.

All right then. How do you find the contact information for your reps and senators? You go to GovTrack (www.govtrack.us) and enter your home address. You're shown the name, photo, phone number, and website link for your two senators and one representative.

The National Resources Defense Council suggests that you enter them into your phone's contacts app "with 'Politician' in front of the official's last name, so you have all three grouped together on your contact list."

You don't have to go it alone. You might find it easier getting started if you join an existing volunteer organization that's already designed to communicate with lawmakers at every level. Turn the page to read about some of them.

Help a Campaign

If you've perused the previous paragraphs, one point about political pressure probably pops out: Those are all *really* indirect methods of attacking the climate crisis. If your elected reps are in the pocket of the fossil-fuel industry, no amount of voter pressure will affect their voting habits.

Here's an infinitely more effective tactic: Help elect somebody better.

Where to start? At votesmart.org, you can look up the records of all state, regional, and national politicians: a history of their votes (and quotes) on the issues; and which special-interest lobbies have funded and endorsed them. It's eye-opening to see what your elected officials have been *doing* all this time.

There are all kinds of ways to help out with a candidate's campaign. You can go door-to-door (with a partner, if you like). Offer to enter data at the campaign headquarters—recording donations, for example. You can deliver yard signs to supporters who've requested them. You can host "meet the candidate" events at your house. You can help raise money.

On Election Day, you can stand outside a polling place, smiling and asking for voters' votes for your candidate. Or you can volunteer to drive voters to the polls; plenty of your fellow citizens people lack wheels, mobility, or bus routes that aren't a nightmare.

Most of these efforts require that you live where the election is taking place—but volunteering time at a phone bank does not.

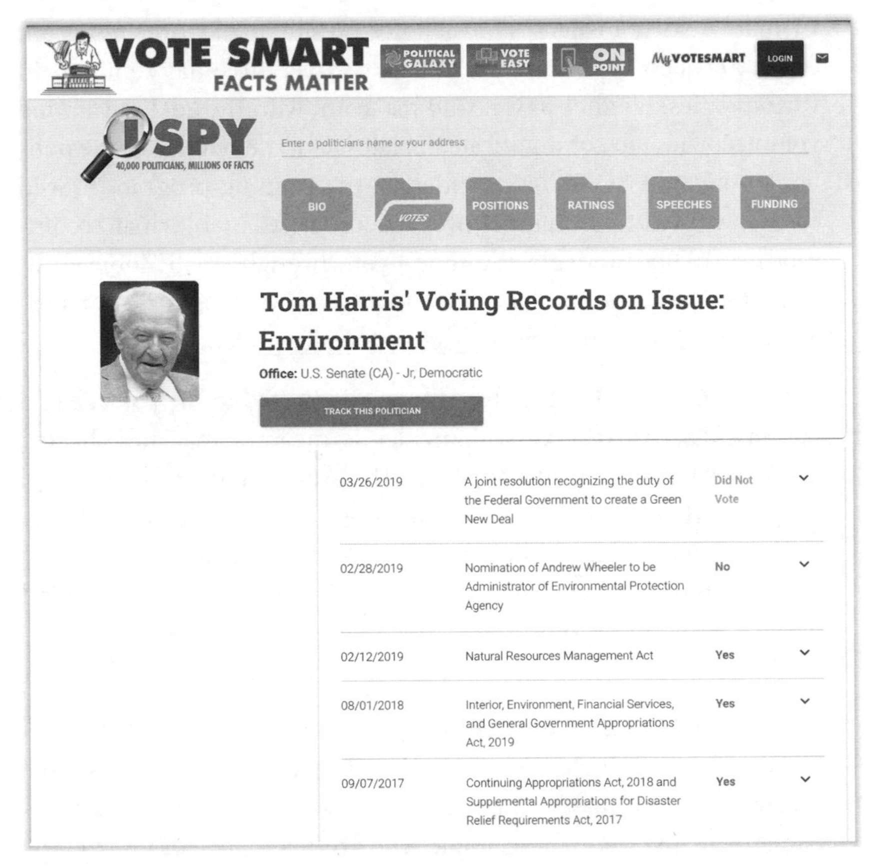

Figure 1-1. On this single, nonpartisan web page, you can look up the voting record, stated positions, and lobbyist funding of any politician on any issue.

Power in Groups

"Advocacy" can mean going door-to-door with petitions on clipboards, or making signs for an Earth Day rally—but it doesn't have to. If you work in an office, it can mean making a few green changes. If you're a teacher, maybe you can begin to introduce a curricular element or an after-school program. "There are endless ways, in every single area of human endeavor," says therapist Leslie Davenport.

The world of sustainability warriors is crying out for your help—and offering to train you for the job. They include:

- **Climate Reality Project** is former vice president Al Gore's nonprofit outfit. Between 2011 and 2019, it held 43 free, three-day seminars in various cities, taught by Gore and his team, with the goal of training volunteers how to communicate the latest climate science to the public. During the COVID-19 pandemic, the training program moved online, with Gore presenting over video. The virtual format accommodates 10,000 students at a time—yet, through small Zoom calls, still offers participants the chance to ask questions and meet fellow advocates who live nearby. www.climaterealityproject.org
- **Citizens' Climate Lobby.** This group is dedicated to the passage of a national U.S. climate tax—with dividends, meaning that the collected tax money will be passed along to us. The group has over five hundred chapters, where you'll learn how to lobby Congress, make your case in the media, and introduce the carbon-pricing idea to the masses. This group has already had some success in Congress; it had a hand in the Energy Innovation and Carbon Dividend Act of 2019, which is working its way through Congress. https://citizensclimatelobby.org.
- **350.org.** This international group seeks to end the era of fossil fuel burning. Its goals include stopping the construction of new fossil fuel projects and cutting off funding and financing for fossil fuels companies—by divesting, for example. The group's name comes from its goal: to keep the carbon levels in the atmosphere below 350 parts per million. https://350.org/.
- **Union of Concerned Scientists.** Founded in 1969 by students at MIT, the Union is now dedicated to "mobilizing scientists and combining their voices with those of advocates, educators, businesspeople, and other concerned citizens." Its work includes reporting, publishing, and mobilizing citizens to pressure oil companies and the government into taking action: www.ucsusa.org/climate.
- **PowerShift. org** is a huge network of groups of young activists all over the world: International Student Environmental Coalition, UnKoch My Campus, Sierra Student Coalition, Rainforest Action Network, and dozens of others. The PowerShift meta-group shares training and skills and even provides funding for individual projects.

Stress Relief

What happens if you've taken the major steps toward addressing your eco-anxiety—you've taken stock of your own flavor of despair, you're talking about it more, you're taking some action—but you still can't sleep?

At that point, it's time to look into stress-relief techniques. The time-honored, well-studied protocols for general stress reduction and mood improvement work beautifully for eco-anxiety. You may have heard these ideas a thousand times, but they've also been *studied* a thousand times—and they work.

Exercise

Let's face it: Your brain is basically a sloshing chemical sponge. Why does sleep feel good when you're tired, food taste good when you're hungry, or a loving touch feel good when you're lonely? Because they release natural hormones, enzymes, and chemicals in your brain. Those physical acts make you feel mentally rested, satisfied, and cherished.

Suppose, therefore, that you had a mechanism for releasing those pleasure hormones *on demand*, whenever you're feeling down, oppressed, or afraid?

You have one. It's called exercise.

You know the "runner's high"? Well, there's also a swimmer's high, a hopscotch high, a tennis player's high, a weight lifter's high. Any activity vigorous enough to make your heart beat faster releases endorphins in your brain—chemicals that relieve pain and boost happiness. (*Endorphins* comes from *endogenous*, meaning "from the body," and *morphine*, which is an opioid.) That's your brain on drugs—the best possible kind.

Even a little bit of physical activity can protect you against depression, regardless of your age or gender. (About 50% of depressed people do no exercise at all.)

A Dose of Nature

It's possible to lower your stress levels, and your susceptibility to stress-related illness—dramatically—just by spending time in nature. Scientists have measured the effect across every age, gender, and wealth level. For example, people who move into leafy neighborhoods experience a long-term decline in mental illness—and people who move to less green areas experience *more* mental illness.

"We live in this tech bubble that's all sped up," says Dr. Doherty. "But the universe is still out there."

If you've ever gone camping or hiking or playing outside, chances are good that it was when you were younger. But as we age, we tend to have less contact with the wild. "People are parenting and working, and they wonder why they're feeling out of sorts and burned out," Doherty says. "They've moved away from the things that actually feed them."

The prescription is obvious: Make regular visits to a park, spend time in a garden, or find ways to walk near forests, fields, lakes, or oceans. As a bonus, time outside is likely to grant you two other therapies for anxiety and despair: exercise and contact with other people.

Leslie Davenport cites a Chinese proverb: "In every moment, there are ten thousand joys and ten thousand sorrows."

"If you've gotten into a habit of tracking the ten thousand sorrows," she says, "what if you took an intentional walk, where the whole point of the walk is looking for things that are pleasant and beautiful? That give you joy, like acts of kindness? It's like some people do a gratitude journal at the end of the day: What were three good things that happened to you? Maybe a stranger held a door for you when your arms were full of packages."

These habits, she says, help balance your negative emotions with positive ones.

News Diet

Among people who are deeply concerned about climate change, there's a common denominator: They read a lot of news.

"It's coming from all these different directions, and they're feeling totally unstable," notes Thomas Doherty.

Often, he'll recommend a news diet to his clients, or even a news fast. The idea is to prune the number of sources you consume—and to balance them out with healthier, nondistressing reading. He often prescribes Thoreau, poetry, or anything that celebrates nature.

Or, after a subjecting yourself to a session of doomscrolling, sample a snack of happier news at GoodNewsNetwork.org.

Deep Breathing

"We know that trauma and anxiety are generated by thoughts and feelings, but they also live in the nervous system in the body," says Leslie Davenport.

That's why working your physical systems—like breathing—can affect your mental ones.

The web is full of tutorials on deep breathing (also called diaphragmatic or abdominal breathing), but here are the basics from the Harvard Medical School:

"Find a quiet, comfortable place to sit or lie down. First, take a normal breath. Then try a deep breath: Breathe in slowly through your nose, allowing your chest and lower belly to rise as you fill your lungs. Let your abdomen expand fully. Now breathe out slowly through your mouth (or your nose, if that feels more natural)." Keep it up for at least two minutes.

Deep breathing encourages greater oxygen exchange (for carbon dioxide). It also slows your heartbeat and can lower or stabilize your blood pressure. If you have a Fitbit band or something similar, you will indeed see that your heart rate has slowed by the end of the exercise.

There are plenty of other techniques that serve the same stress-lowering purpose, including progressive muscle relaxation, meditation, yoga, tai chi, qigong, repetitive prayer, EFT tapping, and guided imagery. (Google 'em.)

If you're just starting out, try out the Headspace phone app, which incorporates an instructor's voice gently guiding you through mindfulness exercises.

And if all that's too woo-woo for you, even chewing gum has been shown to relieve stress.

The Worry Hour

Therapists sometimes recommend a cognitive behavioral therapy (CBT) tool called the Worry Hour. Its effectiveness has been demonstrated in high school students, soldiers with PTSD, and others.

The idea is that you can't tell people to *stop worrying*. They can't. It is possible, however, to *postpone* worrying. Every day—say, at 4:00 p.m.—you sit down for 15 or 30 minutes to write down everything that's bothering you. Really dwell in it. Marinate in all of the "What's the worst that could happen?" thought-experiments.

But that's the *only* time you're allowed to stew, at four every day. (Helpful hint: Your Worry Hour should not be right before bed.)

If worries start to flutter into your thinking at any other time of day, you postpone them. You have an agreement with yourself: Now's not the time. You'll wallow at 4:00 p.m. Over time, you'll get better and better at this worry-postponement process.

At the end of the week, review your worries. Look for patterns and changes in what you've been stressing about.

"I know it sounds odd," says Davenport, "but it actually works."

Creative Outlets

Art, music, and writing are all excellent medicine for times of anxiety and depression. They're proven treatments for pain relief, general hospitalization anxiety, and the mental health of stroke, cancer, HIV/AIDS, and epilepsy patients. "The antidote to depression is, obviously, creativity," says psychologist Renee Lertzman.

The point is to focus your thoughts, express your feelings, keep yourself engaged, and allow less brain space for dismal thoughts about the world.

Write, paint, sketch, noodle on an instrument, take (and edit and organize) photos, shoot and edit a video, learn a magic trick, make a stop-motion movie with your phone, dance, act in a play, knit or crochet or quilt. No masterpieces are expected, and no audience is necessary.

Therapy

For most people in the throes of climate grief, the Big Three—taking action, getting exercise, and using stress-relief techniques—are a magic bullet. You don't stop feeling terrified and angry, but those feeling no longer keep you up at night, incapacitating you with a cycle of obsessive thoughts.

For some people, though, those aren't enough, especially people who have experienced depression or trauma before. In that case, there's one more tool in the box: therapy.

Group Therapy

Hundreds of studies have established the mental and physical health benefits of social contact: Exercising in a group lowers stress levels more than exercising by yourself. Chemotherapy is more effective if the cancer patients are around people. Teenagers are less susceptible to depression if they hang out with buddies. People without much social contact get sicker and die sooner than people who spend time with people. And on and on.

That's why group *therapy* has become a useful resource for anyone who suffers from mental or psychological challenges.

It's not touchy-feely magic at work; it's brain science. Experiencing an event by yourself activates the right parietal lobe of your brain. But when you go through an experience *with other people*, a different part of your brain lights up.

"The minute that you become a member of a community taking action, it moves like magic to the left parietal lobe," says Lise Van Susteren. "The left parietal lobe used to be called the God spot, the spirituality center. Now what we know is that that center helps you stop focusing on your individual pain and instead gives you this uplifting feeling of being a part of something bigger; your brain waves show it, and functional MRIs show it. And when people feel part of something bigger, it has immensely healing benefits. It's incredible."

Environmental-grief groups aren't yet in every town, but they're growing. For example, after struggling with eco-despair themselves, LaUra Schmidt and Aimee Lewis-Reau founded the Good Grief Network. Its regular in-person meetings moved online during the coronavirus pandemic, but its website (www.goodgriefnetwork.org) offers a podcast, training materials, and a start-your-own-group handbook.

In a group, Schmidt says, "people feel less alone, which is absolutely essential. People can say, 'I'm scared, I don't know what to do,' and then somebody else says, 'Me, too. Let's talk about it.' We've found that that is what transforms people."

Websites like ClimateChangeCafe.com and ClimateAndMind.org offer tips on finding or starting a climate therapy group. And sites like www.ecoanxious.ca and Parents for the Planet (a Facebook group) serve as *online* outlets for climate anxiety. If you can't find an eco-therapy group, then at least find *a* group. Book club, sports team, church/temple/mosque, dance lessons, classes, bowling league, choir, community theater, volunteer work, old-friend get-togethers, jogging group, language-practice clubs, game nights, or any of the thousands of groups in every town and city that list themselves on the searchable database at meetup.com. There may not be many eco-grief support groups listed, but there *are* plenty of climate advocacy, action, and education meetups.

Individual Therapy

Eco-grief barely existed ten years ago, but is now a rapidly growing discipline, taught at colleges and programs all over the world.

To find a climate-trained therapist, you might start with a Google search for your area—for example, *eco therapists Minneapolis*. Or search by city at www.psychologytoday.com/us/therapists. If you have health insurance, Medicare, or Medicaid, its website can refer you to in-network therapists. There's a climate-aware therapist directory at www.climatepsychology.us, too.

Once you've got a name or two, visit their websites to research their familiarity with eco-anxiety. Most therapists offer a free initial consultation, where you can assess their expertise, approach, and personality. Don't be shy about interviewing several people in this way—it's very common, you won't hurt their feelings, and research shows that people who shop around get much better results from their therapy.

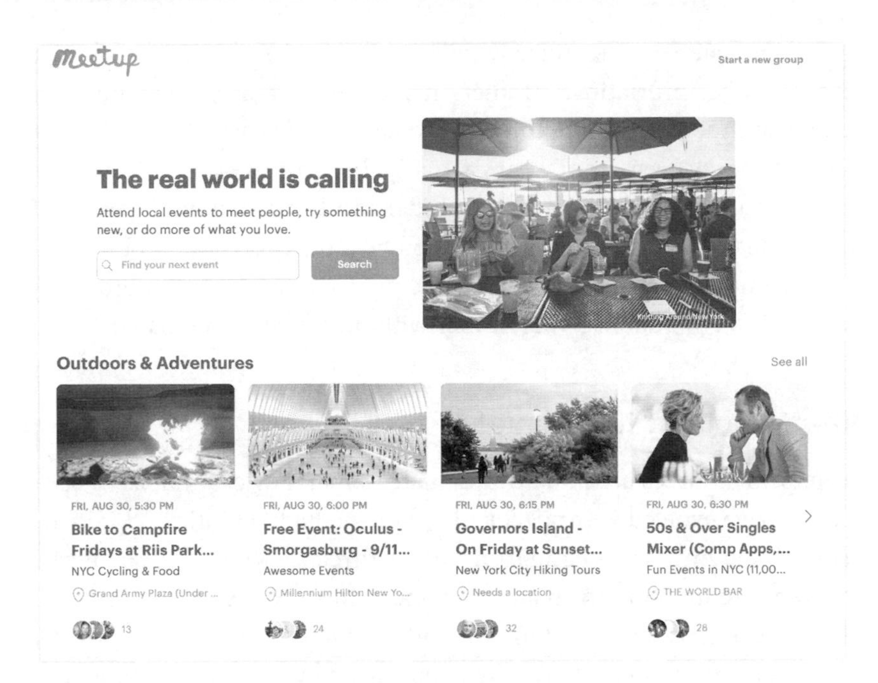

Figure 1-2. You probably won't talk much about your climate grief at your dance lesson or bowling league. You and your brain will, however, get all the health benefits of being with other people, and that's not nothing.

Dealing with Deniers

To a logical person, it might seem bizarre that climate change is controversial. The science is rock-solid and backed up by thousands of studies. What's the controversy?

What we call a denier may be either of two things: (a) somebody who doesn't think the climate is changing at *all*, or (b) someone who simply disputes that human activity is the cause.

Fewer and fewer people are in the first category. The 2020 numbers from Yale's Climate Change Communication program show that 73% of Americans believe that global warming is happening, up 17% from its survey five years ago. Sixty-two percent attribute the changing climate to human activity (up 15%), and 29% think that this is all a natural cycle. Unsurprisingly, people identifying themselves as conservative Republicans make up the bulk of the doubters.

To someone who's familiar with the science, the very existence of a denier can be infuriating. "Deniers make people crazy," says therapist Leslie Davenport. "It can become the focus: 'If only these people . . . !' It can take energy away from what's actually there for us to do."

But it's useful and important to understand why deniers deny—and how to speak to them. Only then can you help make societal progress toward solutions, grow more comfortable in speaking about the problem, and get through Thanksgiving dinner without a screaming match.

What's Behind the Denial

Deep, primal forces are at play in climate denial. For example, as a species, we have evolved with spectacular fight-or-flight abilities. We're fantastic at dodging or defeating imminent threats. When a coronavirus threatens, we mobilize to create a vaccine at historic speed.

But we're terrible at reacting to gradual, long-term, abstract threats, even if they can kill us. That's why people are slow to quit smoking, lose weight, exercise, save for retirement, and so on.

Harvard psychology professor Daniel Gilbert proves the point with a great example. "The density of Los Angeles traffic has increased dramatically in the last few decades, and citizens have tolerated it with only the obligatory grumbling," he writes. "Had that change happened on a single day last summer, Angelenos would have shut down the city, called in the National Guard, and lynched every politician they could get their hands on."

Guilt and accusation are also involved in accepting that we've altered the planet. "The subtext of climate change is 'We messed up,' " says Renee Lertzman. "We've benefited tremendously from all these amazing developments in our lives, but we did not mean for those benefits to lead to the potential demise of so much of what we care about."

The result, she says, is a double bind. On one hand, we know that business as usual—burning petroleum—is unsustainable, a sure path to the devastation of our planet.

On the other hand, condemning our current consumption patterns can feel like a betrayal. "This is a very common theme: People say, 'Well, if I were to acknowledge that it's real, then I feel like I would be betraying my family—my ancestors who worked in the coal mine or the plant, or dedicated their lives to make sure I could go to school. They were part of industries that are now being seen as bad.' ”

We have a hard time with double binds, Lertzman says. Instead, we avoid confronting them. “You find all kinds of strategies: to minimize, to question the science, to demonize the messengers, to make it ideological, to make it political,” she says.

All kinds of other factors make it easier for people to avert their eyes from confronting the climate problem:

- **Self interest.** Lise Van Susteren points out that people in power have a strong motivation to resist change. “Politicians, of course, because they're aligned with a political ideology that they believe will keep them in office,” she says. Or they fight change out of greed: People in the oil, gas, plastic, travel, manufacturing, airline, and car industries may believe that they'll make more money if they play down the news of a climate crisis.

- **Feeling ignorant.** It doesn't help that climate science is complicated. Hearing all the statistics and jargon can make people feel stupid. And when that happens, people ram their heads into the sand, fast and hard.

  “When people feel that they do not understand a domain or an issue, they will disengage from it,” concludes a 2017 study published in the *American Journal of Political Science.*

- **Psychological distance.** There's always been a tragic asymmetry to the climate crisis: The rich countries produce most of the greenhouse gases. The poor countries suffer most of the consequences.

  That effect allows us to take refuge in psychological distance: mentally categorizing climate chaos as something that's happening to somebody else, in some other place, at some other time.

- **Invisibility.** There's a fourth obstacle to recognizing climate change as an imminent threat: Not everybody has *seen* it yet.

  Sure, almost every farmer has; the Inuit people of Canada have; island dwellers have; and the victims of wildfire, floods, droughts, and superstorms certainly have.

But plenty of people, including 56% of Americans, report that they've only *heard* about climate-related disasters. Maybe they've seen some strange weather, but that's easy to chalk up to natural variability. And by this point, you know how people work: They find it hard to believe something if they can't touch it with their own hands, hear it with their own ears, or see it with their own eyes. *Especially* if believing in that something involves revising their entire belief systems, identities, and sometimes livelihoods.

- **Feeling overwhelmed.** You don't need a PhD in psychology to understand that the earth's destruction can be so overwhelming and frightening that many people fall into denial as a defense mechanism.

Put it all together, and you can see why every fiber of some people's psyches are inclined to fight the concept of a changing climate—especially when the change is *our fault.*

How to Speak to Deniers

In speaking to a denier, your natural instinct—everyone's natural instinct—is to offer them facts. Point out the science of the situation. Quote studies. Cite the latest horrifying numbers.

That, as it turns out, is exactly the wrong approach. It's worse than a waste of time; it's actually counterproductive. It's likely to entrench deniers even more firmly into their beliefs.

You're especially likely to lose them if you start jargonizing. At his 2017 TED Talk, psychologist Per Espen Stoknes made that point by offering a sample message in two different styles. He asked the audience to gauge which version had the most impact on them:

- **Style 1:** "We are seeing rising carbon dioxide levels, now about 410 ppms. To avoid the RCP 8.5 scenario, we need rapid decarbonization. The global carbon budget for 66% likelihood to meet the two-degree target is approximately 800 gigatons."

- **Style 2:** "We are heading for an uninhabitable earth: monster storms, killer floods, devastating wildfires, crazy heat waves that will cook us under a blazing sun. . . . We have a three-year window to cut emissions. Three years. If not, we will soon live in a boiling earth, a hellhole."

There's copious brain research that reveals what happens to someone with a tightly held opinion (*not* based on facts) when you offer a factual argument.

"Functional MRI shows the folly of it," Lise Van Susteren says. "It's not the cortex—our gray matter, our rational centers—that lights up. It's the *emotional* centers. It's the amygdala, the limbic system. So don't try to convince people with the facts, which is a debilitating and depleting experience."

So where does that leave you? With an appeal to their emotional centers, of course.

"You say, 'I understand that this is a very difficult topic.' Then you talk about why you, personally and emotionally, might be thinking what you do. Your uncle, whose farm is underwater. Or your child who's worried. Or your own personal losses."

At the very least, Leslie Davenport says, let them speak. "Rather than 'You're crazy!' 'No, *you're* crazy!,' listen to them. 'Well, that's an interesting perspective. Tell me more about that.' There may not be a lot of movement, but it's a softening. It goes a long way."

Therapist Thomas Doherty sums it up with an old saying: "People don't care what you know—until they know that you care."

Chapter 2

Where to Live

YOU HEAR A STRANGE PHRASE FROM CLIMATE SCIENTISTS these days: The world is *shrinking*.

"A period of contraction is setting in as we lose parts of the habitable earth," writes climate author Bill McKibben.

Welcome to the concept of climate migration, where people flee unlivable regions and crowd into the more sustainable ones.

For millions of people, climate migration isn't voluntary. When sea-level rise floods your homeland, or a superstorm flattens your city, or drought dries up the fields, or wildfires turn your region into a wasteland, you move to survive. Droughts and monsoons have already driven 8 million people out of Southeast Asia; crop failures have driven millions more out of the countries just below the Sahara Desert. The IPCC estimates that by 2050, about 200 million people may have to move—1 in every 45 people on earth. (That's a likely number, but the IPCC says that the number could be as high as a billion.) And you think anti-immigrant sentiment in richer, cooler countries is high *now*?

Climate change is hardest on poor countries, but there are climate migrants even in the United States. Hurricane Katrina, in 2005, drove about 400,000 people out of Louisiana for good. In the year following Hurricanes Maria and Irma in 2017, about 50,000 Puerto Ricans moved to Florida. In 2018, the most destructive wildfire in California history destroyed 95% of the buildings in the town of Paradise, sending most of its 26,000 residents scrambling to find new homes. And Flagstaff, Arizona (elevation 7,000 feet), is enduring such a flood of refugees from Phoenix's blistering heat that locals joke about building a wall.

No wonder, then, that a growing number of people are considering relocating *voluntarily*—now, while there's time to do it thoughtfully and calmly.

"People are tired of the heat stress, flooding stress, and—particularly among those who are nearing or at retirement age—the physical stresses," says social scientist Jesse Keenan, who studies climate adaptation and the built environment at Tulane University. "People are on the move, not because their house burned down or because it got flooded; they're on the move because they're trying to preemptively get ahead of this."

Moving preemptively has some overwhelming logic. The sooner you put down roots in a safer place, the sooner you'll be able to stake a spot, set up shop, and escape the ravages of the new climate. There's a self-interest element, too: The early bird gets a better choice of new home options—and a lower price.

Who's Moving?

Moving, obviously, isn't an option for everyone. It costs money. It's disruptive. It requires leaving behind some of your social network. And, to be sure, few people are yet moving for climate reasons alone.

"I don't think climate change is the most important factor, not by a long shot," says Laurence Smith, environmental studies professor at Brown and author of *The World in 2050*. "People can and will live in places that are miserable for other considerations, from getting a job to politics." For most people, climate considerations come into play only if most other factors are equal.

But 40 million Americans do move every year, for reasons unrelated to the climate crisis. Maybe they're graduating, leaving the military, changing jobs or relationships, or retiring.

Maybe you're among them or will be soon, and maybe other factors *are* equal. Or maybe you're among the growing numbers of people who are freaked out enough that moving to a climate haven *is* a primary consideration.

You wouldn't be alone. Florida State University associate professor Mathew Hauer studies the impact of climate change on the distribution of the U.S. population. Figure 2-1 shows the states that will gain or lose

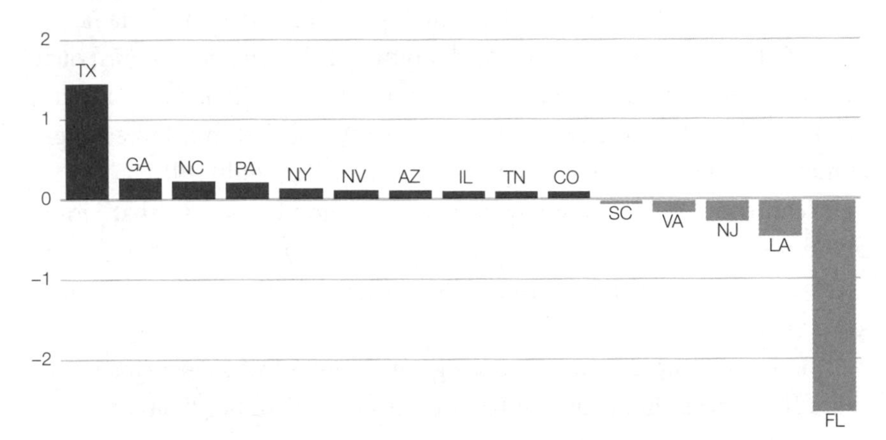

Figure 2-1. This graph shows the states that can expect the biggest population gains (left)—and losses (right), in millions of people. Texas is the big winner—and Florida, at the far right, is the big loser.

the greatest number of residents by 2100, according to Hauer's research, if the six-foot sea-level rise prediction comes to pass. As you can see, Florida can expect a dramatic population crash (far right); Texas (far left) will be the big winner. (Tip: Those climate migrants will largely be settling *inland* in Texas, not along the Gulf Coast, where they'll be smashed by hurricanes.)

Why Texas? Most American climate migrants move no farther away than a four- or eight-hour drive. "They tend to move where their family and friends are, as long as they can maintain or improve their economic situation," says Hauer. That's why most of the climate migrants from heat- and hurricane-threatened regions like Louisiana and Florida move to Texas.

The Finances of Moving

Moving is expensive. But it may also be the investment of the century.

Climate change means real-estate change, and it's happening fast; to many people, buying a place in a climate haven is looking like an excellent investment.

It certainly looks that way to Chinese investors, who, in the last ten years, have spent billions of dollars buying up real estate in Canada. Already, Chinese buyers own 14% of all homes in Toronto, and a *third* of all the homes in Vancouver. For them, owning land in Canada is "a hedge against political, economic, social insecurity, and I think, increasingly, climate change," urban studies professor Andy Yan told NPR.

Meanwhile, in the United States, coastal homes—those that would be flooded by one foot of sea-level rise—already cost 14.7% less than identical homes on higher ground. Researchers expect that discount to keep growing.

There's one more number to plug into your spreadsheet: insurance rates. In general, they drop right along with your distance from the flood zone or the fire zone.

But the biggest reason to consider moving isn't about money; it's about peace. Living through just *one* extreme-weather disaster is traumatizing. It can upend your life for months or years, cost a fortune, and separate you from your home, your job, and everything you own. In the coming decades, some of the world's biggest cities will become unpleasant, dangerous, or even uninhabitable places.

If you're even starting to think about thinking about a move, the following pages can guide you.

The Fundamentals of Climate Havens

Before you hunt for the perfect new homestead, accept that *every* place is affected by climate change. "From Atlantic hurricanes to Midwest tornadoes to Western wildfires, no corner of the U.S. is immune from the threat of a devastating climate-event," says the EPA.

Here's one way to look at it. The maps in Figure 2-2 show the relative climate-chaos effects that each region of the United States can expect to experience between 2030 and 2070.

The second half of this chapter identifies 14 attractive climate-haven cities in the United States. To understand why they qualify, though, here are the four big rules for choosing a climate-safe home city.

Rule 1: Get Away from Oceans

"If I were looking at a blank slate and deciding where I wanted to live," says Alex Wilson, founder of the Resilient Design Institute, "one of the first considerations would be, 'not a coastal area.' "

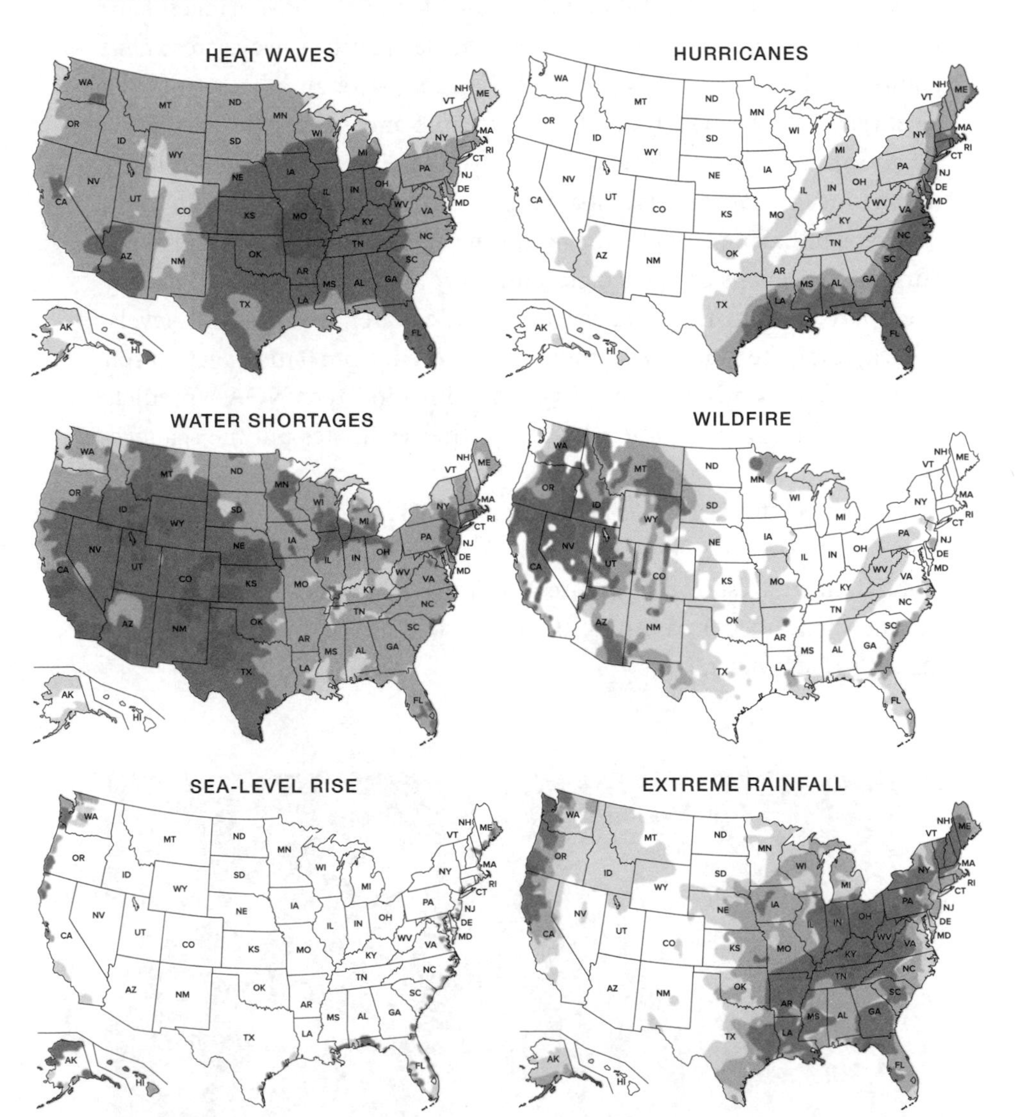

Figure 2-2. These maps show the kinds of climate unpleasantness you can expect in various parts of the country.

About 40% of the U.S. population now lives in counties along the 12,000 miles of coasts: the Atlantic Ocean, Pacific Ocean, or Gulf of Mexico. That's 123 million Americans who should worry about sea-level rise and coastal storms.

And it's not just about flooding of your *home.* "Roads, bridges, subways, water supplies, oil and gas wells, power plants, sewage treatment plants, landfills—the list is practically endless—are all at risk from sea level rise," says NOAA (the National Oceanic and Atmospheric Administration). In other words, just because your living-room rug is still dry doesn't mean your town is still a fun place to live.

If you're having trouble imagining the flooded future, then visit the Surging Seas maps at ss2.climatecentral.org, where you can see *exactly* what parts of the United States will be underwater at various sea levels. You can drag the Water Level slider up or down, from 0 to 10 feet, to suit your inner pessimist. As you drag, keep in mind that NOAA predicts a sea-level rise of up to 8 feet by 2100; other estimates put the sea level even higher.

Of course, the real fun is typing in a particular address and zooming in to street level. You get to see exactly what regions will be flooded and by how much, as shown in Figure 2-3 for San Francisco. Hint for interpreting the shaded area: About three-quarters of the area shown in the image is underwater.

Figure 2-3. At SurgingSeas.com, you can see just how flooded your neighborhood will be in a few more decades. Here, San Francisco's Mission District.

You can also zoom out to see entire cities, like Boston (Figure 2-4). Again, the darker-tinted area represents inundated areas.

Just for fun, have a look at Miami. The nonprofit Union of Concerned Scientists calculates that by 2060, a staggering 58.5% of Miami's inhabitable land will be underwater. By 2100, it'll be more like 94%. Miami is going away.

Living live near the coasts also makes you hurricane bait. As you can read in Chapter 12, coastal storms are getting much more violent and causing far more damage than they used to. One key reason: The higher sea levels provide a much taller launch platform for *storm surges* (towering mounds of water, pushed onshore by storm winds) and *king tides* (freakish "sunny day" flooding about six times a year, thanks to the alignment of the sun, moon, and earth). The seawater flooding from these events is getting deeper, and reaching farther inland, than before.

All right: So what do the projections tell us about flooding risk?

Well, it doesn't take a rocket scientist to guess that the communities that will suffer the most are the ones along the oceans and the Gulf of Mexico. In the big "Where to move?" picture, no factor is more impor-

Figure 2-4. Here's what Boston will look like with a six-foot sea-level rise.

tant. You really, really don't want to live in coastal Florida, Alabama, Mississippi, Louisiana, or Texas. The heat will grow increasingly miserable as the decades pass, and hurricanes are already an annual, life-disrupting nightmare.

Incidentally, not all of the dangers of living on the coast are gradual or occasional, like sea-level rise and hurricanes. *Human* decisions could change your life dramatically in these regions—and abruptly.

"For example, State Farm could suddenly decide not to insure Florida anymore, and boom," says Laurence Smith of Brown University. "FEMA could rezone the maps, and suddenly properties could become very difficult to sell. That's an abrupt change—a human decision, provoked by the steady sea-level rise, that could lead to some very unpleasant surprises for retirees—or anyone living along the coast."

Rule 2: Move North

Just as the Surging Seas calculator lets you see the future of coastal flooding, the University of Maryland's future-climate calculator (fitzlab.shinyapps.io/cityapp) shows you how *hot and wet* a city will be. When you click a city dot on the map or choose its name from a menu, you get to see instantly what *present*-day city it will feel like in 2080.

For example, if you click Washington, DC, you learn that by 2080 it will feel like today's Greenwood, Mississippi, which is 9.8° hotter and 75% wetter than today's DC.

And if you click Jacksonville, Florida, you discover that it will feel like the southern tip of Mexico—practically Belize.

Overall, a typical North American city will feel as hot as though it has moved at least 530 miles south.

By the 2080s, the authors conclude, the climate of cities in the Northeastern United States will feel more humid and subtropical, the way the Southeastern United States does today: warmer and wetter in all seasons. Western U.S. cities, in the meantime, will feel more like cities in the desert Southwest or Southern California: warmer all year long, with less rainfall.

Plenty of Northern cities won't feel bad at all. Future Milwaukee, Wisconsin, will feel like today's Chester, Pennsylvania. Duluth, Minnesota

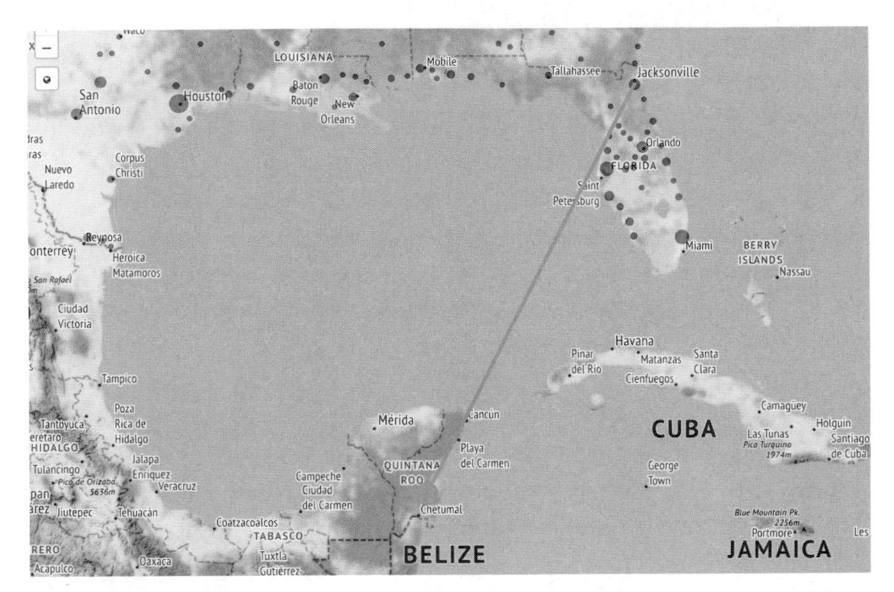

Figure 2-5. This website shows what present-day city yours will feel like in 2080. For example, Jacksonville, Florida, will be as hot as today's Felipe Carrillo Puerto, Mexico.

will feel like today's Cleveland. Seattle will feel like Milwaukie, Oregon, only 150 miles south—about 1.3° warmer.

Now, the biggest part of the "move north" argument has to do with heat, and that's understandable, especially for people who work outdoors. Heat is expected to be the number one killer presented by the climate crisis.

But it's not *just* about heat. Figure 2-6 shows how much economic damage will be caused by climate change by the end of the century—not just from heat, but from coastal storms, agriculture, crime rates, energy demand, death rates, and risk to the workforce. Darker blue means more economic damage; lighter gray means less.

This map puts in stark visual terms one of the cruelest aspects of the climate crisis: It hits poor regions the hardest. The Gulf Coast states will take a 15% to 20% economic hit, thanks to rising sea levels, megastorms, intense heat, and fleeing populations. The "breadbasket" states of the Midwest will suffer, too, as the higher heat, erratic rainfall, and droughts make farming harder and less reliable.

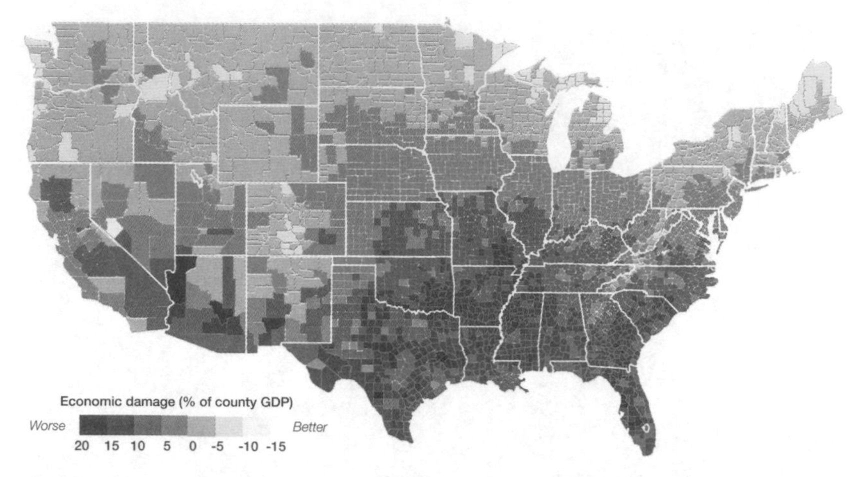

Figure 2-6. How much will it cost you to live in each part of the country? The deeper the blue, the more a county stands to suffer damage from climate-related events.

At the same time, some Northern counties stand to *gain* by a few percentage points. Climate change, in other words, only magnifies the wealth inequality of the country.

There's another tragic irony here: "Many of the jurisdictions that have selected political leaders opposed to climate policy are the most exposed

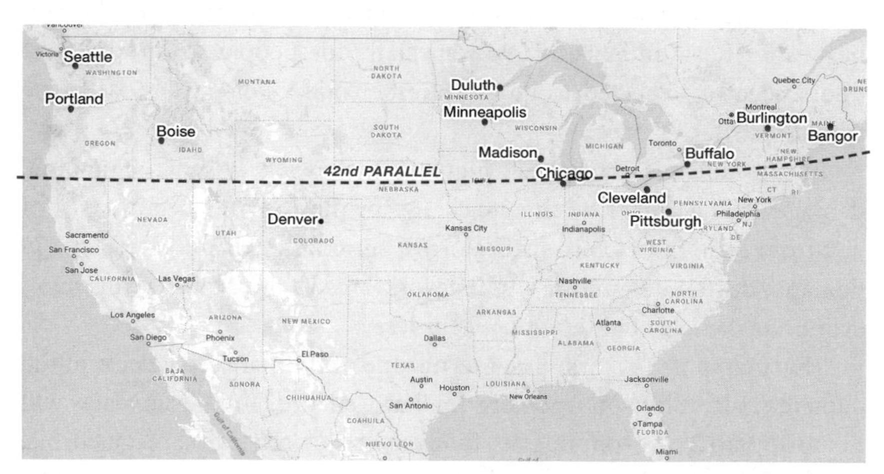

Figure 2-7. The cities that make the best climate havens—with the lowest likelihood of wildfire, hurricane, heat waves, water shortages, drought—are around the 42nd parallel or north of it. (Or at high altitude, like Denver.)

to the harms of climate change," writes the Brookings Institution. "Federal action to curb economically harmful climate change does not necessarily resonate in the places that need it most."

How far north should you move? Look at the band roughly above the 42nd parallel, says Portland State University professor Vivek Shandas, who studies climate change's impact on cities. That's a good guideline both in the United States and in other northern hemisphere countries.

Rule 3: Find Fresh Water

On a planet whose surface is 70% water, you might not think finding water is a tough job.

Unfortunately, 97% of it is *salt* water. It's not drinkable, and it kills crops. And the 3% of water that *is* drinkable is mostly frozen in glaciers.

Over the generations, most towns and cities have worked out ways to tap into water that's drinkable, croppable, and accessible. Some of those sources will be plentiful for centuries. Others, though, won't, including:

- **Snowpack** is the snow that accumulates over months of cold weather, usually in the mountains. For thousands of years, the streams and rivers carrying melting snowpack have been an important source of fresh water for humans. For example, 75% of the water used in the Western states of the United States comes from melting snowpack.

  Over the last 100 years, we've built hundreds of dams to catch and store the water from the snow that melts every spring. Snowmelt makes Phoenix, Las Vegas, Los Angeles, and Denver possible.

  But as the winters grow warmer, there's less snow. And as the winters grow *shorter*, the snow melts earlier. That's part of why California's in a yearslong drought, the worst ever recorded.

  In a study of 65 years' worth of readings from over 2,000 American weather stations, researchers found that *every* region of the United States is affected by the reduction in snowfall. Pacific Northwest, Southwest, Great Plains, the Rockies, the Midwest, the East Coast—all are experiencing less snow at most altitudes. By the beginning of 2018, 38% of the *entire* contiguous United States was in drought.

  For the Northwest (Idaho, Montana, Oregon, Washington), snowpack loss is a terrifying development. If temperatures continue to rise, there won't be any snow-dominant areas left by 2050; by the 2080s, they'll have disappeared completely.

- **Aquifers.** Did you ever dig in the sand at the beach when you were a kid? Dig far enough, and you'd find wet sand, and eventually water.

  That analogy should help you visualize an aquifer: an underground water layer. It's usually a huge pocket of permeable rock, gravel, sand, or silt, all of it soaked through with water.

  We can tap into these natural freshwater tanks by building wells, which supply drinking and irrigation water for 2 billion people on the earth.

  In the United States, 29% of all freshwater comes from aquifers. Ten states get over half of their water from underground: Mississippi (in the top spot, 84%), Kansas, Arkansas, California, Hawaii, Nebraska, Florida, South Dakota, Oklahoma (53%).

  There are also gigantic aquifer systems overseas, affecting dozens of countries in Asia, Africa, and the Middle East.

  In any case, you can probably guess the punch line: As the earth warms and weather patterns change, we're draining our aquifers faster than rain can replenish them. So far, 21 of the world's 37 biggest aquifers have passed that tipping point, according to measurements taken by NASA satellites.

  The biggest aquifer in the United States, for example, called the Ogallala Aquifer, supplies a third of the country's irrigation groundwater. It lies beneath eight states, including Nebraska, Kansas, Oklahoma, and Texas. (It's also called the High Plains Aquifer, but *Ogallala* is more fun to say.) As of 2010, we've depleted about 30% of it. The water table is down 300 feet in some places, and Ogallala wells are completely dry in others. At this rate, the Ogallala will be two-thirds empty by 2060.

- **Reservoirs** are man-made lakes, created by building a dam in a river. They're a critical water source for the Western and Southwestern parts of the United States—and there's not a single major U.S. reservoir whose water level isn't down substantially from its historical levels.

  Here's one screamingly important example: Lake Mead, the

110-mile-long lake created by the Hoover Dam in the Colorado River. It's America's largest reservoir, supplying water to 25 million people in Arizona, California, and Nevada. Las Vegas gets 90% of its water from Lake Mead.

Lake Mead is now just over one-third full. Its surface is 1,083 feet above sea level—a 117-foot drop since the year 2000. If the water level drops 33 more feet, the dam won't be able to provide any more electricity; and if it drops 155 *more* feet, it won't even be able to provide water.

"We're in the nineteenth year of a drought," water-policy expert Robert Glennon told the *Los Angeles Times*, "and it's pretty obvious that climate change is having a devastating impact."

Water shortages are becoming an annual concern in many North American cities—but in some parts of the world, water shortages have become part of life and death. In January 2018, after three years without much rain, Cape Town, South Africa, announced that it was three months away from running out of water entirely. The government took

Figure 2-8. Meet what's left of Lake Mead, America's biggest reservoir. The water level is now 140 feet below the 1983 high-water mark, indicated by the white "bathtub ring."

drastic measures: banning water for use in swimming pools, on gardens, on lawns, and for washing cars; publicly posting how much water each household was using, relative to its neighbors; diverting water from agricultural to urban use; and requiring residents to collect their water, in person, at 200 distribution centers.

Soon, saving water became a point of civic pride. In Cape Town businesses, employees participated in "dirty-shirt" challenges, to see who could go the longest without washing their work shirt.

These practices averted disaster, but the drought continues—and not just in Cape Town. Summer 2019 brought drought and heat waves that almost completely emptied the reservoirs in Chennai, India's sixth-largest city. The government, in desperation, began trucking water into the city's neighborhoods, where hundreds of thousands of residents waited with buckets and vases. They carried home their tiny allowances of water—at least when the water trucks weren't hijacked.

In other words, if you hadn't yet been associating climate change with water shortages, now's the time.

Rule 4: Seek Infrastructure

There's another element of climate survivability that people don't talk about much: wealth.

Richer countries have stronger infrastructure: stable governments, complex food-distribution networks, drinking-water sources, sanitation, medical facilities, emergency systems, communication networks, water purification, paved roads, reliance on imported energy, construction equipment, firefighting teams, agricultural capacity, and so on. That kind of national wealth makes it far more likely that you'll ride out whatever climate chaos dishes out.

The University of Notre Dame maintains a list of 182 countries, ranked by their vulnerability to climate change. The researchers incorporated 74 sets of data into the ranking, seeking to account for both *vulnerability* (that is, extreme-weather risk) and *readiness* to adapt (infrastructure, health care, food supply, government stability, and so on). By this formula, the top 20 countries for climate resilience are:

1. Norway
2. New Zealand
3. Finland
4. Denmark
5. Sweden
6. Switzerland
7. Singapore
8. Austria, Iceland (tie)
10. Germany
11. UK
12. Luxembourg
13. Australia
14. Republic of Korea
15. Japan
16. Netherlands
17. France
18. Canada
19. United States
20. Ireland

All of these countries are far enough from the equator to escape the devastation of heat waves. Many of them have coastlines, but also offer plenty of livable area inland, far from the coastal ravages of sea-level rise and sea storms. And all of them are rich countries, capable of recovering from the worst of what nature dishes out. (This study was completed before Australia's devastating 2019–20 wildfires.)

In the even longer term, migration experts talk about traditionally frozen expanses like Greenland and Siberia. As they thaw, parts of these regions are expected to sustain agriculture—and, because they're vast and sparsely populated today, are capable of accommodating a huge influx of residents.

The American Climate Map

If you're researching the best places to relocate within the United States, you're in luck: You've got good data to work with. At this very moment, several websites show exactly what kind of nasty weather events to expect. For example, you can see your risk of:

- **Sea-level rise.** Visit ss2.climatecentral.org, where you can drag a slider to see the flooding effects of each foot rise in sea level where you live. You can see the effect in Figures 2-2 through 2-4, earlier in this chapter. (NOAA's https://coast.noaa.gov/slr/ is similar.)

- **Flooding.** FEMA's interactive map site (https://msc.fema.gov/) shows the *current* likelihood of any address to get flooded. This tool is essential for figuring out what kind of flood insurance you'll be required to carry. It's also hard to interpret without the guidance in the box on page 201.

- **Storm surge.** The Environmental Protection Agency depicts the depth of storm surges that a certain address can expect for each hurricane strength (Category 1, 2, 3, 4, or 5). The computer model the EPA uses to calculate this data is, hilariously, called SLOSH: Sea, Lake, and Overland Surges from Hurricanes. It's available at https://epa.maps.arcgis.com/apps/MapSeries/index.html; click the "Storm Surge Flooding" tab.

- **Hurricane frequency.** The EPA's map site shows hurricane data, too. Once you're at https://epa.maps.arcgis.com/apps/MapSeries/index.html, click the "Hurricane Frequency" tab.

- **Heat, rainfall.** The Climate Explorer (https://crt-climate-explorer.nemac.org) offers the National Climate Map tool, which lets you split the screen. On one side, the historical average temperatures; on the other, the typical *future* temperatures. Using the pop-up menu at top left, you can change the map so that it's showing rainfall, thawing days, days over 80°F, and so on.

- **State by state.** At https://statesummaries.ncics.org, you can click your state on a map to read a detailed report about how its climate has changed in the last 150 years or so—and how climate change is likely to affect it in the coming years. Detailed graphs cover each state's susceptibility to heat wave, sea-level rise, days of freezing, and a lot more. Brought to you by NOAA, the National Oceanic and Atmospheric Administration.

But if you're just wondering more *generally* where to live in the United States, your government is on the case.

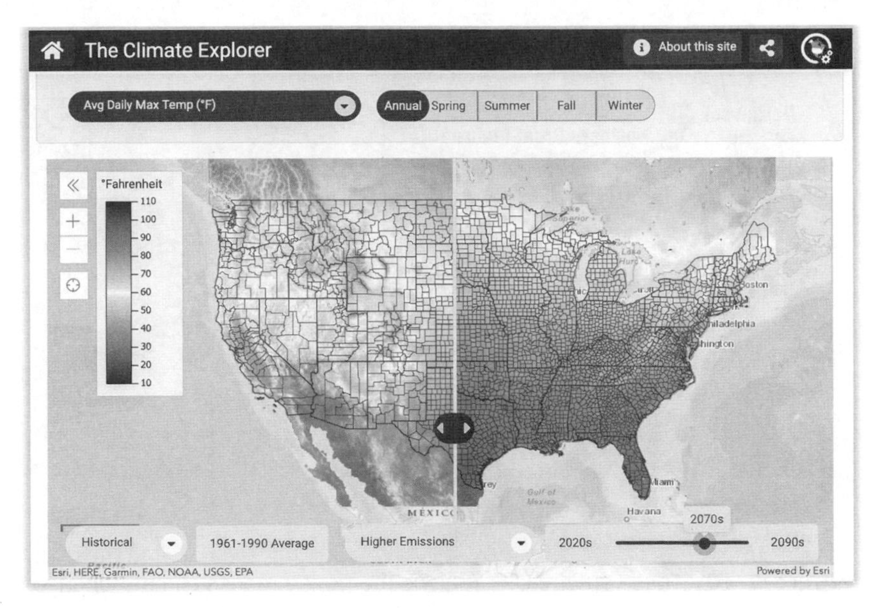

Figure 2-9. The Climate Explorer split screen compares the average temperatures today (left) and in, say, 2070 (right). The same website lets you compare present and future rainfall.

The *Fourth National Climate Assessment* is a massive pair of books created by the Global Change Research Program (GCRP, a group ordered by Congress to coordinate global-change research and funding across 13 federal agencies). This gigantic, multiyear effort was put together by over 300 scientists, resource managers, educators, business representatives, and experts from governments, laboratories, and universities . . . reviewed for accuracy by another army of external experts, the National Academies of Sciences, Engineering, and Medicine (NASEM), and the public . . . and then released by, of all things, the Trump administration.

That monumental research effort includes cheery region-by-region summaries of the weather ahead, as identified on the map in Figure 2-10.

You can read the full governmental write-ups at https://nca2018.globalchange.gov/, or you can just review the following capsule summaries.

Figure 2-10. These are the U.S. regions the government's climate-change assessment report describes.

Northeast

Thes states can look forward to the usual litany of coastal woes: vicious storms, flooding erosion, and ocean acidification In the Northeast, sea-level rise is expected to bc cspecially bad, with a rise as high as *11 feet* by 2100.

And where there's flood there's poop. "Many Northeast cities are served by combined sewer systems that collect and treat both storm water and municipal wastewater," says the GCRP report. "During heavy rain events, combined systems can be overwhelmed and release untreated sewage into local bodies of water." Great.

Ther will also be brutal heat. In fact, the Northeast is expected to warm up by 3.6°F by 2035—a faster warming than any other part of the continental United States.

Th famous fall foliage, once an unforgettable red, orange, and yellow magnet for tourists, is becoming more muted. We have heat, rain, and more clouds to thank for confusing their photosynthetic process.

Southeast

These states can look forward to "sea level rise, increasing temperatures, extreme heat events, heavy precipitation, and decreased water availability," according to the *Assessment*. "Drought and greater fire activity are expected to continue to transform forest ecosystems in the region."

The coastal states here will get the brunt of the hurricane forces every year, causing billions in property damage. Charleston, South Carolina, for example, used to get 11 tidal floods a year (when plain old high tide causes flooding); by 2045, there'll be *180* of them a year.

Meanwhile, the oppressive heat is the most lethal climate-change force—and in these states, it's matched by oppressive humidity. Thanks to the flourishing of mosquitoes and ticks, tropical diseases and Lyme disease are becoming powerful threats. To make matters worse, these states will become poorer as the Northern states grow richer (see page 46).

Midwest

Farming will take a huge hit in these states, thanks to the heat, droughts, thriving insect pests (whose population is no longer knocked out by cold winters), and intense, erratic rain patterns. Flooding from intense downpours near the Mississippi, Ohio, and Missouri Rivers and their tributaries will make life soggy, moldy, and germy for people who live nearby.

On the other hand, the northward and inland positioning of these states will protect their residents from the worst of the heat, and no storm surges exist to wipe out entire towns. And the Great Lakes states are blessed with a reliable, sustainable source of fresh water.

Milder winters and longer warm seasons mean more pleasant living—and, in some places, better crop growing. In particular, the Northern middle states—Minnesota, Wisconsin, Michigan, Ohio, Western New York—are shaping up to be the most climate-proof regions in the United States.

Northern Great Plains

These states—Wyoming, the Dakotas, Nebraska, Montana—are so far from the coasts that they don't enjoy the temperature-smoothing effect of oceans. Wild swings of temperature, rain, and weather will become the norm. There'll be more droughts *and* more torrential downpours. That's all going to make life harder for farmers.

Temperatures all year long will rise, but being so far north means you'll be far more comfortable than you would be in the South.

The biggest challenge for these states, though, will be access to fresh water. Most of the water for drinking and irrigation in this region comes from melting mountain snow, of which there's less and less every decade, and melting glaciers, which are also shrinking away.

Southern Great Plains

Kansas, Oklahoma, and Texas have always experienced dramatic weather. "Hurricanes, flooding, severe storms with large hail and tornadoes, blizzards, ice storms, relentless winds, heat waves, and drought—its people and economies are often at the mercy of some of the most diverse and extreme weather hazards on the planet," says the *Assessment*.

As the century unspools, life will get even harder for farmers. There's the heat, which makes it dangerous for people to work outside. In 2018, Austin had a record 51 days over 100 degrees; by 2100, that number will more than double. By then, 1,300 people will die from the heat every year in these three states alone.

The warmer winters mean that mosquitoes aren't dying off every year. Cases of dengue fever and Zika are already cropping up in Texas.

And then there's the alternation of drought and flooding, which makes it hard for plants to grow. Getting enough water for crops and humans will become increasingly difficult. The *Assessment*'s authors anticipate that farmers will have to contend with growing conditions drier than anything these states have seen in over 1,000 years.

And don't forget about tornadoes. The central corridor of the United States—Texas, Oklahoma, Kansas, North and South Dakota, Illinois, Iowa, Missouri, Nebraska, Colorado, Ohio, and Minnesota—is already called Tornado Alley. During a single week in May 2019, a record 500 tornadoes touched down in a single month. One, in Kansas, was a mile wide. Another, in Dayton, Ohio, created so much devastation that city workers had to use snowplows to clear away the debris.

So far, nobody's proven a link between climate change and the *number* of tornadoes; in fact, Tornado Alley is getting slightly fewer of them, as the American tornado zone shifts eastward into more populated states (see Chapter 12). Indications are, however, that they're growing more powerful.

As for Texas: Well, the Gulf exposure to hurricanes is the big problem. Hurricane Harvey, in 2017, ties with Katrina as the costliest tropical cyclone in history, costing $125 billion in damage in Houston and surrounding cities. One hundred and seven people died, and 30,000 were flooded out of their homes.

And if it's not one thing, it's another: Texas is second only to California for wildfires. In 2018, a whopping 10,541 wildfires burned 570,000 acres of Texas, killing cattle, displacing communities, and costing millions of dollars. "Model simulations indicate that wildfire risk will increase throughout the region as temperatures rise, particularly in the summer, and the duration of the fire season increases," says the *Assessment*.

Southwest

All of these states are facing drought, water shortages, invasive insects, loss of natural ecosystems, and very hot, very dry weather. It's no coincidence that the highest temperature ever reliably recorded on the planet was in the Southwest: 130°F in August 2020 (in Death Valley). Let's not forget that the Southwest was the first U.S. region to discover airplanes can't take off during the worst heat waves—because the hot air is too thin to provide enough lift to their wings.

And, as elsewhere in the United States, it's getting harder to find fresh water. The population is growing, the droughts are getting worse, and the groundwater is getting depleted. In related news, the Southwest can look forward to more megadroughts (those lasting over a decade).

California, as any Californian can tell you, is special. It has a wildfire problem that's devastating, growing, and, so far, unsolvable. Coastal flooding and eroding continue to ravage the coastline; by 2100, two-thirds of California's beaches will be gone.

The California drought has been dragging on for years, despite new laws that make it illegal to hose off your driveway or wash your car with a hose that has no shutoff handle. By 2100, the annual number of very large fires (5,000 hectares, or 19.3 square miles) is expected to triple.

And climate change aside, don't forget about the big one: the Big One. California's San Andreas Fault tends to shake the state once every 150 years. Well, guess what? It's been about 200 years since the last mega-quake. California's overdue.

Northwest

There's a lot to like about the Pacific Northwest—both now (lush forests, clean air, beautiful views)—and in the future. The temperatures will be pleasant. The growing season will get longer, which will help farmers. And while there's less and less fresh water from melting snow in the mountains, melting *glaciers* mean that water shortages aren't guaranteed.

There is, alas, trouble in paradise.

First, hotter, drier weather and even drought are becoming the norm—a perfect setup for wildfires. But since these states aren't *used* to wildfires, they've been developed dangerously, with thousands of homes tucked right up against forests; often, pine trees and other plants grow right up to the sides of houses. There were 194 fires in western Washington State in summer 2019—almost three times as many fires as there were in an average year in the previous decade—and during the catastrophic 2020 wildfires, one-tenth of Oregon's entire population was under evacuation warnings.

Second, there's the bug problem. As the earth warms, invasive and destructive insect swarms move northward. Take the mountain pine beetle (please). Ordinarily, a cold snap of 14 degrees kills them off. But the new, warmer winters of the Pacific Northwest have invited population explosions of these beetles, which burrow into trees and kill them at staggering speed. They can kill 80% of a pine forest in less than four years. In Washington State, that's 235,000 acres of forest destroyed every year, and they're just getting started.

Alaska

Here's a distressing oddity of global climate change: The poles are warming faster than the rest of the planet. Alaska, in fact, has already warmed twice as much as the rest of the United States. By 2080, Anchorage—Alaska's most populous city—will be 24 degrees hotter than it is now, and 3.5 times rainier.

The sea ice and glaciers are rapidly melting. So is the Alaskan permafrost; as it turns to mud, houses and other buildings tip at crazy angles, roads collapse, and sinkholes open up in fields and yards.

Alaska has always had a mosquito problem, but the new warming multiplies its awfulness. And as in the Pacific Northwest, there's now an invasive-beetle problem that has killed tens of thousands of trees—which, once dead, become fuel for wildfires.

Wildfires? In Alaska? Absolutely: 685 of them in 2019, and the problem is getting worse. Since record keeping began in 1939, only 15 wildfires have burned more than 2 million acres each—but 6 of them have occurred since 2000.

Figure 2-11. In Fairbanks, Alaska, thawing permafrost means soft, muddy ground that was solid for thousands of years.

It's probably no coincidence, furthermore, that 2019 was the first time in recorded history that Anchorage experienced a severe drought.

You might think that the prospects for farming in Alaska would be improving—but Alaska still has 22 hours of sunlight during the summer, and 22 hours of night during the winter, which will always make it hard to grow crops.

Finally, as the permafrost melts, travel by ground—whether by wheel or by foot—gets irregular and soggy. (So far, "sogginess" is not a climate-change metric tracked by the U.S. government.)

Hawaii and Nearby U.S. Islands

All of the ocean-related climate-crisis goodies are hitting Hawaii: warmer water, rising sea levels, coral bleaching, disease outbreaks, coastal flooding, erosion of the beaches and the shoreline. Don't forget monster hurricanes, like Lane and Hector in 2018, which dumped 50 inches of rain in Kauai in 24 hours—a U.S. record.

Droughts are tough on Hawaii, too, since you can't exactly divert water from the next state over. In recent droughts, the Marshall Islands actually had to have drinking water delivered by *ship*.

Then there are the wildfires. Every year, fires burn down half a percent of Hawaii's total land area; that's the same proportion as, or a *greater* proportion than, in any other U.S. state.

The Two Great American Climate Havens

The regional summaries you've just read, no doubt, all sound depressing. But not every place in the United States is trending equally toward misery.

Cast your gaze to the regions that obey the four rules of climate havens:

- Look inland, so that you're spared the brutality of hurricanes and sea-level rise;
- Look north, to escape brutal heat and drought;

- Look for reliable sources of fresh water;
- Look for infrastructure that can bounce back.

Where does that leave you in the United States? In two places:

- **The Pacific Northwest.** The upper-left corner of the country doesn't experience the oppressive humidity of the East Coast. Heat waves are less frequent than in most other areas of the country. The economies of the big cities will thrive as climate migrants arrive, and agriculture is getting a boost from the expanded growing season.

  While the shrinking snowpack is bad news, the Pacific Northwest isn't generally worried about water shortages. It's blessed with two backup water resources: aquifers and melting water from the region's 850 glaciers, which computer simulations assure us will flow until at least 2100.

  And what about sea-level rise? After all, Seattle isn't exactly inland.

  First, much of the land there "rises steeply out of the ocean, so it's a relatively small factor," says Ben Strauss, CEO and chief scientist at Climate Central.

  Second, thanks to a quirk of the tectonic plates, the coasts of Oregon and Washington States have actually *risen* slightly over the last 100 years, more than counteracting the sea-level rise.

  Finally, the Pacific Northwest is a big place. Lots of it is *not* coastal.

  You do have to worry about wildfires and those tree-killing beetle invasions. Otherwise, though, the Pacific Northwest is looking very attractive indeed.

- **The Great Lakes.** If there's a climate-change sweet spot in the United States, this is it: the Northern states that get their water from the Great Lakes. They're Illinois, Indiana, Michigan, Minnesota, New York, Ohio, Pennsylvania, and Wisconsin.

  All of these states have vast, clean, reliable sources of fresh water. They won't endure the blistering heat of the South. And except for a corner of New York, all of them are far from the coasts, so they won't suffer from sea-level rise and hurricanes. Because they're cooler and wetter than the West Coast, they're even less prone to wildfires.

  The Great Lakes themselves have recently experienced massive algae blooms, some of which kill fish, birds, and turtles and poison our drinking water. Fertilizer and household cleaning products running into the watershed help to feed these blooms. The affected states have teamed up to study and fix the problem.

Otherwise, though, many of the cities on the Great Lakes are ideally suited to a life in the new, hotter, drier, rainier world. Read on.

City by City

If you have the option of settling in the Pacific Northwest or the Great Lakes region, you have the satisfaction of knowing that you'll experience the least possible climate-related disruption.

Now all you have to do is pick the right city.

And, yes, it will likely be a *city*. "Generally, we're seeing more people move away from rural areas and into cities," says Rachel Minnery, senior director of resilience at the American Institute of Architects (AIA). "We're prepared for more, and bigger, cities in our future."

Today, 50% of the earth's population lives in cities; by 2050, it'll be 75%.

Why a city, and not a suburb or the countryside? After all, cities are responsible for 75% of greenhouse gas emissions. And on hot days, cities are where you get the "heat island effect"—spots much hotter than the surrounding countryside, thanks to dark, heat-absorbing pavement, few trees, wind-blocking buildings, massive air-conditioner use, and so on.

And yet cities are where the jobs, culture, food, and social life are. And from a climate-chaos perspective, the concentration of people in a city means that it will be the first to get protection before an extreme-weather event—and help afterward. You won't be alone before, during, or after a crisis.

Furthermore, moving into a city is also a blow *against* the climate problem. Relative to suburbs and rural areas, there's more walking, biking, and public transportation in a city. Everything's nearby, so there's less driving. Apartments, because they're all attached, use far less energy than stand-alone homes do. And believe it or not, people who live in cities have fewer children than other people.

Finally, cities have the money and the expertise to establish cutting-edge programs for mitigation (minimizing climate change) and adaptation (adjusting to climate change): white roofs, rooftop gardens, solar panels, smart buildings, LED streetlights, and so on.

American city leaders tend to charge ahead with climate-change preparations even when the federal government drags its feet and squabbles. "Cities and mayors are the great pragmatists," Adam Freed, an executive at the Bloomberg Associates consulting firm, told GlobalCitizen.org. "Cities have recognized that there are no Democratic and Republican ways of collecting the garbage."

Nonclimate Considerations

There are plenty of "best climate-proof cities" articles. Almost all of them, however, make their determination *exclusively* in terms of avoiding climate awfulness. But being happy in a new place involves many other factors.

Consider Anchorage, Alaska, for example, which a *New York Times* article pegged as a good climate-refuge city. It's not, actually; Anchorage's climate plate is piled high with wildfires, invasive species, rising sea levels, and drought. But more important, the recommendation ignores some of the *other* aspects of Alaskan life that might affect your decision.

"Before you high-tail it to Anchorage, you might want to consider the Midnight Sun," wrote one commenter on the story. "You might think that nearly 22 hours of sunlight are great—it is unusual for most Americans—but then you have to deal with about 2–3 hours of daylight in the winter." That dark-all-the-time effect may help explain Alaska's high rates of alcoholism—and the highest suicide rate in the country.

So what should you consider?

- **Cost of living.** The Great Lakes states aren't just the most desirable places to be for climate-stability reasons. For now, at least, they represent unbelievable deals relative to the East and West Coasts.

  The median home price in San Francisco, for example, is $1.8 million; in Duluth, Minnesota, it's $178,600. You can spend the 91% you saved on a top-of-the-line Tesla—or a fleet of them.

- **Infrastructure.** How's the city's walkability? How's the airport? Does it have a good public-transportation system, enough parks, good schools, and modern roads?

  And what about medical care? Hate to break it to you, but the result of almost every climate-change effect is *danger to your health.* All of that extreme weather can starve you, bake you, drown you, infect you, and burn you.

  That's why Tulane climate expert Jesse Keenan considers hospitals to be one of the most important factors in a city's climate resilience. "The number one fear that I have, and that many people in the national-security world share, is that there will be bacteria and viruses—they're already thawing out in the tundra—and that mankind will be exposed to those," he said in 2019 (months before COVID-19 emerged). "It's only a matter of time before we have significant public-health outbreaks that are associated with climate change."

- **Future potential.** It's not just about a city's infrastructure *now*, either. If a city's climate-proofness looks good to you, it probably looks good to other people, too. Will it be able to handle the influx? Does it have room to grow? If a city is already reaching the crowding point, its current cost of living, driving-commute time, and building density on Google Maps will let you know.

  And what about the current condition of the city's infrastructure—its buildings, bridges, roads, water, and power systems? Climate change stresses the existing infrastructure. "Most American cities are already at the end of their useful lives in most of their infrastructural classes," Keenan says. "If you've lived in a city from the 1970s through about now, you've benefited from previous generations paying for infrastructure that you are now using. Now we need to build all-new infrastructure, and somebody needs to pay for that."

- **The people.** Climate change has a tendency to push people out of one spot into another, or to mash people from different backgrounds and regions into the same place. Before you move, you should have some understanding of the kind of citizenry you'll be joining. Is it a diverse population? What's the education level? What's the average age? What's the workforce like? What's the culture like?

On a trip to North Carolina, Keenan learned of a startling local demographic phenomenon: In the last few years, a huge number of people have been moving from the Outer Banks inland to Asheville, to escape the monstrous storms that batter North Carolina's coasts.

"Those people were much higher income and were buying up houses and properties—and actually creating cultural conflict," he says. "That culture that they brought with them, which was a certain measure of economic entitlement, was in conflict with this kind of rustic, hipster mountain Appalachia vibe that Asheville has always prided itself on."

Asheville is experiencing a classic example of what Keenan has called climate gentrification: home prices being driven up by wealthier climate migrants.

- **Work.** Unless you're retiring (or life's been very kind to you), you'll probably need a job when you move. What does the job market look like in your prospective city? Does this place lean toward one particular industry, which could make it vulnerable when times change? Could you fit in there?

  If you work outdoors—construction, agriculture, forestry, firefighter, guide, surveyor, gas and oil, utilities—heat is an especially important factor. Already, heat and humidity have dropped our capacity to work outside by 10% since the industrial revolution; it will drop another 10% by 2050, and another 20% by 2100.

  You do *not* want to be an outdoor laborer in the South in 2050.

- **Crime rate.** Of course, you're presumably moving for the long haul, and factors like the crime rate can change over time. But in the short term, it's a factor to worry about.

  From a purely climate-haven perspective, for example, Detroit looks lovely: cooler temperatures, plenty of fresh water, nothing to worry about from hurricanes or rising sea levels. But from a "Shall we move there?" perspective—well, Detroit has the second-highest rate of violent crime in the United States. (On the bright side—for Detroit—St. Louis recently surpassed it as *the* most violent city.)

That's a lot to ponder; in the end, your choice may just boil down to the work you can find or the family connections you have. In the meantime, though, here are 14 cities that are all relatively protected from heat, wildfire, water shortage, sea-level rise, and hurricanes—and, as a convenient footnote, are also great places to live.

For each city, you can see three graphs:

- **Its current population**, so you can get a feel for its size and metropolitanness.
- **Cost of living**, on a scale where 100% is the U.S. average.
- **Average temperature range**, just to give you an idea. This graph shows you the coldest average monthly low and the highest average monthly high, based on NOAA's measurements. You can expect that these ranges will slide to the right—warmer—in the coming years.

Also for each city, you can read about its climate plan. Of course, a city's reduction in greenhouse emissions—its *mitigation* efforts—may not make life more comfortable for you there, except psychologically. But these climate plans are usually coupled with *adaptation* methods, which demonstrate the city's commitment to making life safer during extreme-weather events.

Finally, this is only a starter list. The world is full of northern, inland, water-rich cities, many of them close to the towns on this list. For every Cleveland, there's a Cincinnati and Columbus; for every Buffalo, there's a Syracuse, Rochester, and Albany. Don't be offended if your city isn't on this list. It's intended to prime your planning engine, to teach you how to assess cities in climate terms—not to imply that these are the only options on the planet.

Madison, Wisconsin

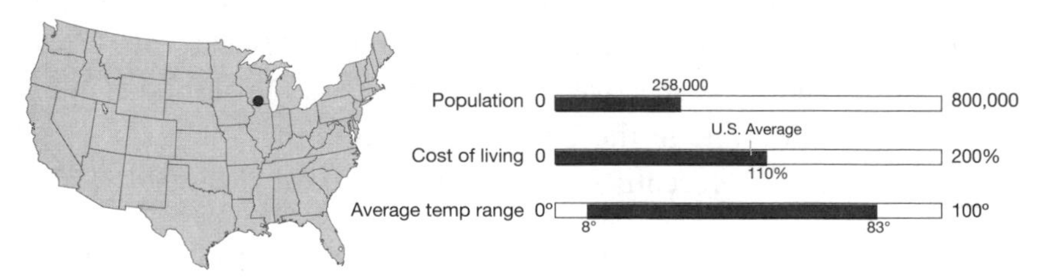

Madison, Wisconsin, is all about the lakes. It has *five* of them, and the city sits on an isthmus between two of them. The views and the recreation opportunities are amazing, of course, but from a future-climate perspective, there's plenty of fresh water for generations.

Like many of the other cities on this list, Madison is rich in outdoorsy opportunities. It's got 200 miles of hiking and biking trails, 11 beaches, and 260 parks (5,000 acres).

Madison is a state capital and a college town, teeming with millennials (27% of the population), young families, and retirees, all of whom enjoy plenty of housing and plenty of jobs. The population skews strongly liberal and progressive, in what's generally a Republican-voting state.

With a diverse population of 258,000 and growing, Madison is a midsize city that's nonetheless as navigable and friendly as a much smaller town. Zoning laws require that no buildings can be taller than the capitol, so Madison never became a concrete jungle of skyscrapers. In related news, Madison is home to nine Frank Lloyd Wright buildings.

Figure 2-12. Madison, Wisconsin, sits in the climate-haven sweet spot of the United States.

Its central spot on the U.S. map means that every city is less than a three-hour flight away. Be aware, though, that there's only one non-stop a day to New York, San Francisco, and Los Angeles. Milwaukee is a 90-minute drive away; Chicago, two hours.

In today's Madison summers, the highs are in the upper 70s, and the lakes invite you to canoe, kayak, sail, swim, and ride paddleboards. The winters can get down to single digits, turning the city into a haven for ice fishing, skating, hockey, cross-country skiing, and downhill skiing a short drive away. Every Saturday, Capitol Square hosts the biggest producer-only farmers market in the country, teeming with locally grown food, artist booths, and live music.

No wonder Madison has been ranked number one in surveys of the fittest cities (Fitbit.com), nicest cities (CheatSheet.com), most caring cities (WalletHub.com), bike-friendliest cities (Livability.com), best cities to raise a family (Zumper.com), best for renters (SmartAsset.com), top cities in tech-job growth (Coldwell Banker), and best quality of life (BusinessFacilities.com).

Climate plan: By 2030, Madison aims to power city operations with 100% renewable energy, become net-zero for carbon emissions, and halve its energy consumption. But its plan also incorporates targets for clean air, clean water, water conservation, landfill reduction, local food use, and natural-habitat preservation.

Already, the bus lines are extensive, but lots of people bike or walk, which the downtown streets make easy. The bike-share program comes complete with bike racks, lockers, and 100 miles of paths. Madison has more bikes than cars.

In 2014, NerdWallet.com named Madison the greenest city in the United States, thanks to its enormous number of parks, clean air, and green-transportation options.

Ann Arbor, Michigan

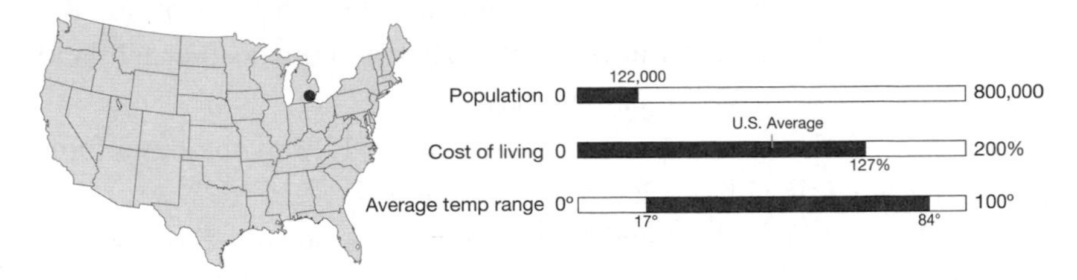

This lovely city is deeply interwoven with the huge University of Michigan, so its residents skew young, brainy, and sports crazy. The school system is excellent, the crime low, the hospital top ranked. There are plenty of parks, museums, restaurants, and cultural events.

Maybe those are the reasons that Ann Arbor has been calculated to be the 38th happiest city in America, and the best place to raise kids.

The worst climate-related vulnerabilities include "heat waves, snowstorms, high winds, and downpours that lead to flooding," according to the city.

Climate plan: The Ann Arbor Climate Action Plan (2012) aims to reduce carbon dioxide emissions 25% by 2025, and 90% by 2050 (relative to 2000). The plan involves 84 steps, including offering a $500 rebate for installing electric-car charging at your home, discounted electricity for car charging (and free charging at city lots), installing rain gardens, funding retrofit jobs in apartments, creating a bike-sharing program, expanding bus routes, and creating a citywide weather-emergency calling system.

Minneapolis, Minnesota

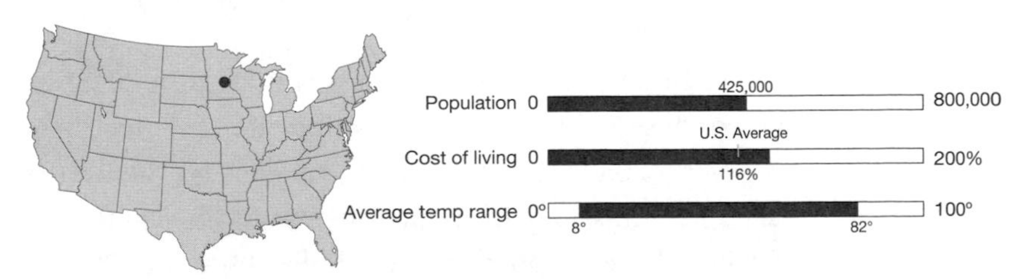

Legend has it that the name *Minneapolis* comes from *mni*, meaning "water" in Dakota Sioux, and *polis*, meaning "city" in Greek—and, boy, is that fitting. Fresh water is Minneapolis's dominant feature. It's built around two major rivers, 22 big lakes, and thousands of nearby smaller ones, plus countless creeks and waterfalls. That, and the northern, inland locale, means good news for climate refugees.

The city's 197 parks constitute an incredible 17% of the city's footprint. They go well with the 200 miles of bike paths.

The Grand Rounds National Scenic Byway is a 50-mile loop around the city that incorporates some of those parks, along with historic sites, art installations, outdoor recreation, shopping, restaurants, the Mississippi riverfront, and the Chain of Lakes (seven of them). You can drive the loop, walk it, or bike it. No wonder Minneapolis has ranked as the number one, two, or three fittest American city every year since 2008, and the sixth-best city for runners—not to mention the sixth-best place to live in America.

Minneapolis has a thriving music scene and has for years (see Prince, Bob Dylan). Because it's a major city (with sister-city Saint Paul, it makes up the "Twin Cities"), it offers all the cultural amenities you'd expect: theater, art, dance, opera, museums, big-league sports teams, a wealth of cuisine options, four hospitals, lots of corporate headquarters, a diverse population, and a major airport 16 minutes from the city center.

Minnesota winters are long and cold, for now. That's why residents welcome the Minneapolis Skyway: nine miles of heated, enclosed pedestrian bridges that connect 80 downtown city blocks, including restaurants and shops.

The cost of living is nowhere near New York's or San Francisco's, but it's 16% higher than the national average. Housing is pricey, too—the most expensive of any major inland U.S. city. Fortunately, jobs are plentiful, and pay is higher than average.

Climate plan: The Minneapolis climate plan was drafted in 2013. Its goals: reducing greenhouse emissions, communitywide, 15% by 2015, 30% by 2025, and 80% by 2050 (compared with 2006 levels). So far, the city is ahead of schedule. Emissions from city operations alone—that is, not including residents—have dropped an incredible 44% since 2008.

The plan also calls for creating 30 more miles of bike lanes and increasing the percentage of bike commuters to 15%, generating 10% of city power from renewable sources, recycling half of all waste, composting 15% of it, improving energy efficiency in commercial buildings by 20% (and residential by 15%)—all by 2025.

Already, Minneapolis has been ranked the second or eleventh greenest city in America, depending on the magazine you ask. The city offers a bike-share system with 170 docking stations, a bus system, two light-rail lines (a third under construction), and a commuter rail line.

Burlington, Vermont

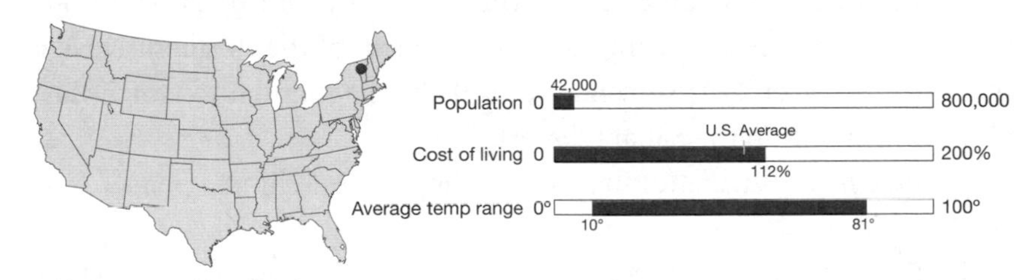

Vermont is said to come from the French words for "green mountains," and that's what you get in spring and summer. "The summers are exquisite," writes a resident on Reddit. "Heaven on Earth. Gorgeous blue skies filled with multi-layered clouds, bees buzzing . . . Ahhh."

In the fall, *Rougemont* would be more like it, because the state offers some of the most spectacular fall foliage in the country. Winters have traditionally been cold and snowy, but they've warmed by 4°F since 1970 and are on track to warm another 13 degrees by later this century. That's bad news for Vermont's famous ski resorts, which are already experiencing shorter ski seasons.

Burlington is on Lake Champlain, an enormous freshwater lake spanning 490 square miles and offering insurance against drought and water shortages. The city has the feel of a seaside town—without being susceptible to the dangers of an actual sea, such as hurricanes and sea-level rise.

The city screams "healthy," both because of the highly rated University of Vermont Medical Center and because the *people* of Vermont tend to be healthy—low obesity rates, high insured rates, few smokers—and live in an area with clean air and fantastic recreational opportunities. It's a liberal, highly educated population, with more than twice the national average of bachelor's and graduate degrees, and maybe as a result, the school system is one of the nation's best.

Vermont also has the lowest crime rate in the country and, maybe in related news, the smallest population—even smaller than Alaska's. Burlington, the biggest city, has only 50,000 residents.

Cheese, maple syrup, covered bridges, and ice cream are points of pride; Ben & Jerry's is headquartered here. The downsides: Housing costs are high, well-paying jobs are scarce, and parts of the state are struggling with poverty and opioids.

"I'm feeling very good being in Vermont. We're sitting at about 300 feet of elevation on our farm," says Alex Wilson of the Resilient Design Institute. "At the end of this century, it's likely to be similar to North Carolina, not similar to the Sahara Desert."

Climate plan: Burlington has one of the weakest climate plans of the cities on this list.

Its 2014 plan calls for a return to 2010 emissions by 2016 and, by 2025, a reduction of 20% *below* 2010 levels. Of course, 2016 has come and gone, and Burlington didn't meet its first target. The plan also aims to reduce driving by 10% per person by 2025, to "work toward" a goal of 100% renewable power, and "reduce" the amount of refuse sent to landfills.

Bangor, Maine

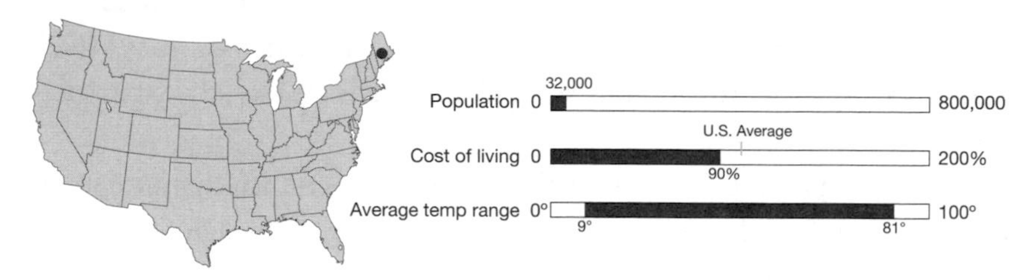

In 2017, the EPA put together a Climate Resilience Screening Index: a color-coded map of U.S. counties showing their climate-change resilience. It incorporates not just vulnerability to extreme weather, but also how prepared and capable each county is for absorbing the impact and bouncing back. When the EPA tallied all of this, guess which state is the most climate-proof of all?

It's Maine.

It's cold, it's Northern, it's got plenty of fresh water—and once you're inland a bit, you don't get the coastal flooding and storm damage that, say, Portland, Maine, does. And that, no doubt, is why Bangor winds up on so many expert lists.

Bangor ("bang-ore," not "banger") is the epitome of a Maine town: chilly in winter, quiet, calm, safe, and absolutely beautiful. It offers two major hospitals, great schools, friendly and supportive residents, and a compact downtown. Everything is bikeable, although bike lanes aren't common. One resident calls it "a series of neighborhoods connected by city parks."

Don't expect the kind of nightlife and excitement you'd find in a bigger city; the adjective *sleepy* comes up a lot. The potholes are prodigious, and there's a homeless population. But overall, Bangor is a sweet, small town that offers the amenities of a much bigger city.

Climate plan: Bangor doesn't have its own plan, but Maine itself does—adopted in 2019. It sets mandatory targets for greenhouse gas reductions of 45% by 2030 and 80% by 2050. It also sets a 2030 target for 50% of all electricity in the state to come from renewable sources—100% by 2050.

Figure 2-13. Bangor's Penobscot River flows through downtown, offering trails, picnic tables, benches, an open space for festivals, and a pavilion for summer concerts.

Denver, Colorado

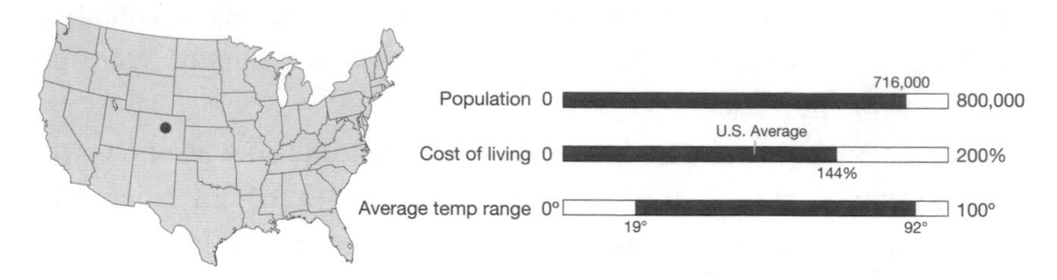

Even before the climate crisis came along, Denver had earned a reputation as the ultimate haven city, at least in popular fiction. Denver and nearby Boulder are the journey's end for the main characters in novels, movies, and TV shows like *The Stand*, *The Man in the High Castle*, *Fear and Loathing in Las Vegas*, *Red Dawn*, and the video games *Horizon Zero Down* and *World War Z*.

The elevation (a mile high) keeps the temperatures cool, the wildfires far away, the humidity low, and the scenery spectacular: 140 miles of mountain views in every direction.

Denver is a young, liberal town—two-thirds of its residents are under 44—and they enjoy an outdoorsy life. Music festivals, climbing, hiking, and biking are big here, and Denver's got over 200 parks covering 5,000 acres, including the Rocky Mountain National Park.

Denver doesn't *actually* get 300 days of sunshine a year, as every article claims—more like 245 clear or partly cloudy days—but the weather *is* very nice.

At least *U.S. News & World Report* thinks so. It declared Denver to be the second-best place to live in America.

On the downside, Denver's population is growing fast. It's up 20% since 2010, which is driving house prices up and creating awful traffic in and out of the city. Note, too, that shrinking mountain snow accumulations mean much shorter ski seasons; by 2080, the EPA estimates, some Colorado resorts' seasons will be 80% shorter than they are now.

Climate plan: Denver's 80x50 Climate Plan, which focuses on buildings, transportation, and power generation, is aggressive and sweeping. The goals:

- ♦ By 2025: reduce greenhouse gas emissions by 30%, and use 100% renewable electricity in city buildings
- ♦ By 2030: increase electric-car registrations 30%
- ♦ By 2035: require all new buildings to be net-zero (produces no carbon emissions—for example, by consuming only renewable energy created on the site)
- ♦ By 2050: Reduce greenhouse gas emissions by 80%

As part of this plan, the city is designating more road lanes for buses only and spending heavily on sidewalks, bikeways, and bike lanes. New light-rail lines make getting around easier and take cars off the road.

Boulder, Colorado

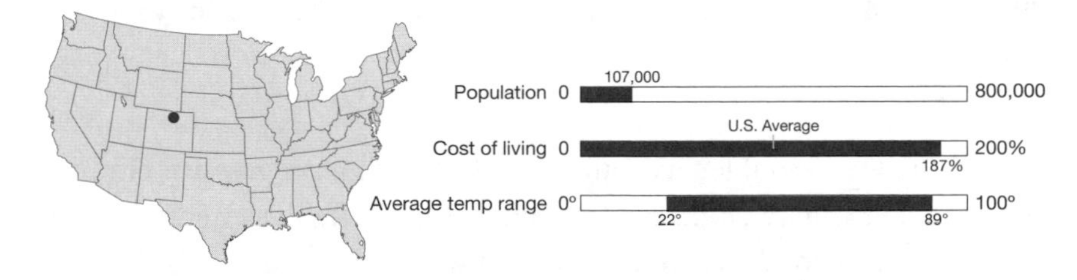

In Boulder, you can have the same climate advantages, views, and healthy outdoor lifestyle that you'd get in Denver. Here, too, the place is teeming with biking and hiking paths, young people, and liberal political views.

The difference, of course, is that Boulder is much smaller: about 107,000 residents instead of 716,000. The schools are better, but the cultural, sports, and travel options are more limited. Boulder may strike you as more earthy-crunchy, New Agey; if Denver is like San Francisco, then Boulder is its Berkeley.

The cost of living is fairly high in Boulder, so the population is heavy on middle-agers or retirees who have the wealth to live here. The University of Colorado Boulder and other schools ensure a steady supply of college kids. Overall, the population is "extremely white and not very diverse," notes one critic, and its city council is noted for its "weirdo ordinances and nanny-state sensibilities."

Figure 2-14. Boulder's mountains form a gorgeous backdrop in every season.

Climate plan: Boulder, as you'd expect of such a progressive city, has been a leader in city climate action since 2002, when it voted to match the Kyoto Protocol. Its current goal: "We will power our city with 100% renewable electricity by 2030, and reduce Boulder's greenhouse gas emissions by at least 80% below 2005 levels by 2050."

By 2031, all new and remodeled buildings must be net-zero, with airtight building envelopes and smart-energy systems. Since 2010, the city has succeeded reducing its own municipal-building emissions by over 40%, saving $700,000 in energy costs and 8,000 metric tons of greenhouses gases every year.

The town even proposed a new tax to support its climate efforts in 2006, which the citizens voted into law—and then voted to extend in 2012.

Chicago, Illinois

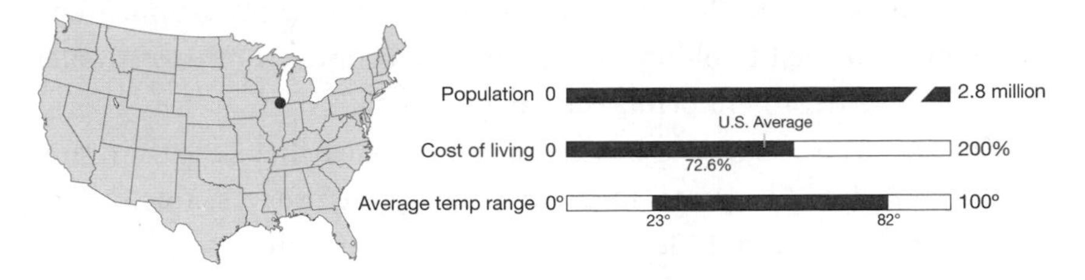

There are so many reasons to love this town. The skyline is gorgeous, the architecture stunning, the art, museums, and food spectacular. It's central, so it's a short flight to anywhere. Lake Michigan is *right there*, its beaches snuggled right up against the downtown. The people have the hustle of big-city-ites, with the friendliness of Midwesterners. And if you're a sports fan, you'll be in heaven.

The transportation system is among the best in the country: *two* big airports, public rail, 165 miles of bike trails, and more bike parking than any other U.S. city.

Chicago still has four distinct seasons, unlikely to blur as they've started to do on the East Coast. The famously cold winters are getting milder.

The only flies in the climate ointment: Heavier rains have given Chicago some flooding headaches, and heat waves occasionally reach this far north. On the other hand, you're free from hurricanes, sea-level rise, wildfires, and the worst of the mosquito-and-tick problem.

Climate plan: Chicago's climate program is called Sustain Chicago. Its goals for 2025: reducing greenhouse emissions by 26% (relative to 2005)—to align with the Paris Agreement—and 100% of public buildings powered by clean energy. The city is well on its way, having cut its emissions by 11% in the decade ending 2015, even as employment and population grew.

Chicago has 5,000 miles of sewer pipes, which *also* collect stormwater runoff. The problem with that design is that during floods, residents get sewage backing up into their toilets and basements; untreated sewage pouring into the rivers and the lake; and beach closings. On average, it rains often enough for that kind of overflow once a *week*.

The city is now midway through a massive construction project designed to solve that problem, which is slated for completion in 2029: the Tunnel and Reservoir Plan (TARP). It's a series of deep rock tunnels and reservoirs, capable of holding 20 billion gallons of overflow stormwater until the city's treatment plants can catch up with it.

Chicago is also investing heavily in technologies that keep rainwater from flowing into the sewers in the first place: green spaces, green roofs, and green alleys (permeable pavement that absorbs rainwater), planting trees, and offering rebates for rain barrels and tree plantings to citizens.

The city's programs offer huge incentives to residents to join in the crusade: free energy inspections, efficiency products, and installation; rebates for efficient appliances; offering discounted solar installations; and so on.

Meanwhile, the city is replacing 1,500 buses with low-emission models, rewriting zoning laws to promote development near public-transit stations, expanding its 248 miles of bike lanes to a massive 645-mile citywide bike-lane network; rejiggering stoplight timings to minimize car idling; installing car-charging stations; and adding 180 acres of new parks.

Cleveland, Ohio

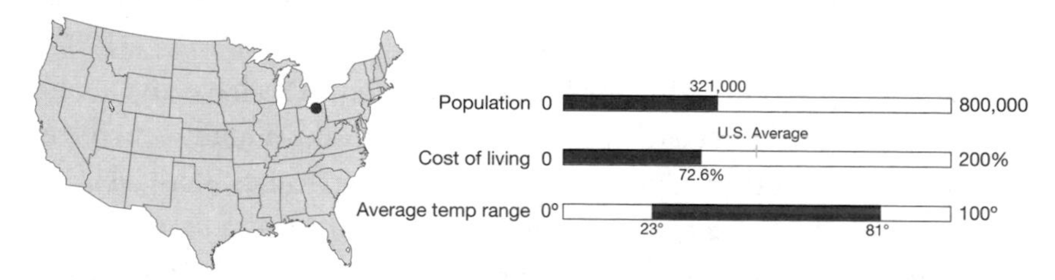

Cleveland got its share of bullying that was well deserved—in the seventies and eighties. You know, the river catching on fire, the mayor's hair catching on fire, becoming the first U.S. city to declare bankruptcy, the world's largest balloon release shutting down the airport with popped latex.

A lot has changed since. Living in Cleveland means access to Lake Erie, complete with sandy beaches and cool little islands to visit; a world-class art museum, symphony, and zoo; major-league baseball, football, and basketball teams (the Cavs!); an airport only 15 minutes from downtown; 18 city parks, 8 of them on the lakefront, featuring 300 miles of trails through 23,000 acres; some gorgeous suburbs (including—full disclosure—the one I grew up in); and an unbelievably low cost of living. One dollar in New York City, for example, buys you $2.71 of stuff in Cleveland.

Oh—and if you get sick, Cleveland is where you want to be. It's home to both the massive, internationally famous Cleveland Clinic network (18 hospitals, 4,200 doctors, 6,000 beds) and University Hospitals (1,900 doctors serving 1,000 beds on a 35-acre downtown campus).

"Cleveland is going to have its heyday soon, because their weather conditions and their climate will become much more appealing in the future," says Rachel Minnery of the American Institute of Architects.

Climate plan: The city has made great progress on its 2013 climate plan (updated in 2018). Its goal is to reduce greenhouse gas emissions from city operations 40% by 2030, and 80% by 2050 (from 2010). It has installed over 70 miles of bike infrastructure; started a bike-share program; funded 50 neighborhood climate projects; scaled up stormwater drainage systems; installed more solar panels; updated building codes to greener standards; and so on.

The Rapid Transit light-rail line is cheap and covers an impressive amount of the city and its suburbs. The downtown trolleys are free.

Boise, Idaho

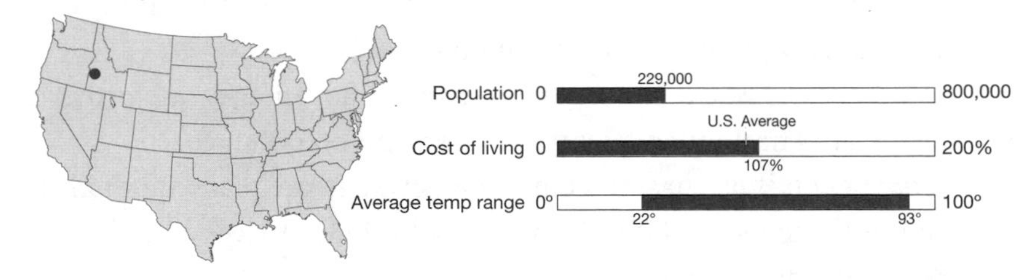

For pure, rugged American natural beauty, it's hard to beat Idaho's capital city. The place is made of rivers, mountains, canyons, lakes, and deserts, all of which invite outdoor recreation by the backpackful. The 180 miles of trails in the foothills invite anyone who likes to hike, bike, or run, and the Shoshone Falls is taller than Niagara Falls.

The cost of living is low, especially compared to other Northwest cities like Portland and Seattle. The airport is only ten minutes from downtown. The population is diverse, ranging from retirees who might, before the country began heating up, have settled in Arizona, to families drawn

Figure 2-15. Boise offers no big-league sports, but there are opera and ballet companies, an orchestra, an art museum, and mountains.

by the low crime rate and inexpensive housing. Unlike most of the city populations on this list, this one reliably votes Republican.

The city's climate assessment points out that it expects to suffer a major drought once every three or four years—four times as often as in the past—and by 2050, wildfires will be four times as likely. Idaho temperatures can be extreme, too, and Boise is the 13th-fastest-warming city in America. (Dry places tend to warm up faster.)

But otherwise, Boise has some key climate advantages: a Northern site that escapes heat waves, plenty of water, and no ocean coasts.

Climate plan: Boise has no comprehensive climate plan, although it has committed to using 100% clean energy by 2035. There's a bike-share program and a modest bus system.

Portland, Oregon

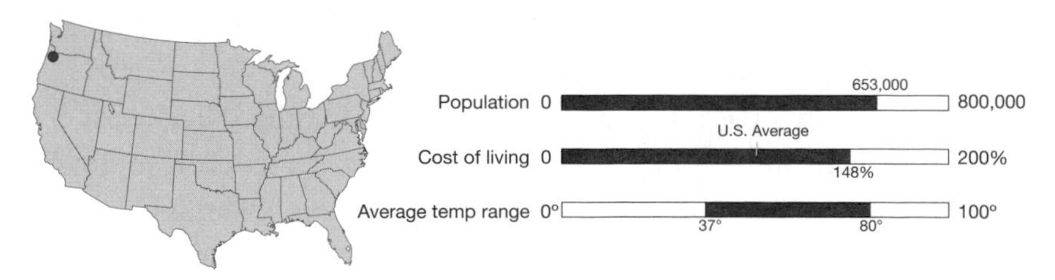

The beauty of Portland is, of course, the *beauty* of Portland. You've got the Columbia and Willamette Rivers, mountains in the distance, lush vegetation everywhere you look, and 11,700 acres of parks, with 70 miles of trails for running, biking, and hiking.

Is it really rainy as much as they say? Cloudy, for sure; the rest depends on whether you define all-day misting as rain. The winters are mild and getting milder; it rarely snows. In summer, the sun doesn't set until 9:00 p.m.

Be prepared for the people of Portland. They're dog friendly; *friendly* friendly (in Portland, people sit in the *front* seat in their Ubers); progressive politically; heavy on the hipsters; and very outdoorsy. Many raise chickens or keep bees, when they're not participating in the annual World Naked Bike Ride. Plastic bags are outlawed, everyone recycles, and Portland doesn't add fluoride to its water.

The food scene is alive with everything that's locally sourced, organic, free-range, artisanal, and craft brewed.

Portland's social problems have a lot in common with San Francisco's: The tech boom is pushing rents up, driving less affluent people out, making traffic a nightmare, and creating a huge population of homeless people. The school system is only average. Portland isn't a very diverse city, either.

Figure 2-16. Because it's a fairly big city, Portland offers every activity, cultural event, museum, zoo, and civic facility you could dream of. Nearby: mountain hiking, skiing, and wine country.

The Northern latitude will keep Portland cool as the earth warms, its elevation will keep it away from the rising seas, and its weather keeps it safe from droughts. Wildfires (and the sooty air that results) are the biggest major climate-change dangers that affect Portland—the devastating 2020 wildfires turned the air brown and gave Portland the worst city air quality on earth. Flooding and landslides aren't unknown, though, and the big West Coast earthquake is always lying in wait.

Climate plan: Portland created the nation's first climate plan, way back in 1993. Its most recent update was in 2015—a detailed, broad, and ambitious program. The goals: a 40% reduction in carbon emissions by 2030 (from 1990) and an 80% reduction by 2050.

Portland will have no problem getting there. By 2016, city operations had switched entirely to renewable energy, and already the city's greenhouse gas emissions have declined 41% since 1990, despite a population that's grown by a third. It helps that the city has a great light-rail and bus system, streetcars, one of the biggest bike-sharing programs in the country, and the highest percentage of bike commuters in the country.

By 2030, the city has a few more targets:

- Net-zero emissions for all new buildings; 50% of all building energy from renewable sources; reduction of all older buildings' energy use by 25%
- Reduce per capita driving by 30% (from 2008 levels), and increase average fuel efficiency to 40 mpg
- 90% of all waste recovered or recycled
- Leafy-tree canopies to cover 33% of the city
- Reduce city carbon emissions by 53% (over 2007 levels)

Not only has the city divested itself of investments in petroleum companies, it has enacted zoning rules that make it harder for oil and gas companies to move their products through the city!

Spokane, Washington

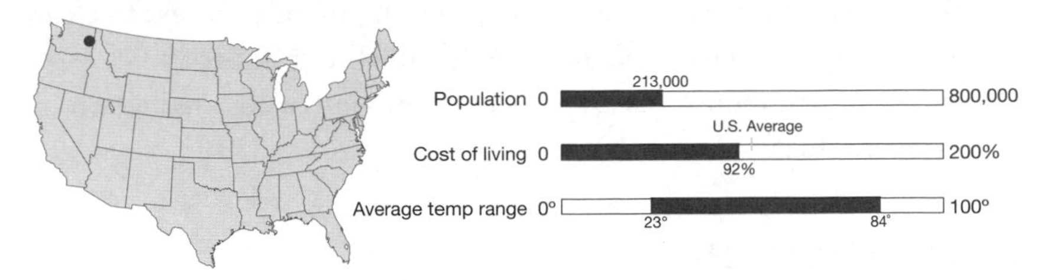

Spokane's marketing slogan is "Near Nature, Near Perfect." From a climate perspective, that's not a bad characterization. The city is surrounded by mountains and majestic forests, bisected by a river, blessed by lakes, and dotted by plentiful green spaces.

And it's got the essentials of climate-proofness: It's northern, inland, and freshwater rich. Wildfire smoke often fills the sky in the summer, and the city is concerned about the water levels of the Spokane Valley Aquifer, which supplies all of the city's fresh water. But you'll escape hurricanes, storm surges, and most heat waves and tornadoes.

Riverfront Park is a highlight, featuring 40 miles of walking, running, and bike trails; the Skate Ribbon (a long, loopy oval for ice-skating in winter, roller-skating/scootering in summer); gondola rides; reverse bungee jumping; pedal-cart rentals; a hand-carved 1909 carousel (54 horses, 1 giraffe, 1 tiger, 2 Chinese dragon chairs, and a brass ring); and the roaring Spokane Falls. But really, you can go just about anywhere to find hiking, camping, skiing, running, biking, boating, and snowmobiling.

Washington doesn't have a state income tax. That, and Spokane's lower-than-average cost of living, makes your dollars go a lot further. You'll find a mix of political views—liberal influence leaking from Seattle and Portland, conservative leaking from Idaho to the east—and a cultural Native American influence. Other pluses are good schools, a well-run airport, and good hospitals.

Some residents describe Spokane's downtown as run-down, although the influx of young people, driven here by sky-high costs in Seattle and Portland, means that there's also a lot of renewal and development. Citizens often mention the homeless population, too.

Climate plan: With the support of a huge range of environmental, religious, and business organizations, Spokane has committed to switching to 100% renewable energy sources by 2030. That effort includes the creation of an inexpensive solar-panel program, broader access to clean buses, and engaging power plants to switch to cleaner sources.

Spokane's greenhouse gas target is more modest: a 30% reduction by 2030 (relative to 2005).

Duluth, Minnesota

It was the *New York Times* headline heard round the world: a top Harvard climate expert declared Duluth, Minnesota, the ideal place to settle in the new climate era.

"By 2080, even under relatively high concentrations of carbon dioxide emissions, Duluth's climate is expected to shift to something like that of Toledo, Ohio, with summer highs maxing out in the mid-80s Fahrenheit," the story went. Duluth is also cool and wet enough to minimize wildfires, far enough from the coasts not to worry about sea-level rise, and close enough to Lake Superior—the biggest freshwater lake on earth—that it will never have to worry about water.

That expert was Jesse Keenan (now at Tulane), and he'd been commissioned to create a marketing campaign by the University of Minnesota Duluth.

Even when he's not being paid, however, he says that Duluth *is* the very model of a promising post-climate-change metropolis. "Duluth has all of the metrics that one would want, in terms of architecture, urban design, public administration, tax base, culture, education, access to health care, fresh water, fairly diverse economic base, investments in R and D—all the standard metrics of urban economy," he says now. "If I had to live up there in my waning years, that would not be a bad thing."

There's a lot of beauty in Duluth: rivers, falls, forests, state parks, and, of course, the majesty of Lake Superior. But for now, Duluth winters are still bitterly cold and long; a wind shift from the lake can drop the temperature 30 degrees in half an hour. There are no pro sports teams, no light-rail system, little diversity. There's not much of either hustle or bustle, either. "The pace of life here is much, much slower," says a resident on Reddit. ("Boring," says another.)

Climate plan: Duluth's Energy Plan, as it's called, aims to lower greenhouse gases from city operations 80% by 2050 (from 2008 levels). The program has so far spent $4 million to install LED street and park lighting and begun improving energy efficiency in the city's buildings. Future plans may include installing solar panels on the city's reservoir covers, replacing some city cars with electric vehicles (EVs), installing EV solar-powered charging stations, and overhauling the town's steam plant.

Buffalo, New York

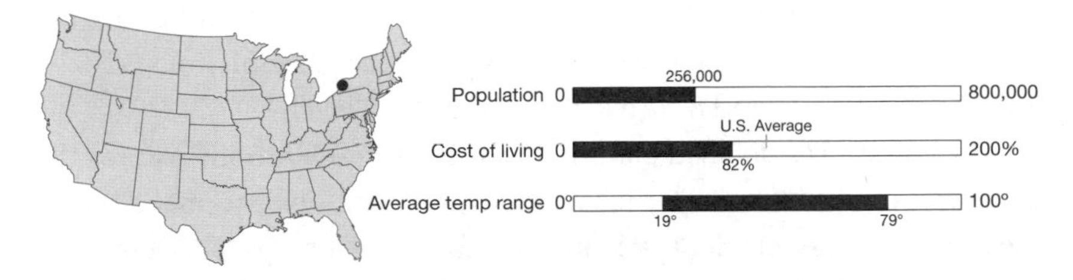

Buffalo's nickname is the City of Good Neighbors, a reference to the community spirit—and willingness of people to help dig you out of the snow. They're also big drinkers, especially on Saint Patrick's Day and Buffalo Bills game days.

The city's prime spot on Lake Erie means that fresh water will never be a problem, that wildfire risk is minimal, and that water sports, fishing, and lovely parks are all part of the portfolio. Niagara Falls is 25 minutes away, and Canada is the next town over.

The city boasts seven cool Frank Lloyd Wright buildings, city parks designed by Frederick Law Olmsted (who designed New York's Central Park), a well-developed waterfront, and top-notch medical facilities. Yet its cost of living is well below the national average.

Figure 2-17. Buffalo enjoys all four seasons ("Sometimes we can experience all four in one day," writes OnlyinYourState.com). Winters are traditionally cold and snowy, but, as everywhere else, they're getting milder.

Jesse Keenan expects that Buffalo will soon be the recipient of a lot of investment. Land is cheap there, with a lot of room for a lot more people, and the labor force is well educated and skilled. And, yes, the chicken wings are really good.

Climate plan: Erie County (Buffalo is the county seat) abides by the Paris Agreement. Its goal is to reduce greenhouse gas emissions 32% by 2020, 50% by 2025, and 80% by 2040 (from 2005 levels).

The county has detailed plans both for its own operations and for the citizenry: discounts on solar installations, tax incentives to make buildings more energy efficient, a loan fund for clean-energy projects, programs to install electric-car chargers in parking lots, and so on.

Moving Day

No city on earth will escape the changing climate. Life will get warmer, more expensive, and more dangerous no matter where you go.

Knowing the four factors that make a great climate haven, though, give you two benefits. First, it helps you assess where *you* might want to live. Second, it makes you aware of where everybody *else* might soon want to live.

Both pieces of insight, in their way, will help you prepare.

Chapter 3

How to Build

MOST PEOPLE DON'T CREATE MANY THINGS THAT WILL LONG outlive them, other than babies. But buildings will. They stand, on average, for 80 years. Same as babies, come to think of it.

Having read Chapter 2, you have some ideas about *where* to live. But if moving to a new place isn't in the cards—actually, even if it is—you should consider how to make changes to an existing home or think about a new one in your community. You should incorporate climate-crisis resilience the next time you build, buy, renovate, or expand your home.

The idea is to construct the building in such a way that you can *shelter in place.* That's emergency-speak for staying put until the weather danger has passed, rather than leaving the house and running to an emergency shelter. Imagine the peace that would come from knowing that the safest place to be during a superstorm or wildfire is *your own house.*

You may find, as you plan, that your contractors aren't up-to-date on the latest resilient building practices. You may also find that they're not especially enthusiastic about adapting these newer methods. "Nobody likes change," says Rachel Minnery, senior director of resilience, adaptation, and disaster assistance at the American Institute of Architects (AIA). "Builders and contractors like doing things the way they've always done them, and they don't want to do them differently, especially if they might be perceived as adding additional cost."

Nevertheless, persist. For reasons of safety, comfort, convenience, generational stability, community, and, yes, saving money, fortifying your home is worth the attempt.

The Costs of Resilient Building

You might guess that a resilient building costs more than a traditional one. You'd be right.

How much more depends on which features you incorporate, and *that* depends on where the building will stand. If you're settling down in the Arizona desert, you need features that protect you from heat waves and water shortages. If you're constructing a home in Miami, New Orleans, or the shelter islands of New York, your worries are flood, wind, power failures, and hurricanes. (Also: Are you nuts?)

But here's the thing: You're investing this money, not burning it.

At the request of Congress, the National Institute of Building Sciences did a little math. Suppose you spend $10,000 making your building climate-change resilient to the standards of the 2018 International Residential Code (IRC) and International Building Code (IBC). Those are thick books of minimum construction requirements that serve as the basis for most states' building codes. They cover things like materials, ventilation, exit points, room sizes, foundation strength, electrical safety, drainage systems, and so on, all designed to protect your safety and health.

Once you take into account the property damage you're likely to experience, what you'd spend while you're temporarily displaced, the interruption in your work, the discounts you'll get on your insurance, and other factors, your investment will ultimately save you $110,000.

That's right: an 11-to-1 payout.

What you're creating, moreover, is a lasting asset. You're creating a valuable inheritance that you can pass down to your kids, if you have them. You're building a house that will survive not only your lifetime, but theirs.

Building Codes

You're not the first person in your community who's ever considered the importance of designing homes to stand up to nature. That, after all, is what building codes are for.

But in poor or rural areas, they may not be meaningful—or even present. Some communities don't have the funding to develop and enforce building codes.

"Probably one of my biggest aha moments was when I was volunteering after Hurricane Katrina," says Minnery. "The majority of structures I was evaluating for safety in Mississippi weren't built to *any* building code and didn't have *any* building inspectors involved. It's shocking."

Even where building codes are in place, they're guaranteed to be out-of-date with the latest environmental data, thanks to a long, slow adoption process. After 2012's Hurricane Sandy caused $70 billion of damage in 24 states, for example, FEMA recommended 47 improvements to aging building codes—and many of them *still* have not yet been adopted.

In other words, when it comes to preparing for climate chaos, building codes are a bare minimum, nothing more. You'll ride out global weirding much more successfully if you follow the advice in this chapter.

Energy Backup

People lob all kinds of statistics at you about how many more floods, hurricanes, heat waves, tornadoes, and droughts we have these days. But according to Alex Wilson, founder of the Resilient Design Institute, what they all have in common is power failures. "No matter what the disaster," he says, "a typical direct or indirect impact is *loss of power*, and there are ways we can design or retrofit buildings to keep occupants much safer if they lose power."

Since the 1980s, we're recording power outages *ten times* as often. We owe some of that rise to improved reporting standards implemented in 2000. Even so, we're blacking out twice as often now as we were in 2003. In fact, the United States now enjoys the distinction of having the most power outages of any developed country.

Take storms, for example. The worst part isn't your house getting pounded by rain and wind directly; it's the aftermath. Your home may sit there, dark and wet, for days or weeks. Without power, you can't dry out the interior. Mold and rot may doom the entire structure.

And don't forget the refrigerator full of food that goes bad, the air

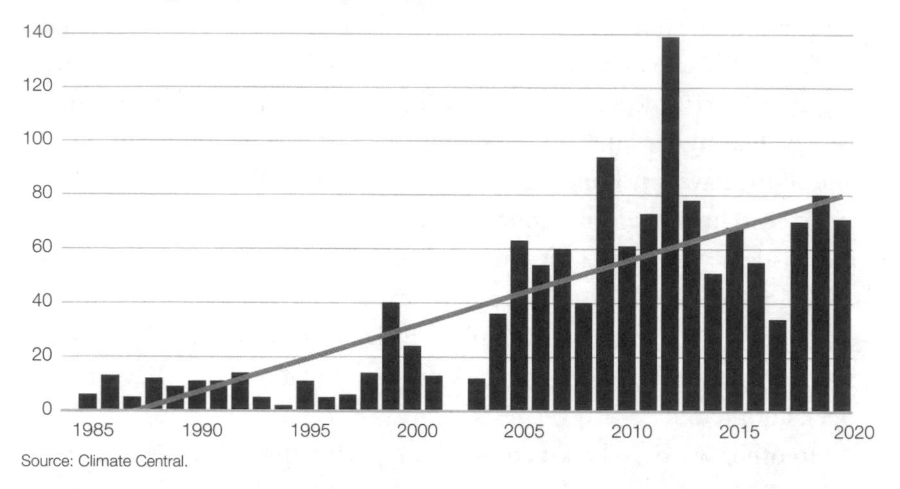

Figure 3-1. Crazy weather is giving us many more blackouts, defined as an event that leaves at least 50,000 people without power for an hour or more.

conditioners you can't run, the phones you can't charge, the pipes that freeze. No power may mean no water, too, if your home relies on electricity for water pumping. And what if your survival depends on a medical machine?

The price of a backup power system depends primarily on how much power you need. Every generator is made to provide a certain number of watts of electricity, which determines how many electrical machines it can power simultaneously. The question is, then, what do you want to be able to run during a blackout? Add up your wattages like this:

- Phone charger: 5 watts
- Lights: 6 watts (per LED bulb), 60 watts (incandescent)
- Laptop charger: 65 watts
- Fridge: 700 watts
- Gas or propane furnace: 800 watts
- Portable heater: 1,500 watts

- Dishwasher: 1,500 watts
- Electric range: 2,000 watts

As you do the math, keep in mind that air conditioners, refrigerators, and sump pumps draw a power *spike* at the moment they kick on. For this reason, most generators identify not only how may watts they can supply, but how big a surge they can handle.

Here's the universe of generators:

- **Portable generators.** These things look like tractor engines on wheels. They're not pretty, clean, or quiet; you have to run them outdoors, away from the house, with extension cords. But when the power's out, you can fill one with gasoline, start it up, then plug things into its standard outlets.

  A 3,600-watt generator from Home Depot might weigh 300 pounds and cost $450. A small one lets you charge your phone, keep your fridge running, and have a couple of lights on; the bigger ones can keep your AC going, too.

  If you have a bigger generator—say, 5,000 watts or more—an electrician will be happy to hook it up to your circuit-breaker panel (using a *transfer switch*), so that you can also turn on overhead lights, central air-conditioning, well pumps, and so on. That job will cost around $700.

Figure 3-2. A portable generator (left) and an inverter.

The primary headache with these things is the fuel. Generators gulp a gallon of gasoline every couple of hours. Fuel can be hard to come by during a local disaster; you have to buy, transport, and store the gas safely ahead of time. If your power is out for days or weeks, fueling these generators can get to be a real nightmare.

- **Inverter generators.** These machines are smaller, lighter, and much quieter than traditional portables. They create DC power and *invert* it to the AC power your appliances need. As a result of that design, they can vary their output according to demand—and as a result of *that* design, they use a lot less fuel and make a lot less noise. They're also more compact, looking more like a suitcase than a naked car engine. You can string them together for more juice, and most let you monitor them with a smartphone app.

  As you probably guessed, there's a catch: They're much more expensive. A 3,500-watt inverter model might cost $840.

- **Standby generators.** These are larger, permanently installed generators that sit next to the house and kick on *automatically* when the power goes out, usually within 15 seconds. Your lights pop right back on.

  A generator big enough to power your whole house, with installation, costs $4,000 to $10,000, depending on the capacity. The big

Figure 3-3. A standby generator can power the entire house, or big parts of it, automatically. No gas-pouring required.

ones can handle 20,000 watts, which, unless you're in Jeff Bezos's tax bracket, should do quite nicely for your home.

On the other hand, you may not need the entire house to spring back to life. In an emergency, powering the kitchen and a couple of bedrooms may be plenty. In that case, you can get away with a smaller, less expensive generator.

If your house is already connected to a natural-gas line, you're in luck; you can hook the generator up to it and never have to think about manually fueling the generator. If you have a diesel or propane generator, keeping the thing refilled is your responsibility. Its tank probably holds enough fuel to last for a few days.

- **Battery backup systems.** Those fossil fuel generators are noisy, they burn petroleum, they emit fumes and smoke, and they require periodic testing and maintenance. There's a newer option, though: home batteries.

  These, too, are available in a range of sizes. On the very, very low end: lunch-box-size, single-outlet, $130 "power stations" with a 167-watt-hour capacity, suitable for charging phones and plugging in a lamp or two. (Total watts provided: 100.)

  On the high end: the Tesla Powerwall, a refrigerator-size, wall-mounted, whole-house beauty that costs $5,500 plus about $1,000 for

Figure 3-4. A Tesla Powerwall is a battery backup for your entire house. No noise, no fumes, no fuel to pour in.

installation. It can supply, for example, 1,000 watts for 14 hours on a single charge.

Battery-power backups are totally silent. They don't burn any fuel on-site and don't produce any exhaust. If you have solar panels, the power is free. It's also available to your home at night, because the battery has stored the solar power during the day. If the Powerwall is recharged from solar, you also get a tax credit from the government.

And during the times of the year when you're *not* experiencing a blackout, you can use your battery to time-shift your power use from the grid. That is, you can buy your power during the day at daytime rates, which are lower in many cities, then *use* the power in the evening, when everybody in the world is home, using juice, and driving up the price of electricity.

Solar Panels

Solar energy, if you hadn't heard, is heating up. Since 2009, the price of installing solar panels on an average house has dropped from $51,000 to about $14,000, after tax credits. That, in part, explains why about 3 million Americans have now installed solar panels, and about half of the rest are thinking about it.

Solar panels are a fantastic idea for climate preppers like you because you'll still have power when the grid goes down. As a bonus, they give your house free, clean, unlimited energy, reducing or eliminating your electric bill. They also add an average of $15,000 to your home's resale value.

And you can deduct 26% of the installation costs off your taxes. This is not some lame tax *deduction*, which lowers the amount of *income* you use to calculate your taxes. This is a tax *credit*, which subtracts real money from your final tax payment. A tax credit is much juicier.

That windfall, known as the Residential Renewable Energy Tax Credit, will drop to 22% on your 2021 taxes, and to 10% for 2022. Then it goes away—unless Congress renews it, which it's done several times before.

Depending on the size of your house and the quality of the system, solar panels may cost $10,000 to $30,000 to buy and install. But *you* don't have to pay all that. Various financial aids are available:

- **Solar-panel loans.** They're available from banks, credit unions, power utilities, solar-panel makers, and sometimes towns and cities. Even state and federal governments offer loans, through programs like FHA Title I or NY-Sun. (Solar panel installers may offer loans, too, but their rates and fees are usually pretty high.)
- **Rebates and tax incentives.** Thanks to these government programs, you can get back nearly half of your outlay. There's that 26% tax credit, of course, but your local government, state government, and utility company may offer further rebates. To find out what's available in your neck of the woods, enter your zip code at the Database of State Incentives for Renewables & Efficiency, www.dsireusa.org.
- **Sell your power.** Once the panels are in place, you can *sell* any excess power you generate to the local power company. They're required by law to generate some of their electricity from renewable sources, so they're delighted to take some of that cheap sun-powered goodness off your hands.

  Every 1,000 kilowatt-hours of power generated by your roof earns you something called a Solar Renewable Energy Certificate. The power company buys that from you for $70 to $200, depending on where you live (and supply and demand). If you sell 10 of those annually, that's $2,000 a year in your bank account, courtesy of Mr. Blue Sky.

Even without that selling-back business, solar panels on your roof or in your yard save you a lot of money on electricity. They pay for themselves in as little as six years.

There's another approach, though, that requires no financial outlay at all: you can *rent* solar panels. Companies like Sunrun and SunCity install solar panels on your roof at no charge. They do all the paperwork, permitting, repairs, and maintenance. But they, not you, get the rebates and incentives, and *they* get to sell the Solar Renewable Energy Certificates that your roof generates.

You don't get free electricity, either. You pay a small amount to the panel-rental company, courtesy of a contract called a PPA (power purchase agreement). Fortunately, it's less than you'd pay to the local power company, and it's also a fixed rate.

This online calculator lets you see how much it'll cost to install solar panels, and how much you'll profit from them after 20 years: www.energysage.com/solar/calculator/.

And this one shows you whether buying or renting your panels will save you the most money: https://sunmetrix.com/solar-buy-vs-lease-calculator/.

Not everyone can do solar. Your roof or backyard may not have enough room or may not get enough sunlight. Maybe you rent your home. Maybe you can't afford the investment.

But if solar *is* an option, you should pursue it. In all of climate preparation, there are few better win-wins.

Backup Water and Drought Protection

Besides power, *water* is the most important backup you can have. As the earth warms, more rain and flooding wash fertilizer, pesticides, and other pollution into our waterways, shut down water-treatment plants, and contaminate water lines.

During a wildfire, treatment plants may once again shut down; later, ash, nitrates, and organic carbon wash down burned-out slopes into the water supply.

And then there's drought. In the era of climate change, there's no state in the United States, nor any country on earth, where drought is unknown.

For millions of people, "emergency water" means whatever bottles of water are left in the grocery store when a disaster is approaching. That tactic is better than nothing, and the FDA says that unopened store-bought water can remain drinkable indefinitely. If this is your backup plan, buy the water *now.*

But gallon for gallon, bottled water is massively expensive and environmentally disastrous. And storing enough to last you and your family a decent number of days means filling (and weighing down!) an entire closet with bottles of water.

Emergency Tanks

You can buy big plastic backup water tanks ($100 for 55-gallon capacity, for example) to fill in readiness for a drought or disruption. Here again, those are fine *if* you know that an emergency is coming. They're basically more sanitary versions of filling your bathtub before a hurricane.

You don't want to leave water stagnating in them for more than a few months, however. In other words, those tanks won't help you if the water disruption is sudden and unanticipated.

For that reason, you can also consider an *inline* water tank, which looks something like a water heater. In calm times, your household water flows through this tank constantly, always circulating, unnoticed and forgotten. But when there's a disruption or contamination, you flip

Figure 3-5. An inline backup water tank (right) gives you an always-ready, always-fresh supply of clean drinking water, no matter what's going on outside your home.

a lever to cut the tank off from the world. The last 80 or 120 gallons of clean, drinkable water is now trapped inside, ready to see you through the disruption. These systems, called things like LifeTank, ReadyMade-Water, and Constant Water, cost around $1,300.

If you're a do-it-yourselfer, you could probably rig up something similar yourself.

Rain Catchment

Rain catchment sounds like an awkwardly made-up term, like *delicious eatage* or *teenage annoyification*.

But at least it conveys its meaning: preparing for drought by storing whatever rain falls the *rest* of the year. In most cases, they're basically barrels under your gutters.

These systems are fantastically effective and dirt cheap. The City of Chicago, for example, offers its residents a 50% rebate on any rain barrels they buy. ("Why take from the lake what you can get from the sky?") In Australia and Germany, people in rural areas use these systems as their *main* sources of water.

Figure 3-6. A typical rain-catchment setup is an excellent source of fresh water—for free.

For many people, a single 55-gallon barrel from the hardware store, or even plastic trash can, is plenty. Park it behind your house, direct your gutter's downspout into it, and boom: 55 gallons of emergency water, ready at all times.

Then again, the roof water from a single mild shower can pour 500 gallons of water into that 55-gallon barrel. You'll have to consider where the overflow will go—onto the lawn, for example, or to thirsty plants and trees. You also have to make sure it doesn't run toward your foundation or your next-door neighbor's property.

Some people install a *series* of barrels connected by pipes, or a huge cistern designed for rain catchment. Keep in mind that water weighs a *ton*, literally; 240 gallons of water—about four barrels' worth—weigh one ton. You may have to build a foundation for your catchment system.

There's some maintenance involved. You have to:

- Seal the barrels to prevent mosquitoes from breeding inside during the warm season.
- Drain the barrels if freezing is a risk.
- Clean out leaves and stuff that flow in from the roof (or install a screen to keep that out). Most people also install a *first-flush diverter*, a $50 mechanical device that diverts the first gush of water after a dry spell—the flow most likely to contain crud from the roof—away from the barrel.

Otherwise, though, these systems are pretty great:

- Because the water doesn't have chlorine or other additives, it's better for your plants.
- The water you use for watering the lawn and garden is now free, so your water bill goes down.
- These systems cut down on storm runoff, which is *another* hazard for cities in the climate-chaos era.

Above all, though, these systems give you backup water during droughts and disruptions. And if the city lays down water-use restrictions, which happens often in places like California and Colorado, you can laugh and go right on watering your lawn.

Most people who have these systems use the water for watering their plants, washing the car, and bathing the pets. If you've invested in a bigger cistern to hold the rainwater, you can connect it up to your house's plumbing for use in washing machines, toilets, and even showers.

Preparing the Water for Drinking

What you can't do with the water from your barrels is *drink it*, at least not directly. It's got pollen, dust, bird poop, and bugs in it. It's run through gutters, down downspouts, and across asphalt-shingle roofs, all of which contain chemicals that you're better off not ingesting.

You *can* drink the water from your rain-collection barrels, cook with it, and brush your teeth with it, as long as you clean it first, as described in Chapter 8.

But if you'd like to drink and cook with collected rainwater all the time, you can install a reverse-osmosis purification system under the sink (about $300). Reverse osmosis removes all kinds of nastiness from your water, including impurities like chlorine, nitrates, nitrites, PCB, lead, radium, and arsenic, as well as bacteria and viruses. (Also fluoride, so brush your teeth religiously!)

Note, however, that this process is very water wasteful, requiring between two and five gallons of water to produce *one* gallon for drinking.

Whole-house purification systems, which include reverse osmosis and UV virus-killing, cost $7,000 and up. For most people, they don't make sense—why purify water you're going to use for watering the lawn, flushing the toilet, and washing clothes?

Probably better to *filter* your whole-house water system, and *purify* only your drinking water.

Flood Resistance

If you're even reading this paragraph, the first question is, Why are you planning to build in a flood zone at all? Every year, the seas will get higher, the superstorms will deliver more water per second, and the storm surges will get higher. Every year in a flood zone, property values will go down.

Bottom line: It's probably easier to find a higher-elevation lot for your house than spend a lot of money trying to make it floodproof in the face of an uncertain future. (To find out how likely you are to get flooded as climate change intensifies, use the interactive visual SurgingSeas maps at ss2.climatecentral.org. Or look up the FEMA flood maps for your neighborhood at msc.fema.gov. For instructions on interpreting these maps, see pages 200–203.)

But if you must build your home in harm's eventual way, or if the idea is to flood-prep the house you already own, here are some of the ways to proceed.

Building Materials

Choose the materials for your house wisely. Some can recover from a flood, some can't.

The materials that FEMA describes as flood resistant must be capable of withstanding 72 hours of contact with water without requiring any repair (beyond cleaning). Here's the basic rundown:

- **Acceptable materials:** Waterproof stuff. You know, brick, stone, steel, concrete, two-by-four solid-wood beams, plastic lumber, epoxy, glass blocks, sprayed polyurethane foam insulation, latex and oil paint, most rubber, vinyl, silicone, ceramic and porcelain tile (set with mortar, not adhesive), and so on.
- **Unacceptable materials:** Plaster, carpeting, fiberboard, particle board, wood (including standard plywood, trim, doors, floors), paper-faced gypsum board, asphalt tile, any kind of insulation that's not sprayed, any floor tile glued down, wallpaper (paper, cloth, vinyl, plastic), laminate floor coverings. All of that stuff rots in water.

  Hope you're not a fan of natural wood.

Raise the Electrical and Mechanical Systems

In a flood, wet wiring and electrical panels mean no power, sometimes for a long time. Wet heating and AC systems mean you'll shiver or swelter.

And that's what's wrong with the usual location of your electrical panels and HVAC machines (heating, ventilation, air-conditioning): in the basement.

If you're building new, great: Set your house's electrical and mechanical systems higher than the highest possible flood level—the second floor, perhaps.

If you're retrofitting, an electrician will be delighted to move your electrics up higher—in exchange for a couple thousand dollars.

Keep Out the Sewage

When all hell is breaking loose, and everything in your house is soaking wet, the last thing you need is raw sewage spurting *out* of your toilets, sinks, and floor drains, flowing in the wrong direction.

But that's what happens in floods. For decades, many cities were designed with sewer pipes that carried *both* sewage and all the rainwater from our gutters, downspouts, and storm drains. In a heavy storm, those municipal systems become overloaded, pushed beyond capacity by millions of gallons of stormwater. That combined sewage-plus-stormwater brew has nowhere to go but back up.

These days, some cities are decoupling the storm and sewage lines, but it's a slow, expensive process.

It's not just gross. Raw human sewage is teeming with microbes that can bring you such delightful gifts as cholera, typhoid fever, and meningitis. It's also not great for your carpets, walls, and electrical systems. You have to hire a professional cleanup company to take care of that.

Backflow, as it's called, also contains everything else society intended to flush away: fertilizer, pesticides, chlorine from people's pools, soap and shampoo from their showers, detergents from their dishwashers and washing machines, and so on.

The solution: a backflow valve. A plumber installs it on the pipe that carries clean water into your house (between $600 and $1,400 installed). The one-way valve closes if any liquid starts to flow the wrong direction.

While you're recovering from that sticker shock, you can buy plugs from Home Depot to seal your basement floor drains and standpipes. Those things cost about 10 bucks. Do it now; if you wait until a megastorm is approaching, they'll be sold out.

Wet Floodproofing

When a flood happens, massive amounts of water press inward on the foundation walls of your house. It's called hydrostatic pressure, and it's sometimes enough to make those walls collapse, which will cost you dearly in repair and cleanup.

Therefore, FEMA recommends, and some local building codes require, something called wet floodproofing, which means installing flood vents (also called floodgates or flood ports). They're wide rectangular holes cut in your foundation, so that floodwater can slosh in and out without building up any pressure. To calculate how many of these vents you need,

Figure 3-7. If you have a crawl space, your flood vents usually have louvers or bug screens; if the flood vent gives direct access to your first floor of living space, it's usually solid and engineered to open only when it feels water pressure.

FEMA proposes a formula: one square inch of vent opening for every square foot of your house. If the crawl space totals 2,000 square feet, then you need 2,000 square inches of flood-vent opening (about 16 vents).

Your bank account recommends flood vents, too, because they have an immediate lowering effect on your flood-insurance cost.

Anchor the Tanks

Post-flood inspections often reveal another preventable hazard: fuel tanks beside the house or in the basement. In a flood, they become torpedoes. Water lifts them up and waves smash them into your walls, or other people's. If they break open, a spark can actually blow them up.

Propane tanks, for example, release highly flammable clouds of gas as they float along. "Propane is a nasty little devil that always finds an ignition source somewhere," Fire Chief Neil Svetanics told the *St. Louis Post-Dispatch*. "You have a flash, and a huge fireball powerful enough to blow roofs off buildings and walls down in the immediate area."

Getting a thousand-gallon tank bolted down to a concrete pad, usually with metal straps, costs $300 to $500. But in a pinch, if a big storm is coming, you can also just make sure it's full (it's less likely to float), then rope or chain it to something immovable.

Raise the House (New Construction)

Here's the ultimate flood protection for new construction: Start the ground floor above the projected floodwater level. The idea is that floodwaters, when they come, can surge away all they want, underneath the house, while you remain cozy in your elevated living spaces. (Local building codes may *require* this kind of elevation, in fact.) If elevated houses worked for our ancestors in the Neolithic Age, 12,000 years ago, they can work for you.

Elevated houses like this also remain cooler in hot weather because breezes can pass beneath your floor. They keep critters out and provide a lot of underhouse storage.

The trick is to elevate the house higher than you expect—maybe even higher than the local zoning rules dictate (the FPE, or Flood Protection

Figure 3-8. This house is ready for a storm surge. Once you've built the front stairs to the front door and planted some bushes, nobody on the street will know by looking that there's anything weird about your house.

Elevation level)—because, as sea levels rise and superstorms deliver more water per second, storm surges are getting higher.

Just ask the residents of Oak Island, North Carolina. They built their houses on seven-foot pilings—and then Hurricane Florence came along with a *nine*-foot storm surge.

Usually, you can situate the "ground floor" just a few feet off the ground.

Raise an Existing House

As flooding becomes a more imminent threat, more people are choosing to raise their *existing* homes—a complex, labor-intensive procedure that's performed by specialized contractors. And it's a luxury: Costs range from $50,000 to $300,000, depending on a house's size and shape, condition, foundation type, number of floors, landscaping, and the urgency of the procedure.

On the other hand, elevating an existing building is usually far less pricey than knocking it down and building a new one. You'll get a break on your flood insurance, too. The higher you raise the house, the bigger the insurance discount, especially if the house is in a high flood-risk area. FEMA calculates that raising a flood-prone average single-story house could save you more than $90,000 in insurance costs over ten years. That should make you feel a lot better about the cost of the raising.

Before you begin, get multiple quotes. Choose a contractor that offers at least $200,000 in liability insurance. Check references.

The steps for lifting a house depend on what it's made of and what kind of foundation it has. But in general, the contractor will get building permits, turn off and disconnect utilities and plumbing, cut evenly spaced holes in the foundation, then insert massive steel I beams through the holes—first one orientation, then a second, perpendicular set.

At that point, the crew uses enormous hydraulic jacks to raise the building. In a time-lapse video, you can see the house rising like magic.

Finally, they build the foundation up to the new bottom level of the house, using posts, columns, piles, concrete blocks, or poured concrete.

Figure 3-9. A modern-day house-raising, partway finished. If you had a basement, well, it's gone now. The contractors will have to fill it in—another offering to the flood gods.

Then they lower the house gently onto the new foundation. They reconnect all the utility wires and pipes, repair plaster cracks and the other inevitable minor damage, install a stairway up to the new entrance height, relandscape . . . and voilà.

Just be aware that the new foundation supports are intended to provide only *vertical* support to your house, to get it up away from flooding. They may actually make your house *less* resilient to *horizontal* forces—from wind, mudslides, earthquakes, and, yes, floodwaters. An engineer may recommend steel-bar reinforcing or some other solution for horizontal stabilization.

If You Can't Elevate

Not all houses are elevatable. If your home is a row house, town house, or brownstone—that is, attached to other buildings—you'll have a hard time raising it.

Fortunately, there are all kinds of other ways to floodproof your house. They all cost money, but not nearly as much as elevating your house. Meanwhile, they all lower your flood-insurance rates.

- **Fill in the basement.** The idea here is that you sacrifice everything below ground level by filling it in. You'll have to move your water heater, furnace, and other utilities to a higher floor.

  If there's any basement left *above* ground level, you're supposed to install flow-through flood vents, described earlier, so that floodwaters can slosh on through without caving in your walls.

  Of course, filling in your basement means sacrificing a lot of living space to Mother Nature. That's why people often add a floor to their homes as part of the surgery.

- **Sacrifice the first floor.** If you have no basement, you can perform the same kind of surgery by sacrificing the entire "ground floor" to floodwaters. Once again, this means installing flood vents and relocating the utilities.

  In dry times, you can use that underhouse area for parking and storage. Your living space "begins" on what's technically the second floor.

- **Raise only the floor.** This sneaky option is available only if you have high ceilings at the moment—ten feet, for example. The idea is that you can elevate the *floor* two feet, keeping it up out of floodwaters' way. The resulting rooms will have only eight-foot ceilings, but you may have saved a lot of money and effort.

  Once again, you'll need to install flood vents and move your utility gear.

FEMA notes that if you're creative, you can also build a reinforced-concrete flood wall, or an earthen levee (a berm, or ridge), all the way around your building, with a sump-pump backup for water that seeps in. These are complicated—you have to worry about what effect you're having on the *neighbors* when there's flooding, as well as how you're going to drive in and out—and at the moment, they don't produce a lowering of your national flood-insurance rates.

Paying for Modifications

These kinds of changes to your house are expensive. They're speculative, too: You're making enormous changes to your home on the premise that there *might* be a flood.

But your government would like you to consider these changes anyway, because ultimately they'll save *it* money. Therefore, FEMA has created the Flood Mitigation Assistance Grant Program, which gives you money to pay for floodproofing your house. (*Mitigation*, in this sense, means "structural changes that will protect your house from flooding.")

To apply for one of these grants, you submit an application to your State Hazard Mitigation Office. Google it to find its website—*oregon state hazard mitigation office*, for example. That office then forwards all the grant applications it has received to FEMA, which then decides who gets the money.

The Flood Mitigation Assistance piggy bank isn't bottomless; they have about $160 million to give out each year. Therefore, they give grants only to people who have National Flood Insurance policies (see Chapter 6), and even then, not everyone gets approved. So make it good.

Wind Resistance

In the nor'east, you get nor'easters. In Tornado Alley, you get tornadoes. In the South, you get hurricanes. In the Plains states, you get high winds barreling across the fields, unimpeded for miles.

High winds are an increasing risk in every part of the country, and wind causes more damage than any other weather element. Its damage accounts for 25% of all insurance claims, and winds sometimes destroy houses completely.

Wind can exert three destructive pushes on your house: shearing forces (which make the building tilt and sway), lateral forces (which push the building sideways off its foundation), and uplift forces (which rip off your roof).

Preparation begins with awareness of the usual direction of the wind in your region: from the south in Oklahoma and Texas, from the west in California, and so on. The risk varies depending on how many tall, wind-blocking structures are around your house, such as trees, hills, and other buildings. They all reduce your house's vulnerability by acting like wind-breaking shields (assuming that they're far enough away that they won't *fall* on the house). A farmhouse in the middle of flat, unobstructed land is a sitting duck.

Roof Design

The weakest part of your house are the joint points: where walls meet the foundation, where the first and second floors connect, and where the roof meets the walls. That's why, when the Weather Channel shows hurricane devastation on loop, roofs blowing off make frequent appearances.

Losing your roof is often the first domino to fall in creating even more devastating damage. The framing of your house, and its contents, are now vulnerable to the storm.

Plenty of studies, and too many case histories, have established the right way to build wind-resistant roofs:

- **Roof geometry.** Hip roofs (slopes on all four sides) perform better in high winds than the more common, less expensive gable roofs (see Figure 3-10). Hip roofs deflect storm winds, no matter which direction they come from, up and over your house. Gable roofs, on the other hand, leave a triangular wall patch exposed at each end of the house—and their roof overhangs act like beer-can pop-tops for your house.

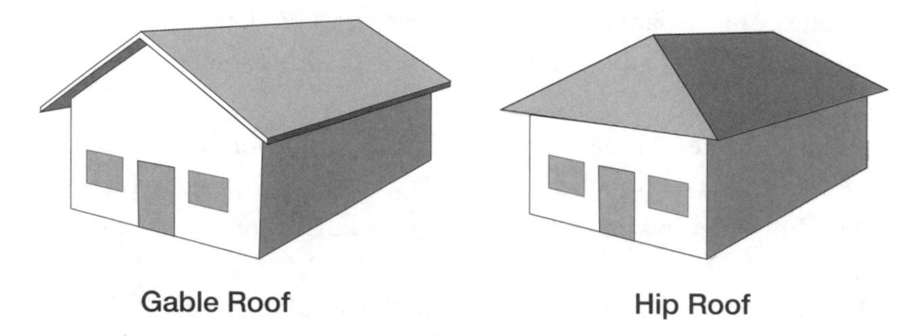

Figure 3-10. A hip roof channels wind up and over your house much better than a gable roof.

If you need any further convincing, consider this: A hip roof gets you an instant discount on the wind-coverage part of your insurance policy.

- **Steeper angles.** If your roof is flat or flattish, it can create an airfoil effect. High-speed wind passing over it exerts a tremendous lifting force, just as it does on an airplane wing. If that force builds high enough, your roof becomes Flight 165 to the next county.

  By the time the roof's angle is sloped at 30 or 40 degrees, that effect goes away.

- **Deeper soffits.** Soffits are the eaves—the underside of the roof's overhang. The wider they are, the more surface area they provide to massive updrafts of wind and water, which is bad.

  Unfortunately, shallower eaves mean less shade. "Sometimes you have to look at these different strategies and balance conflicting goals," says Alex Wilson of the Resilient Design Institute.

You'll also impress your contractor or roofer by asking what materials they intend to use:

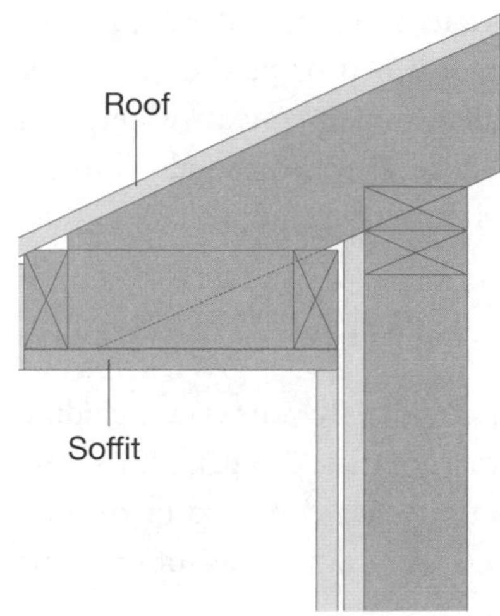

Figure 3-11. Your soffits, shown here as a cross section of your house, make a difference during storms.

- **Connectors.** Nails and staples can pull away; bolts, screws, and metal fasteners do much better.
- **Tiles** come in varying degrees of strength, rated Class 1 to Class 4 (the toughest). Metal roofing may be the best of all, because each sheet runs continuously from the peak of the roof to the eaves, giving no entry points for wind or water.

Figure 3-12. Metal roofing is a blessing in the new climate.

- **Waterproofing.** Roofers have various tricks to keep water out of your house. They include flashing (sheets of metal or plastic that cover seams or cracks in your roof and walls), underlayment (waterproof sheets underneath the shingles), taped decking seams (adds protection even if the shingles and underlayment have blown off).

The Debris Problem

In a windstorm, all kinds of airborne missiles can fly your way, including gravel, trash cans, fencing, grills, branches, roof tiles, and patio furniture. For decades, the classic prehurricane activity has consisted of rushing around nailing sheets of plywood over your windows. But as superstorms become more frequent, that task gets old.

Instead, consider installing hurricane shutters; they're available in either manual or electric models. You can get rolling shutters that store themselves above the window, accordion-fold versions that slide shut horizontally on tracks, colonial models that look like traditional window shutters, and so on.

There are two problems with hurricane shutters. First, they're no help if you're not home when the storm hits; you have to be present to close them. Second, they make the house incredibly dark in a power outage.

A more expensive alternative is "impact" windows (and doors). They have heavy-duty frames, two panes of glass bonded together with a plastic layer, and silicone glazing (sealing putty) that helps keep the glass in the frame. If some errant chaise longue does fly into this glass, the window may still crack, but it doesn't shatter.

Impact windows also block sound, insulate your house better, make life miserable for burglars, and usually land you a discount on your insurance.

Your trash cans, grills, and patio furniture are a danger in a hurricane, too—both to your neighbors and to your own house. It's worth looking into how they can be anchored, bolted, chained, or cabled down.

You have to worry about trees falling, too. According to FEMA, 75% of all home damage caused by falling trees is predictable and preventable. It usually doesn't take a tree doctor to spot a tree that's too close to your house or is dead or has shallow roots. Take care of those trees *now*, before they try to get into your house in a storm.

House Design

For high-wind and hurricane protection, here are a few other design principles:

- **Connect everything to everything.** A wind-resistant home is one designed with a continuous *load path*, meaning that every piece of the house is securely connected to every other one, from the roof to the foundation. This principle ensures that the force from a huge gust gets transferred from one part of the house to the next, all the way to the ground.

  For example, consider hurricane straps: strips of metal, often nailed to the rafters and to the roof, connecting the two. For maximum protection, each part of the house should be strapped to the next.

- **Low, single-story homes** fare much better in high winds than taller ones.

- **Garage doors** can be a weak spot in your building's "envelope." The problem is that once they've blown in, the wind can blast directly into your garage and home, building up pressure on the *inside* and exploding outward.

  You can buy pressure-rated garage doors designed just for this problem, or you can reinforce existing ones with braces across the back of the doors and strengthened wheel tracks (see page 395).

- **Design a safe room.** The United States is the tornado capital of the world: 1,200 twisters a year. Chapter 12 covers tornado protection in delicious detail; for the purposes of your building or renovating a home, consider designing a "safe room"—a reinforced room where you and your family are protected from the violence and debris that might strike the rest of your house.

- **Chimneys** are tall, relatively fragile structures, and sitting ducks when the big storms come. Unless they were originally built with rebar (reinforcing steel) or otherwise braced, they can collapse or snap right off.

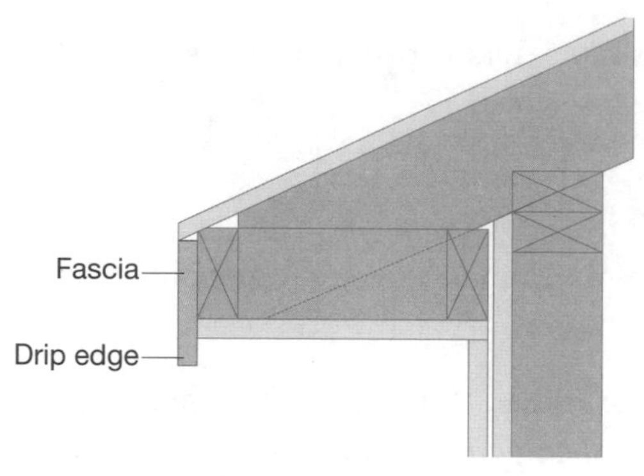

Figure 3-13. This fascia has a drip edge.

- **Extend the fascia below the soffit.** The fascia ("FASHa") is the *vertical* face of your roof overhang. Usually, it forms a tidy L where it connects to your soffit. But if you make the fascia tall enough that it extends *past* the soffit, you create a drip edge. That way, storm winds can't drive the rain horizontally across the surface of the soffit and into the eaves of your house.

The Fortified Home Standard

Since local building codes are usually old and insufficient, you might wonder why somebody doesn't sit down with scientists, architects, and engineers to come up with a *realistic* building code for the climate-chaos era.

Actually, somebody has—several somebodies, in fact. There are several alternative rating systems, but the one backed by the most data and testing is called Fortified Home.

It's brought to you by the IBHS—the Insurance Institute for Business & Home Safety, a nonprofit research group funded by insurance companies. The IBHS has come up with a complete set of climate-considered building and engineering codes called Fortified Home (https://fortified home.org). It's an upgraded, modernized set of building standards that specifies every detail, right down to the spacing of screws and the thickness of panels.

The program offers three levels of certification:

- **Fortified Roof** specifies improvements to the roof deck, sheathing, and ventilation system. "Tremendous amounts of damage can be avoided by focusing on the roof," says Alexandra Cary, market devel-

Figure 3-14. At the Insurance Institute's torture-chamber campus in South Carolina, engineers subject full-scale structures to 130-mph winds, with rain, hail, and other elements.

opment manager for the Fortified program. "That's a very low-cost option because you're already up there doing work; it's little as fifteen hundred dollars over the usual cost."

- **Fortified Silver** adds strengthened doors, windows, gable ends, soffits, and attachments such as porches.
- **Fortified Gold** also specifies a continuous load path (see page 115) and reinforced chimneys, walls, windows, and doors. At this level, the IBHS says, you'll pay 3% to 5% extra for the construction—but you'll get a home that's twice as strong as a non-Fortified building.

When it's all over, you can have your house certified by a Fortified-trained inspector, which costs about $500. It may be a good investment; in one study, Fortified certification raised homes' resale values by 7%. Your insurance premiums will probably go down, too.

But here's the thing: You don't have to get the certificate. The IBHS publishes all the details for the Fortified building standards on its website, free to all. If your main objective is to build a resilient house, you may not care about the official piece of paper.

Keep in mind, by the way, that the Fortified program is geared toward storm, wind, and rain protection. It doesn't include standards for wild-

Earthquakes and Climate Change

In the era of global weirding, you might think you have enough to worry about without taking earthquakes into account. How could they even be related to climate change?

Yet huge earthquakes do sometimes seem to occur directly after gigantic hurricanes. That was the sequence in Nepal in 2015, where an earthquake killed 9,000 people, or in Haiti in 2010, where the earthquake killed 220,000. Could it be that intense storm rains washed away thousands of tons of mud, releasing the weight from a fault?

Or what about California, where we've been pumping water out of the ground for decades—a water loss accelerated by the yearslong California drought? As a result, millions of tons of water weight are no longer bearing down on the earth's crust. Could the removal of all that water weight, lifted off the San Andreas Fault, mean that California is now freer to shift and quake?

"Yeah, no," says earthquake expert Lucy Jones, author of *The Big Ones: How Natural Disasters Have Shaped Us*. "You do not need a chapter on earthquakes."

No studies have shown a convincing relationship between climate change and earthquakes. "In earthquakes, we so desperately want *prediction* that a lot of people try to get into it," she says. "You've got to do your statistics and know the difference between coincidence and causality."

She does note, however, that being prepared for climate change *also* prepares you for earthquakes. Building your house with a continuous load path—an excellent preparation for high winds and hurricanes—*also* protects you against earthquake damage. Having backup power, food, and water supplies can *also* save you after an earthquake.

But you *don't* need to worry about climate change making earthquakes stronger or more frequent.

fires, earthquakes, flood, high heat, or drought. But a companion website, DisasterSafety.org, does include suggestions for these kinds of resilience.

Finally, remember that the IBHS is funded by the insurance industry, which has an ulterior motive in promoting these building standards: to pay out fewer insurance claims.

But just this once, your interests are precisely aligned with theirs. Given the choice, *you'd* probably rather not have to file insurance claims, either. It's much more pleasant to ride out a storm, flood, or fire in your own, undamaged home.

Wildfire Resistance

Something is going on with wildfires. In 2018, over 49,000 fires took out 4.5 million acres of California; the Camp Fire alone—then the deadliest and most destructive fire in California history—killed 85 people and destroyed 19,000 buildings. Everyone thought that was a horrifying fire season—until 2020, when the worst West Coast wildfire season ever recorded burned over 5 million acres, wiped out endangered species, and filled the air with so much smoke that it made the sky hazy in New York City, 3,000 miles away.

Chapter 13 describes how wildfires work. But from a "how to build" perspective, you can stack the odds of survival by thinking about four factors: the site of your home, the safe zone you create around it, the materials you use, and the room you create for fire and rescue trucks to maneuver.

"You can make your house the one that's left standing, if there are houses left standing," says Linda Masterson, author of *Surviving Wildfire: Get Prepared, Stay Alive, Rebuild Your Life.*

Siting for Wildfire

When you're moving to a new neighborhood, you may have no clue what wildfire risk you're stepping into. There aren't exactly signs that say WELCOME TO PARADISE, CA—HOME OF THE DEVASTATING INFERNO. And real-estate agents aren't about to volunteer that information.

In every town, however, there is someone who'll speak to you frankly about the risk: the fire department. Masterson suggests asking the fire department about the location of the home you're considering. "They'll say, 'That community's pretty well mitigated,'" she says. "Or 'That community's in a terrible position.'"

You can also have the property inspected for risk by someone from the fire department, the state forest service, or the local building department.

Beware homeowner associations, too. Sometimes, their covenant *requires* fire-friendly building materials for the home you're going to build (wood siding, cedar shingles) or forbids you to cut down trees to create a fireproof zone around your house.

If things look promising, work with a builder who knows about fire mitigation—and about the massive trove of building guidance available at Firewise.org, run by the nonprofit National Fire Protection Association. Your architect or builder also knows about the local building codes (right?), which usually specify fire-safe materials and components.

FEMA offers some important considerations for choosing where to build your house:

- **Build on a level spot.** Fire almost always races uphill; the steeper the slope, the faster it moves. Remember, too, that the path of the sun generally makes vegetation on *south- and west-facing* slopes drier and hotter than north- and east-facing ones. Fires are more likely to start there and to travel faster.

  The moral here: For best fire resistance, don't site your house at the top of a slope.

- **Beware the canyon effect.** A canyon acts like a chimney, channeling superheated gases and heat. Avoid building a house in a narrow canyon.

- **Orient the house** so that the side facing the likeliest fire path is the narrowest. Twist the house on an angle, too, so that the wind (and therefore embers and debris) doesn't build up against one wall.

Defensible Space

This is the big, screamingly obvious one: Create defensible space around your home. That means setting up a kind of moat between your home and any vegetation that surrounds it. This 30-to-100-foot nonburnable buffer is the best chance you have of saving your home in a wildfire.

In this safety gap, remove not only flammable trees, bushes, and dry grass, but also leaf piles, gas mowers, motorboat engines, propane tanks, RVs, and woodpiles. Clear out all the leaves and needles in your gutters and on your roof.

The idea is to deprive a wildfire of the fuel that would let it reach your home directly. You're also lowering the chances that flying embers can reach you, and you're providing some open space for firefighters to defend your house.

It's okay to leave trees that are right next to the house, as long as the defensible space begins *past* them, and as long as they don't overhang the house.

The defensible gap can include your driveway, stone walls, a patio, or plants that don't catch fire easily. ReadyforWildfire.org has a great list of fire-retardant plants, bushes, and trees.

Woodpiles should be uphill and downwind from the actual house, and at least 30 feet away. If the woodpile catches fire, you don't want it to act as a giant fuse for your home.

Figure 3-15. This California government graphic illustrates 30- and 100-foot "defensible space" zones.

In California's wildfire areas, you're not *allowed* to build a home without at least 100 feet of defensible space, split up like this:

- **Zone 1** is a 30-foot gap without any dead plants, grass, leaves, or pine needles—not on the ground, not on the roof, not in your gutters, not near or under your deck. No woodpiles, either. Tree branches have to be at least 15 feet from other trees' branches, can't overhang your roof, and have to be 15 feet away from your chimney.
- **Zone 2** extends another 70 feet beyond Zone 1. It can contain plants, but they're subject to horizontal and vertical spacing minimums. Fallen leaves, needles, twigs, and so on are okay up to a depth of three inches.

Defensible space is a big, big deal. After every wildfire, aerial photos show that the houses still standing are usually those with these cleared zones in place.

Truck Turnarounds

Linda Masterson notes that your home's ability to survive a wildfire may come down to the shape of its driveway. "Fire engines can't turn around in your steep little, cute little driveway," she says. "And where they can't turn around, they won't go." Rescue and fire vehicles need broad, straight roadways, and a parking area big enough to turn around in.

How much does it matter? Fire departments don't even *try* to save houses that don't offer easy access because it's not safe for the firefighters. "They're going to save people who have done their part," she says.

Building Materials

What you use to construct your house has a big effect on its ability to withstand fire, too:

- **Exterior material.** No material is completely fireproof, but metal, brick, stucco, tile, adobe, and concrete block take a lot longer to burn than wood.

  Drywall (gypsum board) underneath forms an excellent fire barrier, too; it's usually treated with fire-slowing chemicals. Some contractors use more than one layer of gypsum board for added fire resistance.

The *undersides* of things—roof soffits, decks, floors—are especially vulnerable. That's where flames get trapped, and temperatures go sky-high. You can protect them by adding a second layer of protection and beefing up the structural reinforcements.

- **Windows and frames.** The intense heat of a wildfire can shatter your windows from a *distance*, even when the flames themselves haven't reached your house. In fact, things *inside the house* have been known to combust before the fire has even reached the building.

  You can buy fire-resistant windows—tempered or double paned. Don't forget about the window frames, either. Steel is best, wood and aluminum are just okay, vinyl is bad. For windows that don't open, you can get windows made of fire-safety glass, or wire-reinforced glass, which may crack in heat but won't shatter.

  Plastic skylights: no. They melt and let the firestorm in.

  The best protection of all is metal fire doors. Most of the time, they sit there hidden, rolled up in roof overhangs or the recessed sides of the windows. But when their sensors detect intense heat, they unroll automatically. (Fold-down panels and self-closing shutters are also available.) That way, they protect your house even when you're not home, and even if you left the windows open.

- **Vents.** Automatic fire dampers are also available for the other areas of your house, including garage doors and big vents in the attic and the undersides of your floors. The vents in your kitchen, bathroom, and laundry room should have automatic back-draft dampers, which prevent burning embers from being sucked in. Eave vents and attic vents should have, at the least, one-eighth-inch mesh screens to keep the embers out.

  And don't install plastic or fiberglass vents. They melt and burn.

- **Chimneys** are intended to vent hot gases and smoke *out* of your home, but they can also serve as a handy *inlet* for sparks and embers. The solution is a spark arrestor: a cylinder of double mesh, part of your chimney cap, that prevents sparks from getting in *or* out. Keeps critters out, too.

- **The roof.** You want what Underwriters Laboratories calls a Class A roof: "effective against severe fire." (There are also Class B and C materials.) As long as your roof tiles are tightly interlocked, any fire-rated shingles work fine: metal, slate, asphalt, and so on.

  Wooden shake shingles, such as cedar-shake roofs, are bad news. They're pure fire fuel and have lots of exposed edges to catch embers. Your insurance company may not even insure a house with wooden shingles.

  Spanish-tile roofs—those rounded, tubey, reddish tiles so popular in California and other wildfire hot spots—aren't so great, either. They allow the wind to blow hot embers straight into your attic.

- **Decks.** Even fire-resistant wood can burn. Deck material sold as fire*proof* is a safer bet. Oh, and don't store burnable stuff under your deck. No wooden boats, woodpiles, furniture, and so on.

  It helps to seal in the area beneath the deck with metal screens to keep embers (and leaves, twigs, and animals) out.

- **Sprinklers.** You can also install sprinklers on your roof, or on poles over your patio or deck, that turn on when they sense a fire's heat. It's not a simple proposition, though; in a wildfire, your house is likely to lose both power and water. You'll have to hook up your sprinklers to a generator and a backup water source—like your pool, if you have one.

- **Protect the soffits.** Remember: Embers wafting through your soffits (the undersides of your roof overhangs) are one of the most common ways that homes catch fire. You want ember-resistant soffits, which have metal screens over the vents.

- **Your address sign.** Masterson points out a small thing that becomes huge when fire trucks are trying to find your house: Make sure that your house's street number is big, clear, and illuminated at night. Firefighters shouldn't have to decipher your font to fight your fire.

  If you have gates or live in a gated community, confirm that the fire department has the entry code.

Air Quality

Air pollution isn't just outdoor air. These days, the *indoor* air, which you breathe 90% of the time, is a fresh worry, for all kinds of reasons:

- In drought conditions, you get more dust sneaking into your home, sometimes piggybacked with pollen, bacteria, and fungal spores.
- In wildfire conditions, it may be the smoke that gets you, not the fire.
- The warmer, wetter world of climate change also helps explain the rise of Legionnaires' disease, a nasty, bacteria-borne kind of pneumonia. For 10% of the victims, it's fatal. From 2000 to 2017, reported Legionnaires' cases *quintupled* in the United States.
- Greater humidity indoors—another climate-change effect—"will in turn increase indoor mold, dust mites, and bacteria," says the government, "as well as increase levels of volatile organic compounds (VOCs) and other chemicals resulting from the off-gassing of damp or wet building materials. Dampness and mold in U.S. homes are linked to approximately 4.6 million cases of worsened asthma and between 8% and 20% of several common respiratory infections, such as acute bronchitis." Lovely.
- As we use more air-conditioning, we're pumping more pollutants around inside our buildings.

Bad indoor air quality isn't good for anyone, but it's especially dangerous for the very old, the very young, and people who already have asthma, allergies, or lung diseases. You have three options for clearing the air:

- **Better HVAC filters.** Your house probably already has air filters. They're part of your air-conditioning/heating systems. You're supposed to change their filter cartridges every three or six months, however—more often if there are smokers or pets in the house, less often if you don't run the AC or heat very much. Are you doing that?

  Unfortunately, these filters are intended to protect your HVAC system, not your lungs. The standard one-layer filters remove only the biggest particles—mostly dust—and not pollen, bacteria, and other nastiness.

Accordion-folded filters cost a little more and do a better job. Their effectiveness is measured on the MERV scale (Minimum Efficiency Reporting Value), a rating system dreamed up by ASHRAE (the American Society of Heating, Refrigerating and Air Conditioning Engineers), which may be a case of AFMATWRN (A Few More Acronyms Than We Really Need).

The higher the MERV number on a scale of 1 to 16, the better the filtration.

By the time you're at MERV values of 8, you're filtering out pollen, dust mites, dust, mold spores, and pet dander. To filter out bacteria, you need at least MERV 13; for viruses, you need MERV values over 17.

But here's the problem with MERV: As the filter's openings get finer, they make it harder for air to pass through. In effect, they're clogging your air system. Bottom line: Install the highest-numbered MERV filter that your fan can accommodate (you may have to consult an HVAC pro).

Figure 3-16. Every air filter lets you know its MERV value—what size stuff it filters out of the air.

♦ **Portable room air "purifiers. "** These machines cost from $50 to $1,000, depending on the size and capacity.

No single technology gets everything out of the air:

Activated carbon filters pull smoke, smells, and gases out of the air—but not particles or germs. Also, you have to replace the filter periodically.

HEPA filters ("high-efficiency particulate air") remove 99.97% of particles bigger than two microns. That gets rid of dust, pollen, spores, mold, and some bacteria and viruses—but not volatile chemicals. For most units, you have to replace the filter every few months, which costs $40 or more. (Some let you just wash the filter and reuse it.)

Electrostatic precipitators are electronic, charging the air particles so that they stick to a collection plate, which you have to clean in soapy water every few months. Gets rid of particulates and some gases, but does nothing for bacteria or viruses.

UV irradiation is an ultraviolet lamp that kills most bacteria, viruses, and some volatile chemical gases—but not particles or smells.

Ionizers make particles clump together and drop out of the air. Does nothing for fungus, bacteria, or most viruses, and not nearly as effective as a HEPA filter.

Ozone generators pump out an oxygen molecule (O_3) that kills odors and makes the air smell better. (They don't remove particles or

Figure 3-17. Portable air purifiers come in various designs.

germs.) They are not, however, intended for occupied living spaces. "Avoid portable air cleaners and furnace/HVAC filters that intentionally produce ozone," writes the EPA. "Ozone is a lung irritant."

If you're going to buy a room air cleaner, *Consumer Reports* recommends buying a bigger model and running it on a lower speed, rather than buying a small one and cranking it to full power. You'll save money on power and get a much quieter room.

- **Whole-house filters.** You can get these units professionally installed on your furnace or in your ductwork, so that they filter all the air of your house. Some ("extended media filters") are boxes that stack multiple accordion-fold filters together. Others are larger versions of the electronic filters described above.

  As you'd guess, this approach costs the most—up to $1,000, plus another $500 or more for a UV lamp to kill viruses. And, of course, they clean the air only when your heat or cooling is actually *running*, which is usually less than a quarter of the time. Running your HVAC longer eats massive amounts of power.

Heat-Wave Resistance

Heat is the deadliest force of the climate-change era, and it's getting worse. In the developed world, it kills more people than hurricanes, floods, extreme cold, or drought.

If you're indoors, well hydrated and air-conditioned, you're not at much risk. Trouble is, everyone *else* has their air conditioners cranked up, too, and AC poses problems of its own. As Chapter 10 makes clear, running it is costly to you, to the city, and to the planet.

The thing is, air-conditioning didn't become standard in homes until the 1960s. There was a time before air-conditioning. How on earth did those people survive?

Through *passive cooling*, also known as clever design.

Picture those beautiful antebellum Southern homes. They had wraparound porches, which shaded the body of the house and gave the residents a cool, breezy place to hang out. The rooms had tall ceilings, which

Figure 3-18. Tall ceilings, wraparound porches, shady trees: Older Southern houses used passive cooling instead of air-conditioning.

allowed heat to rise above human height. The layout featured a long central breezeway forming the spine of the building, a channel for cooling breezes to run through the house.

"We kind of forgot about those passive design features when air-conditioning came along," says Alex Wilson of the Resilient Design Institute. "As a result, when we lose power, those newer buildings are more vulnerable."

These days, those old technologies—and a lot of newer ones—make it possible to design buildings that stay much cooler as it's getting hotter outside.

In warmer states like Arizona or Florida, the average air-conditioning bill is $750 a year and climbing; in the next 20 years, Northern states will start to see similar bills. Designing a house with passive cooling means that money can stay in your bank account.

Building Orientation

No matter where you live, the sun rises in the east and sets in the west. Our giant celestial heat source crosses the sky in a straight line, day after day. If you're smart, you'll situate your house so that as few of those rays blast your walls and windows as possible.

The solution: Lay out the house so that the longer sides face north and south. The sun's path, as it crosses overhead, is parallel to the house, and therefore its rays don't strike as much of it. Your eastern and western walls get the brunt of the sunshine—but they've got smaller surface area, because you were clever.

Another clever move: Build deep roof overhangs on the sunny sides of the house, to keep those heat rays off the walls.

Walls and Windows

A north-south building orientation also means that most of your windows are parallel to the sun's path. That's good, because windows let in a lot more sun and heat than walls do.

The thing is, you generally don't want to install the *same* glass in all your windows. How they face the sun makes a big difference.

- **East and west windows.** The glass in today's windows comes in a variety of formulations, with varying degrees of infrared-heat blocking. On the window's sticker, you'll see a rating for its *solar heat-gain coefficient.* The scale goes from 0.0 (no heat passes through) to 1.0 (all the heat and light passes through).

  On the east- and west-facing windows, you want glass with *low* heat-gain numbers, such as 0.2 or lower, to keep out the heat, since these are the windows that catch the most of the sun's energy.

  Shade is also good on those windows: roller screens, vertical louvers, awnings that you can open and close, plantings, and so on. Shading contraptions on the *outside* of the windows keep out more heat than interior ones do.

- **South windows.** In the summer, at least in the northern hemisphere, the sun's angle is high enough that it doesn't hit your south walls much, so you don't have to worry about the heat gain. Therefore, for your home's south-facing walls, you want windows with *high* heat-gain numbers such as 0.5 or 0.6, which give you slightly clearer views and, if you need it, warmth in the winter. (The sun's angle in winter is lower in the southern sky, more likely to transmit heat through the glass.)

- **North windows.** For north-facing windows, the heat-gain number doesn't make much difference, since direct sunlight almost never strikes these windows. Instead, you want to look at how well these windows insulate. You want glass with a high *R-value*—which means a low *U-factor*—as declared by the sticker. Those related numbers may have a low comprehension value and a high confusion factor, but at least now you know what to shop for.

The Roof

No surface of your house is more exposed to the sun's heat than the roof. But most roofs are dark, and dark colors absorb heat, for the same reason that dark leather car seats in summer are sometimes scorching when you first get in. A typical roof can be 90 degrees hotter than the air around it.

But your roof can serve as a giant solar shield if you play your cards right—by making it *white.*

A white roof can drastically reduce a building's heat gain. Since the

Figure 3-19. A white roof is cooler to anyone inside.

mid-1990s, Walmart has been painting the roofs of its 4,800 stores white for this very reason. Sure, that's a noble environmental step. But it also saves the company 8% a year in cooling costs.

(Physicist Art Rosenfeld, former California energy commissioner, once calculated that turning all the world's roofs white would save the equivalent of 24 billion metric tons in CO_2 emissions. That's two-thirds of *all* the carbon dioxide the entire world currently produces each year.)

The EPA lists over 3,000 cool roofing materials that can be any color or even reflective, including white asphalt shingles and metal. As a handy basket of bonuses, metal also resists fires and storms, it's ideal for harvesting rain (page 100), and it makes solar-panel installation easy. Look over the certified options at the government's Energy Star roof program site (http://j.mp/2ZeIueO) or CoolRoofs.org (yes, there is a Cool Roof Rating Council).

You can also get a coating put on your roof to improve its reflectiveness.

The roof's shape matters, too, especially on the south-facing sides of the house. You want an overhang that shades the side of the building in summer, when the sun's path crosses overhead—but lets the sun shine in during the winter, when it passes lower in the sky.

You don't have to do the math yourself. Believe it or not, there's such a thing as a roof-overhang calculator that lets you play with its angle and depth at different times of the year. Here's one example: http://susdesign.com/overhang/index.php.

In any case, a cool roof confers a long list of perks. You can save as much as 40% on your AC bill. A cool roof generally lasts a lot longer than traditional roofs. You're also doing a favor to the planet by lowering your contribution to the heat island effect (page 64) and pumping fewer emissions into the air by using less AC. You may even get a rebate from your utility company as a little thank-you.

Shading

Any heat that you can prevent from striking your house is heat you don't have to counteract with air-conditioning. That's a good argument for growing trees, vines, and tall shrubs around the walls of your house.

Awnings and shutters can do that, too. Awnings aren't cheap—a single awning over a door or window can cost between $300 (for canvas) to $3,500 (for a top-of-the-line, remote-controlled, motorized retractable one), plus installation. But they're exceptionally good at keeping the building cool. On the south- and west-facing sides of your house, awnings can eliminate three-quarters of the sun's heating effect.

Insulation

Architects and builders often get surprisingly excited about insulation. They're sometimes baffled at why it gets so little love from the general public. (Hint: because insulation is incredibly boring.)

Yes, we know that we're supposed to insulate our homes well to keep them warm in winter. But great insulation in the roof and walls also keeps the heat from *entering* the house in the new, hotter era.

You can buy insulation with various effectiveness ratings, on a scale that begins at R-1 (the equivalent thermal protection of one inch of wood) and goes—well, as high as you want, because you can make your walls ever thicker and stack layers of insulation.

So how much insulation do you need? It depends on whether you live in a hot or a cold climate, what kind of heating/cooling your house has, whether you're building a new house or updating an old one, and even which parts of the house you're talking about. The walls, floors, and roof don't all get the same R-values.

But here's an example: In a hot climate, in a new house, you might go for R-20 in your walls and R-60 in the attic. You can look up the recommended R-values on the web or just trust an insulation contractor.

There are so many kinds of insulation! There's the foam sprayed into the walls with huge hoses; there are rigid slabs of the stuff; and there are the traditional *batts*—fluffy fiberglass rolls that they stuff between the studs. And then there are all the different materials: the old standby fiberglass, but also cellulose, polystyrene, gas-filled panels, wool, cork, and many others.

Insulation can't do its job if the house is leaky. It's worth checking the place for airtightness, whether it's a new house or the one you're living in now. Air leaks out around chimneys, out the eaves, through the windows, out gaps around your doors, and anywhere else the "building envelope" isn't perfectly sealed.

Finally, if you're building a new house, there are two final insulation-related points to ponder:

- **Windows.** A piece of glass is pretty terrible as insulation; heat goes right through. That's the argument for installing double-paned windows, which, thanks to the air gap between the panes (or, more commonly, argon or krypton gas), are far better insulators.

- **Thermal mass.** If you're planning to build in a region with big temperature swings from day to night—say, 15 degrees or more—heat-protection wonks might also bring up the topic of *thermal mass.* That's architect-speak for how much heat the *walls themselves* can store—and flooring, ceilings, and chimneys. Massive stuff such as bricks, plaster walls, slate or tile floors, and concrete holds a lot of heat. Lighter materials, such as wood, don't.

  A house with a lot of thermal mass can time-shift heat. During the day, it absorbs heat from the air, keeping you cooler. Then, at night, it releases that warmth to the cool outside breezes and open sky. (Or, in winter, it gives that warmth back to you inside.)

If you're still bored by the topic of insulation, consider this: A well-insulated house isn't just protection against heat in summer and cold in winter. It makes your house more livable when the power goes out. It can save you a huge amount on your heating and cooling bills—and even on your insurance rates.

Layout

You may not want your house to look like an 1850s plantation house. Even so, there's nothing wrong with stealing some of the design ideas that kept those houses cool in hot weather:

- **Breeze orientation.** The best way to cool a house is to let air flow through it. If you're about to build a house, then you'll want one of the long sides of the house facing the direction of the prevailing winds, with lots of doors and windows that you can open at night.

- **Sun orientation.** The coolest house is one that exposes the fewest big surfaces (walls and roof) to the south, which gets blasted the most by the sun during the day in the northern hemisphere. (That's also why a good roof overhang on that south face is a great cooling idea; it shades the southern walls of your place.)
- **Vegetation.** Plants and trees block sunlight and, thanks to the cooling process of photosynthesis, literally cool the air that blows into the house when you open it up.
- **Pass-through channels for breezes.** The "dogtrot" designs of older Southeast U.S. houses featured a gigantic front-to-back open hallway that let breezes blow through. Any time you create openness on directly opposite sides of the house, you help with the cooling. Open floor plans help a lot (and are trendy anyway). So does positioning windows on opposite, unobstructed walls.

Opening the House

Most passive-cooling strategies work without any effort on your part. But longtime Southerners can tell you about another one: closing the house during the heat of the day, and opening it when the sun goes down, for a little "night flushing."

"Opening" and "closing" in this context means whatever screened windows and doors you can open, as well as the shades or drapes over them. (For bonus coolness: Choose *light*-colored shades or drapes.)

That way, you're blocking the sun's heat and retaining the cool air inside as long as possible—then, once it's cooler outside than inside, letting out the warmed-up indoor air. It's as though your house is exhaling and inhaling once every 24 hours.

Usually, just opening windows on opposite sides of the house creates a cross-breeze that cools the place down fast. But whole-house attic fans, which suck hot air up and out, speed the cooling along. Or just crack one window on the ground floor and open a window wide on the second floor, preferably with a window fan aimed outward. Heat rises, so you'll create a two-story breeze tunnel that cools the place off quickly.

A few more manual-operation tips:

- **Launder at night.** Appliances like dishwashers, washers, and dryers throw off huge amounts of heat. Run them at night.

- **Open the inner doors during the day.** Keep the *inside* doors—to the rooms of your house—open during the day, to encourage air movement.

- **Ditch the incandescents.** Incandescent light bulbs are not just energy hogs; they're basically high-wattage heaters. (They're banned for sale in the United States for these reasons.) Replace them with LED bulbs. LEDs require one-tenth the energy, last 25 years or more, are hard to break, contain no mercury or other bad stuff—and they're cool to the touch, even after they've been on for a while.

Ceiling Fans

About 30% of American homes don't have air-conditioning. It's a pretty good bet—or at least a fervent hope—that they at least have ceiling fans.

And why on the ceiling? Because hot air rises. Upon reaching the fan blades, it gets pushed back down again. And presto: You've got a continuous *current* of air, cycling low/high/low.

The temperature of the room doesn't change; this isn't air-conditioning. But it *feels* cooler to you because of the windchill effect. When air moves past your skin, your sweat evaporates—and evaporating moisture, anywhere in nature, makes the temperature drop.

A ceiling fan's effect amounts to about a five-degree temperature drop. If you don't have AC, that's a huge assist to whatever passive-cooling features you have. And if you do have AC, you can use a lot less of it, saving money and slowing down climate change.

More tips:

- Turn off the ceiling fan when you leave the room. As the government puts it, "Fans cool people, not rooms."

- Fans work best when the blades are 7 to 9 feet up, and 10 to 12 inches below the ceiling.

- Fans bearing the Energy Star sticker move air at least 20% more efficiently than other models.

The bottom line (or, rather, top line): Ceiling fans create a lot of cooling for very little money.

Heat Pumps

For decades, we've built houses with furnaces for heat, and air conditioners for cooling.

But these days, there's an alternative that saves a lot of power, and therefore money: a *heat pump.* It replaces both your furnace *and* your air conditioner and dramatically reduces your heating and cooling bills.

You may already be familiar with how a refrigerator works. Its tubing contains a refrigerant—a liquid that can convert back and forth to a gas, absorbing or releasing heat each time.

"Your refrigerator's sitting in your kitchen, taking heat from inside your refrigerator and dumping it into your kitchen," explains Alex Wilson. "It's taking heat from a colder place and dumping it into a warmer place—which, on the surface, would seem to violate the laws of thermodynamics. But that's the secret of refrigerant-cycle air-conditioning."

Heat pumps cool your house using the same technique. Think of your house as the fridge, and the great outdoors as your kitchen.

Better yet, in *cold* weather, heat pumps can move heat the other way, thereby *warming* your house. Heat pumps just shuttle heat around instead of actually generating it; as a result, it can heat and cool your house for as little as a *quarter* of the cost of furnaces and air conditioners.

Heat pumps come in two types:

- **Air-source heat pumps** are the most common. For heating, they use only half as much electricity as furnaces and baseboard heaters; for cooling, they dehumidify much more effectively than air conditioners, so you feel cooler and use less power. Average cost, with installation: $5,100—but the government offers some juicy tax credits, such as the Renewable Energy Tax Credit, that can soften the sticker shock.
- **Geothermal** heat pumps don't exchange heat with the air; instead, they exchange it with the ground, or with a nearby body of water. The beauty of this system is that, from day to night and season to season, the earth and water maintain much more even temperatures than the

air does. As a result, you get far more efficiency, and spend very, very little on heating and cooling your home.

Geothermal heat pumps confer some substantial benefits. "They can reduce energy use by 30%–60%, control humidity, are sturdy and reliable, and fit in a wide variety of homes," says the EPA. "Ground-source or water-source heat pumps can be used in more extreme climates than air-source heat pumps, and customer satisfaction with the systems is very high."

On the other hand, geothermal systems cost a lot more than air-source heat pumps; the average price is over $8,000.

Heat pumps are already popular in Europe and Japan. They're catching on slowly in American homes—but considering that they save you money *and* spare the environment, they deserve even greater popularity.

Building for the Future

This much is clear: Safety, structural integrity, and economy in building means something different in the new climate than it did 100 years ago. Fortunately, we now have the knowledge, materials, and technology to build much more resilience into our homes.

But this much is also clear: Neither the government's building codes nor the building contractors themselves are as up-to-date as they should be. If you plan to move, build, or renovate, the driving force in disaster-proofing your home will have to be *you.*

Chapter 4

What to Grow

GARDENING WAS THE MOST POPULAR HOBBY IN AMERICA even before the pandemic lockdown, which gave it a huge boost. But in the climate-crisis era, gardeners have the same problems that farmers do: Every element of growing plants is *changing*. The heat is different, the rainfall's different, the seasons are different, the pests are different, the invasive plants are different, the pollinators are different (bees, beetles, birds, butterflies).

The techniques and timings we've used for thousands of years no longer work as well as they used to. You simply can't plant the same stuff you used to and expect it to grow the same way.

But *how* to grow is only the first question. The second one is *what.*

The hallmark of global weirding is periodic disruption: of power, of heat, of communication, of normalcy, of the ability to shop for groceries.

There's a certain comfort, therefore, in maintaining a *survival garden*: a garden that can produce food for you and your family to live on when things really get bad. In fact, a survival garden gives you a satisfying sense of self-sufficiency whether the food system breaks down or not.

Of course, most people don't grow anything during the winter. So an important part of survival gardening is preserving your crops through the off-season—a process that gardeners call *putting up.*

"People need to learn to put food up," says garden writer Nan Fischer. "With climate change, we're going to have so many crop losses in places like Florida and California. Growing your own food is a buffer from those, and from price fluctuations."

There's a secret bonus feature of growing your own food, too: It's adaptation *and* mitigation. You're protecting yourself against food shortages *and* limiting greenhouse gas emissions—by eliminating the step of flying or trucking your food.

The Survival Garden

You don't want to grow just any old vegetables in a survival garden. The best crops are easy to grow (they're tolerant of bugs, weird weather, and your mistakes); they keep well through the winter; they're not water hogs; and they're calorie- and nutrient-dense, meaning you get a lot of fuel with every bite.

Here's a sampling of good options that fulfill those requirements:

- **Peanuts.** As any trivia fan knows, peanuts aren't actually nuts. They're legumes, just like peas and beans. They're also weird: The flowers are above the ground, but the *peanuts* grow underground.

  You can grow peanuts almost anywhere, although they like sun and warm weather; if you're up North, plant a variety that matures quickly, like Early Spanish. Even then, you might want to grow them for a few weeks indoors, in peat pots, then transplant them.

  Almost all peanuts you buy are roasted; you've probably never eaten raw peanuts right off the plant. But they're fine to eat raw—and will, in their shells in an airtight container, keep for a year.

  You can also boil them or roast them yourself. Roasting them requires brining (soaking in salt water) for six hours, drying them overnight, then putting them in a 350° oven for 20 minutes.

  Caution: Your house will smell amazing.

- **Dry beans.** "Dry beans," like pinto, navy, kidney, and black beans, are extremely easy to grow. They're incredibly good for you, delivering fiber, protein, and antioxidants, and they're also good for your garden because they restore nitrogen to the soil.

  For survival/storage purposes, wait until late in the season, when the bean pods begin to turn dry and brown and start to crack open. If you were to just ignore them at this point, they'd open all the way, and the beans inside would drop to the ground, planting themselves for the next season, perpetuating the great bean circle of life.

You, however, have other plans.

Pull the pods off the plants. Store them in grocery bags for a few weeks, stirring them by hand occasionally. Let them dry out until they feel crisp.

At that point, you have to *thresh* the beans, which means getting them out of the pods. If you've got nothing but time (and few beans), you can do that by hand. But the internet is full of more efficient techniques, like folding them inside a tarp, taco-style, and then stomping on them with your kids. The brittle pod husks crumble; the beans inside are hard enough to withstand the pummeling.

Then comes the *winnowing*: Turn on a fan to blow away the pod fragments as you shuffle through the beans with your hands. Let the newly freed beans dry for another week, then seal them into jars or airtight containers. They're ready to soak, cook, and eat when you need them.

"You can eat them all winter, and they last several years," says Fischer. "I'm a huge beans fan."

- **Other pulses.** Chickpeas, lentils, and peas are, like dry beans, members of the *pulse* domain, meaning the subset of legumes that produce dry, edible seeds inside pods.

  Talk about easy to grow: Pulses are self-fertilizing (they pull nitrogen from the air) and don't need much water. "Just put them in the ground and leave them alone—piece of cake," says Fischer.

- **Green beans.** Easy to grow, easy to eat, and easy to put up: rinse, slice off the stems, pull out the dental-flossy thread down the middle, blanch (boil for three minutes, then plunge into cold water), pack into airtight bags or bowls, and stuff into the freezer.

- **Tomatoes.** One way to preserve your tomato crop, of course, is to can them. Actually, that's a misnomer; you're actually going to *jar* them.

  Canning is a hobby unto itself, but the general idea is to core the tomatoes, blanch them (boil until the skin splits), skin them (the skin slips right off), put them into sterilized, heated mason jars with a little lemon juice and salt, seal them up airtight, and then boil the jars in a big pot, or the *special* big pot known as a tomato canner.

An easier method, best for smaller amounts, is to core, blanch, and skin them, chop them up, pour into a labeled ziplock bag, squeeze the air out, zip, and freeze. You now have them ready for making pasta sauce, pizza sauce, stews, soups, and chili all through the winter.

- **Root vegetables.** Things that grow underground—beets, carrots, turnips, potatoes, onion, garlic—don't need any special treatment to store for the winter. Just dust off the dirt, cut off the tops, then store them in boxes or crates in a cool, dry place. That might be the garage, crawl space, mudroom, or anywhere else that stays cool but won't freeze. In Northern climates, an unheated basement works great. Most root crops are fine that way for up to four months for eating; potatoes, eight months.

- **Winter squash**, like butternut squash, acorn squash, and pumpkins, grows easily and makes a great survival food. It fills you up and it's full of vitamins A and C.

  Once you've picked the squash, it can just sit there, as picked, for three months or so. To ward off mold, wash each one, dry it, and buff its skin with a little vegetable oil.

  Or you can cube the squash—raw or cooked—and freeze that. (If it's raw, freeze the cubes solid, spaced out on a baking sheet, and *then* put into the freezing container.)

- **Broccoli, squash, everything else.** You can always use your fresh veggies in a cooked dish, and then freeze *that*—a stir-fry, for example. In the dead of winter, you can dump some of that into chicken broth and, presto: soup!

It's actually fairly difficult for a family to survive entirely on what they grow. (It also gets culinarily boring, reports Nan Fischer, who tried it one winter.) It helps if you live somewhere where you can raise chickens, so that eggs supplement your garden's output.

But if you're on your own for food for more than a few days, you'll likely supplement your homegrown produce with canned emergency rations you've stashed away as described in Chapter 8.

The Urban Gardener

The phrase *urban gardener* may sound absurd. How can an apartment dweller have a garden?

In containers, that's how. Your empire could consist of two pots: one growing a cherry-tomato plant, and one growing cilantro. Add a hot-pepper plant, and you can make your own salsa.

But pots are only the beginning. Just about anything that can hold soil can be the basis of your garden: a bucket, a basket, a wooden or plastic box, a washtub, a big food can, whatever. It just needs holes in the bottom for drainage, and probably a catcher tray beneath, to avoid leakage all over your apartment.

What else can you grow in container? All kinds of stuff: lettuce, cabbages, tomatoes, beans, peppers, carrots, radishes, and even potatoes. If

Figure 4-1. A little terrace is all the land you need for a miniature farm.

you can set up some kind of fake trellis or fence for the plants to climb, you can also raise cucumbers, squash, and pumpkins in pots.

When you're buying seeds, look for "compact," "space saver," or "bush" varieties of your plants. They're bred for container life.

Now, plants' requirements don't change just because you're growing them in your apartment. They still need:

- **Sunlight.** Most fruits and veggies need at least six hours of sunshine a day. If you don't have a windowsill or a window box that can handle that, you can look into using a balcony, or the building's rooftop. In that case, take some steps to keep the wind from blowing over your pots or plants.

  If the roof thing doesn't pan out, you can keep the plants indoors and just buy a "grow light"—a full-spectrum LED bulb—which acts as a stand-in for the sun.

  Or just scale back your ambitions to growing herbs and salad greens, which don't need as much sunshine.

- **Soil.** If you're growing in pots or containers, traditional garden soil doesn't do the job well; it gets compacted, and water doesn't flow through. Instead, buy potting soil, which is lighter and fluffier, which both the roots and the water will appreciate.

- **Water.** Containered plants require a lot of watering—much more than backyard gardens. Keep that in mind when you're choosing a spot for them, so that you don't spend your life carrying water back and forth to the sink. (To see if they need watering, poke in your finger. If the soil isn't still damp a couple of inches down, it's time to water.)

  Keep in mind, by the way, that wet soil is *heavy*. Your pots or containers should rest on something sturdy.

- **Fertilizer.** Sometimes, potting soil is already fertilized, so your job is done. Otherwise, you can mix in some slow-release fertilizer at planting time and forget all about it; it will keep nourishing your crop all season long.

Gardening in the Hotter, Wetter, Drier Era

The good news is that the growing season—the frost-free season—is getting longer in most places. You can plant sooner, keep growing later, and expect more plants to survive the winter.

Sara Via, research scientist and University of Maryland climate professor, suggests that you plant a couple weeks earlier than you might have in previous years. "Just keep your eyes on weather," she says. "If there's a frost forecast, toss a sheet over it and it'll be fine."

The Pollinator Mistiming

But the earlier warming has an unfortunate side effect: Different species "wake up" after the winter at different times. As a result, the insects that would ordinarily pollinate your plants arrive too soon or too late to overlap with the plants' blooming, like guests showing up after the party's over. "I see these butterflies flying around, but there's no host plants for them," says Via. "That's a disturbing trend."

You have only one weapon against pollinator mistiming: diversity. Plant a range of different flowers and crops (a theme you hear again and again in climate-change conversations). The idea is to plant as many different species as you can, with all-different bloom times. That way, if one thing isn't blooming, something else is. "The more stuff you have, the more habitat you've created for the native pollinators and all the beneficial insects that you want to bring in," says Nan Fischer.

Oh—and don't pick the dandelions. "I don't care if they're on your lawn," she says. "They're the very first thing that the bees come and eat, so they're really important."

Crop Selection

Your garden's new climate also means that you have to reconsider what to plant.

Many gardeners, for example, are in the habit of consulting the "hardiness zone" labels on seed packages when choosing what to plant. These color-coded tables match up with the USDA's hardiness-zones map at

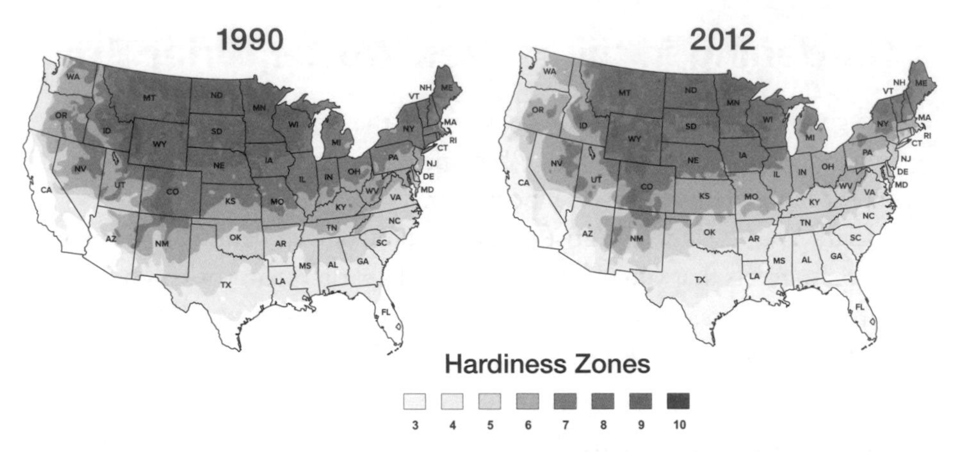

Figure 4-2. The zones of the USDA planting hardiness map shifted northward in 25 years.

https://planthardiness.ars.usda.gov. The colored bands indicate the coldest average winter temperature for each spot in the United States—your guide to survival for permanent shrubs, perennials, and trees. This system is intended to let you know if a certain plant will be able to survive your winters.

Keep in mind, though, that the most recent hardiness-zone maps were drawn up in 2012, and the world is steadily warming. In your head, you should shift those zone colors upward by yet *another* zone or so to account for the warmer climate.

(This effect, notes the National Wildlife Fund, is producing a peculiarly American loss: "Many states across the country may lose their official State Trees and Flowers. Imagine Virginia without the flowering dogwood, or Ohio without the Ohio buckeye!")

False Spring

So far, the notion that you have a longer growing season sounds like it's good news. It means that we get our upstate New York apples eight days sooner, our grapes six days sooner.

Unfortunately, earlier spring warming also poses a new danger for growers: False spring. That's when a warm stretch fools dormant plants into thinking that it's time to wake up and blossom—only to be slammed with a return to freezing temperatures.

That's especially problematic for fruit trees, whose blossoms are the precursors to the fruit. A freeze kills the blossoms. They all fall off, and no fruit is forthcoming.

In the last ten years, false spring has struck the U.S. agricultural industry broadly three times: once in 2007, causing a $2 billion loss of winter wheat, corn, and legumes; again in 2012, causing Michigan to lose 95% of its peach crop, 90% of its cherries, and 88% of its apples; and again in 2017, when a devastating false spring killed off 90% of South Carolina's peach crop and North Carolina's entire blueberry crop.

The Pest Problem

It's not just your plants that are enjoying a milder winter and an earlier thaw, by the way—the bugs are, too. The earlier spring means that bugs hatch sooner, grow up sooner, and lay their own eggs sooner. As a result, many regions are getting two generations of pests per season instead of one.

Worse, the second generation is a far bigger swarm because there was no winter to kill off many of the eggs. In the later parts of the summer, "you get massive amounts of insects," Via notes.

To make matters even happier (for the pests), the warmer climate means that the bugs that were once confined to Southern states are now free to spread northward.

So what's a gardener to do, especially a gardener who'd prefer not to spray insecticide?

- **Plant flowers nearby.** Even if your goal is vegetables, planting a range of nectar-bearing flowers nearby is an ingenious nature hack. Plant marigold, zinnia, verbena, borage, black-eyed Susan, sunflower, hollyhock, lavender, phlox, aster—the broadest array you can. The nectar of these flowers attracts and feeds not only bees and other pollinators, but also *good* insects—the ones that *eat* the bug pests.

- **Don't clear up the mulch.** Especially in the era of global weirding, the wise gardener uses mulch.

Mulch is covering material that you spread over the ground of the garden. It keeps the roots cool, keeps the soil moist, minimizes temperature fluctuations, holds down the weed population, and looks good. You can even buy organic mulches, which decompose and fertilize the soil.

People use all kinds of stuff as garden mulch, like straw, compost, or grass clippings (although grass clippings get slimy and stinky as they decompose). Three or four layers of dampened newspaper or brown package-wrapping paper make an excellent weed barrier. As a stack dries, it gets hard and ugly, but you can always toss some straw on top for esthetics.

Fallen tree leaves make spectacular mulch, too. If you have a mulching lawn mower, run it over your leaf piles, and boom: fantastic, organic, free mulch.

"I don't get this: People put their leaves on the curb—but then they go to Home Depot and buy topsoil!" says Via. "I often see big bags of leaves that people put on the curb. Sometimes I just pick them up and take them home and use them."

If you don't have a lawn mower, pile the leaves up in a corner of your yard, or stuff them into a chicken-wire cylinder, and let them compost themselves. In the spring, it'll all be a little slimy and decomposed, but it's perfect mulch. Wait even longer—a year or two—and you've got the world's most excellent soil.

In the fall, dump the leafy mulch onto your garden rows. Then, in the spring, when it's time to plant, just nudge that mulch aside. Install your seeds or transplants, re-cover them with the mulch, and you're done.

There are also plastic mulches. They're reusable from season to season; but then, do you want to be putting more plastic out into nature?

In any case, back to the pest problem: When the season ends, leave your mulch in place. Hidden inside are ground beetles and wolf spiders, whose favorite food is—hey, cool!—the pest bugs.

"And don't squish spiders," says Via. "That spider's going to eat a lot of stuff that could damage your plants! The more spiders you have in your garden, the better."

- **Inspect the plants.** Get down on your knees and look closely at the leaves. Look for holes in patterns. Turn the leaves to look for caterpillars on the undersides. If you spot them early, you can control their population with a chemical or organic spray. Choose the right stuff for the job; soft-body and hard-body bugs, for example, require different insecticides.

  Or just pluck the bugs off yourself. "I'm a hand picker," says Fischer. "I just squish them between rocks."

- **Put out row covers.** Row covers are strips of very thin fabric that you lay on top of the transplants or yet-to-germinate seeds. The tiny holes let the heat out, let the rain in—and keep the bugs and rabbits out.

  If you use row covers on top and a weed-suppressing mulch below, like straw or grass clippings on wet newspapers, you can increase the yield of your peppers, strawberries, and cucumber-family crops by more than a third.

  You can use heavier row covers as frost protection, but for pest control, use lightweight ones, to maximize sunlight—or even tulle (wedding-veil stuff).

Figure 4-3. Row covers in action.

One drawback of row covers is that they also keep the pollinators out. Lots of vegetables require pollination—those that sprout a blossom before growing the actual vegetable, like cucumbers, tomatoes, peppers, and squash. Once the blossoms appear on those plants, you'll have to remove the covers.

Greens and underground stuff (carrots, onions, garlic, beets, potatoes) are okay to keep covered all summer long.

The other drawback is the visual element: Row covers turn the natural gorgeousness of your garden into what looks like strips of dirty bed linen. But you can't have everything.

Heat Stress

You, as a gardener, may find the new world of heat upsetting. But it could be worse: You could be the tomato.

Some crop species are known as warm-weather crops: beans, corn, cucumbers, eggplant, melons, peppers, zucchini, squash, pumpkin, sweet potatoes, tomatoes, watermelon. They thrive in the heat; in fact, many of them *need* a stretch of 80°-to-90°F days. (And a lot of water.)

But it's a different story for the cool-weather crops: asparagus, beets, broccoli, brussels sprouts, chives, cabbage, carrots, cauliflower, Swiss chard, kale, leeks, lettuces, onions, parsnips, peas, radishes, spinach, and turnips.

You can plant these early in the spring, even before the final frost. You can even get away with planting them *again* in the fall, as long as they'll mature before winter sets in for good.

But cool-weather crops don't like extreme heat. Days of 80° or 90° heat stunt or kill them—and we're getting a lot more of those days these days. "Three years ago, the highs of the day used to be eighty, and eighty-five was a heat wave," says gardening columnist Nan Fischer. "Now, *ninety-five* is a heat wave, which is frightening."

The problem is with fertilization. Even if a plant has successfully been pollinated, a few more steps have to happen before it's *fertilized*. And hot weather interferes.

The result: You get misshapen tomatoes or none at all, and your peppers fall off and die when they're still babies. Lettuces, spinach, broccoli, and peppers may "bolt," where the plant, weirdly, sprouts a stalk that flowers and then goes to seed. The leaves, which would have been your salad someday, come out small, tough, and bitter.

How do you combat heat stress?

- **Avoid the worst heat.** Plant the cooler-weather crops earlier in the spring and again in late summer.
- **Choose your species.** You can plant heat-resistant varieties of many vegetables.
- **Keep the soil cool.** If you use drip irrigation hoses and add a thick layer of mulch, you keep your plants cool from the bottom up.
- **Lay out some shade.** Here we go again with draping cloth over your garden. "Covered gardening is the future," says Nan Fischer.

  Here, the idea is to provide some shade from the hot sun. You can buy shade cloth expressly for this purpose, or you can use the same lightweight row cover described above. Drape it right over the cabbage and carrots; for tomatoes and peppers, drape the fabric over hoops—for example, cheap electrical conduit piping from Home Depot—that you've shoved into the ground. Clip the shade cloth to those supports, or drape the row cover over them.

  The shade helps the plants retain their moisture and prevents sunscald, which is basically sunburn for crops. In one study, 40% shade cloth doubled the survival rate of transplanted lettuce.

Not Enough Water

It's a hallmark of the climate crisis: You get massive, weeks-long downpours, which can delay planting, wash out the garden, stunt or drown the plants, and compress the soil—and *then* you get weeks of drought. "This is very symptomatic of climate change: You get these prolonged, excessive weather periods," says Via. "It's really hard on gardeners and farmers."

For the drought periods, come up with a smart watering strategy like this:

- **Water early in the morning.** Watering early, while it's still cool, lets the water trickle through the soil to the roots. Now the plants have enough moisture to face the heat of the day.

  The problem with watering once the sun is up, of course, is that most of the water will just evaporate. You're wasting it.

  The next-best time to water is late afternoon. Here again, you're avoiding the stretch of daylight that will evaporate your efforts. (Don't water at night—that's inviting mildew and mold.)

- **Use drip irrigation or soaker hoses.** These are hoses or flat tubes with holes that let the water trickle directly into the ground. The water's going onto the roots, where the plants need it, and not onto the top parts, where it will mostly evaporate. You're directing water right where you want it—not on the weeds. And the water's released slowly, for hours, for better spread and penetration of the ground. You can set one up on a timer, if you like.

  If your garden is small and level, use a soaker hose, which is basically a garden hose with holes that ooze water all day. Soaker hoses are inexpensive and idiotproof: You just hook one up to your water spigot.

  Drip irrigation systems are a little more elaborate. You can snap them together so that the main line has branches sprouting off like the letter *E*, for example. These setups are better on sloping ground or long rows.

  In both cases, adding some mulch or straw on top makes the system much more effective; once again, you're preventing evaporation. Just connect your garden hose to the soaker hose once every few days until the soil is completely wet. It's more effective *and* lot easier than watering with a hose.

  (But if you do water with a hose, aim at the base of the plants. Water the soil, not the leaves; anywhere else is wasted water and effort.)

- **Mulch.** Here, once again, mighty mulch makes its appearance. The same mulch that's good for suppressing your pest population is also excellent for preventing the soil from dehydrating.

  And no tilling, Via says. Tilling smashes up the structure of the soil, makes it hard for the soil microbes to survive, and ruins the soil's health. "The whole idea of going out in the spring and rototilling your garden—that is so 1960s," says Via. "No-till gardening is the way to go."

Too Much Water

Climate craziness means both baking-dry periods *and* relentless downpours. Too much rain can stunt, wilt, or even kill the plants, primarily by depriving the roots of oxygen. Wet weather can also leave your garden with a range of fun-sounding diseases like powdery mildew, fire blight, and apple scab. Rain also puts something of a damper on pollinator activity.

After a big rain, try not to walk around in your muddy garden, which compresses the dirt. Maybe put down some boards if you have to move around. Put some more mulch down to further fight compaction.

Your sole defense is paying attention. Remove diseased leaves and fruit as soon as you see them—even from the ground. Let everything

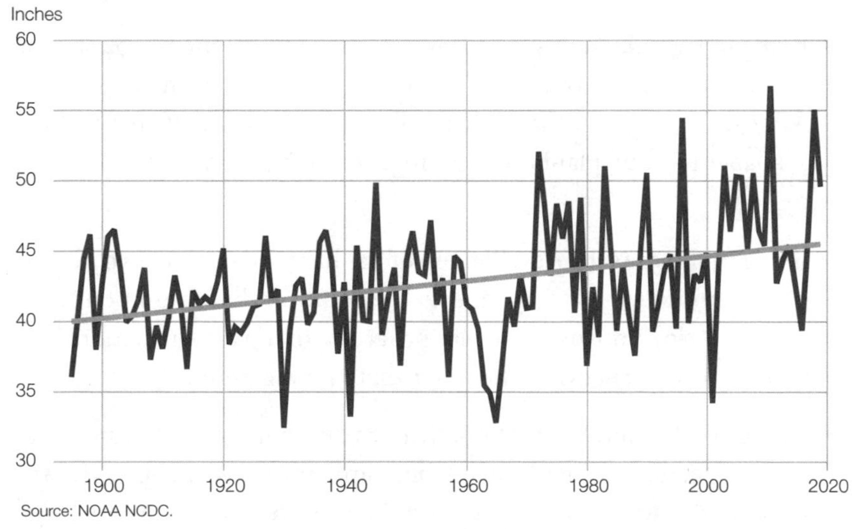

Figure 4-4. Rainfall amounts always swing crazily—but in general, the seasons are getting rainier. The straight line shows the average trend.

dry before you prune the plants, too, because your pruning shears themselves can spread the disease from one plant to another. Watch for, and remove, slugs, or set up a slug trap; slugs come out after a rain and treat your garden as a free buffet.

Both during and after the rain, you also have drainage to worry about. If the water's got nowhere to go, it can flood your seed beds, rot the roots, erode the soil, and kill fragile plants.

Your options are:

- **Fix the clay.** A lot of clay in your soil makes it hard for water to drain. Clay soil also gets compacted easily if someone walks on it, meaning that it can't absorb water as well as sandy soil, and that means more runoff, erosion, and flooding.

  Fortunately, you can fix the makeup of your dirt with organic "soil amendments" like compost, rotted manure, and leaf mold. Spread this stuff out a few inches deep and then work it into your soil, six inches deep.

  Or cover the clay soil with organic mulch like sawdust or wood chips, which will gradually break down and leaven your clay soil.

  Be patient with all of this; it can take a few seasons before you've fixed your soil problem.

- **Bring in the mulch.** If you spread a thick layer of mulch around your plants before a big rain, you'll protect their roots from soil erosion. After the rain stops, the mulch will help keep some of that wetness available to your plants—preparing, in effect, for the *next* drought period.

- **Bring in the gravel.** Gravel on top of the bare soil confers two degrees of rainproofness. First, it protects the soil from hard rain pounding; second, it slows down the water's progress, giving the earth more time to absorb and process all the water that's falling onto it.

- **Healthify the soil.** *Soil* isn't just dirt. It's dirt combined with air, water, organic matter, decaying plant bits, and maybe sand or clay. And *healthy* soil holds 20 times its weight in water—far more than unhealthy soil—which goes a long way toward solving a standing-water or runoff problem.

To be healthy, your soil needs a lot of organic matter (compost, decaying leaves, decaying straw, and so on), which brings with it nutrients and microorganisms—billions of them per inch—that do wonders for your plants.

You can also healthify your soil by growing lots of different plant species (rotate the crops, use cover crops); keeping the ground planted as long as possible; and covering the ground with plant residue year-round. And don't use a tiller.

Cover crops are what you grow after the vegetable-growing season is over, to serve as a kind of living mulch. They reduce erosion, keep weeds in check, and, once they die in the winter, serve as "green manure" to feed the soil in the spring.

You can buy all kinds of seed mixes for cover crops—annual ryegrass, oats, rapeseed, winter wheat, winter rye, and buckwheat, for example. But cover crops from the legume family, like clover, hairy vetch, beans, and peas, offer a huge added benefit: They convert nitrogen in the air into a form that the roots of your plants can use. Once the cover crop dies, that nitrogen becomes available to your other crops, acting as the world's best fertilizer. Nitrogen is a key ingredient in fertilizer; if you use a cover crop, you may not need to add fertilizer at all.

If it looks like your cover crop is going to seed, chop or cut it back. You don't want the stuff to become a weed farm unto itself.

After the frost has killed your cover crop, you can turn it under into the soil—but if you mow it at your mower's lowest level or weed-whack it, you can leave the cut-up bits in place to act as mulch. That way, the corpses of your cover crop continue to control weeds and contribute nutrients to the soil.

You remember, no doubt, what Franklin D. Roosevelt said in 1937: "The nation that destroys its soil destroys itself."

- **Raise the bed.** Raised gardens have a lot less problem with rain overload; gravity does a great job of pulling the excess water out of your little farm. If you've ever played in a sandbox, you get the idea: You build a wall around a rectangular space, maybe six or twelve inches tall, using wood, concrete, cinder blocks, stones, or bricks as the walls, then fill in that frame with dirt.

Just keep in mind that raised beds don't just dry out faster in wet weather; they also dry out and heat up faster when it's *not* raining. Water accordingly.

- **Swales.** A *swale* is a depression that directs groundwater in a certain direction—for example, down a hill or across a slope. Creating one is usually a job for a landscaper and a Bobcat digger, although for a small swale between growing beds or tiers, you and a shovel might suffice.

  "It can look like a natural part of the landscape," says Nan Fischer. "A wide, shallow depression with the same grasses as the surrounding area is hardly noticeable. If it's filled with decorative landscape rocks instead, it becomes a part of the design and can mimic a meandering stream. The line of it can also define areas of the yard, as a fence might."

- **Trenching.** Dig shallow trenches between rows of your garden—or do it in reverse: build up the rows of plants with mounded soil. That way, flash-flood waters have an exit strategy. You can create deeper trenches for the collected runoff. Just make them wide enough that people won't trip on them.

- **French drains.** If you have hard-core drainage problems—if, for example, you get standing water after even a medium rain shower—you may have to install a more hard-core solution, like a French drain. That's a perforated plastic pipe, surrounded by gravel and then covered over, which gives the water someplace to go.

- **Rain gardens.** Another idea to be filed under "Give the water a place to go": set up a rain garden. It's a small depression in the ground, usually on a downslope, that you plant with shrubs and flowers. The idea is to provide a temporary holding tank for water runoff from your roof, patio, and yard—and to filter out fertilizer, chemicals, oil, and bacteria that would otherwise have flowed directly into the groundwater. The makeup of the soil is important, so Google this project before you dive in.

- **Rain barrels.** Rain barrels connected to your house's downspout keep the water from flooding your yard and garden in the first place. Much more on this topic in Chapter 3.

Figure 4-5. Permeable pavers let water slowly soak back into the earth.

- **Permeable paving stones.** The perpetual problem with pavers is that they're impermeable. When it rains, the water has no place to go except to the lowest point, where it pools or pours into already full storm sewers. *Permeable* pavers look the same as regular pavers, but they let rainwater slowly filter through, downward into the soil. They let rain replenish the groundwater, but they reduce runoff, erosion, and ponding.

Getting Help

You're not alone. The world is full of people who can help you. You can find out what to plant these days, how to recover from extreme weather, how to design your garden, how to fight crop diseases, and so on.

You can ask at home-and-garden stores. You can join a gardening Facebook group for your area—great for asking questions about some weed or bug. Somebody there will know the answer.

Or you could exploit your own tax dollars at work: an agricultural extension service.

In 1862, in a masterstroke of forward thinking, the U.S. government donated free land to colleges that would be dedicated to teaching agriculture, science, and engineering. These are now known as land-grant universities, and there are 100 of them—at least one in every U.S. state.

In 1914, scientists from the land-grant colleges started making house calls in rural areas, bringing their knowledge to individual farmers and growers. Today, these *extension offices* are thriving, and they exist to help out anybody who's trying to grow anything. You won't believe how much knowledge they're willing to share with you—for free.

"You can call, you can email. You can take a picture of a pest or something on your plant and send it to them. They'll tell you what it is and what to do about it," says Sara Via. "People pay for them with their tax dollars—they should make use of them!"

You can look up your own state's extension services online (Google, for example, *texas extension service*), or you can use the government's directory at http://j.mp/2qdmQLy.

Your Landscape

Your home may also be flanked by all kinds of foliage besides your garden—plants grown for shade, runoff control, environmental diversity, or, above all, looks. But these plants, too, require different care as the weather intensifies.

Native Plants

You get a long list of benefits from growing *native* flowers, bushes, grasses, and trees on your property—plants that grow in your region naturally. They've adapted and evolved to thrive in *that* soil and in *those* weather conditions. You can find out what they are in your region at, for example, www.plantnative.org.

Native plants don't require any fertilizer; need less water and pesticide than "imported" plants; tend to have deep roots that fight erosion; and attract exactly the right kinds of native bugs, birds, and bats for pollinating your garden.

Diversity

These days, there's a second consideration in choosing what to plant around your home: diversity. Those poor native plants may soon discover that the weather is far less predictable and far more extreme than what they grew up expecting.

The moral: Plant native plants that have always done well where you live, but *also* plants that do well in hotter or wetter climates. When you examine the USDA growing-zone map, don't look only for plants that are supposed to thrive in *your* zone; also look at plants from the next zone south of you. Those are likely to be the new winners in the hotter era. Try them out and keep notes.

Diversifying—planting a wider range of plants than you used to—is especially important when you're planting trees, because they can live for decades, well into the worst of climate chaos.

The Lawn

Oh, man, the lawn thing. Environmentalists, botanists, and horticulturalists cannot *stand* mowed lawns.

"Let's lose the lawn," says the University of Maryland's Sara Via. "Mowed grass is like an infant: You have to provide it with everything. It can't get anything for itself because it doesn't have very big roots. It's introduced, so it's an instant monoculture. It provides no habitat. It's the worst thing ever."

What's so wrong with those beautiful carpets of green grass? Let scientists count the ways:

- **They're water hogs.** Their roots are only one inch deep. That's as far down as they can stretch to get water, so they require constant irrigation to avoid turning brown. A typical lawn requires the equivalent of a 1.5-inch rain every week. (Some native plants' roots, on the other hand, grow down 12 *feet*.)

- **They require chemicals.** You can't have a great-looking lawn without applying fertilizer, pesticide, and herbicide. Fertilizer is a carbon-emissions nightmare, and the chemicals can wash into our drinking water, or the fishes'.

- **They have to be mowed.** That primarily involves loud, smelly lawn mowers that burn fossil fuels and emit CO_2. And mowing takes either time (yours) or money (to hire someone).

So what are you supposed to grow instead of grass?

- **Clover.** If you want a sea of green that won't raise the eyebrows of the neighborhood association, here you go. No watering, no mowing, no fertilizers, and you can walk on it. It's incredibly cheap, on the order of $4 for a 4,000-foot yard.

  You can even buy seed mixes (for example, see www.earthturf.com) that combine microclover with low-maintenance grass. Since clover is nitrogen-fixing (gets nitrogen from the air, feeds it to the roots), the clover fertilizes the grass, and the entire thing looks just like a regular lawn.

- **Ornamental grasses** are fancy, lovely grass breeds that grow in splaying tufts or shimmering sprays. They're drought resistant, bug resistant, and weed resistant, so they don't require any fertilizer or chemicals. They're almost zero maintenance.

Figure 4-6. Microclover mixes can look exactly like grass—but they're self-fertilizing and don't require mowing.

The downside: You can't walk on it, so there's an art to choosing where to put it.

- **Moss.** No mowing, no watering, no chemicals. You can choose from a huge array of species. It grows superfast, and you can walk on it. It does not, however, like direct sunlight.

Of course, it's not *really* realistic to suggest that Americans should rip up and replace their lawns. What would become of the suburbs? It's a ludicrous suggestion—like asking people to give up beef.

But now, at least, you've heard the arguments. You know that there are cheap, easy alternatives. Maybe someday you'll have an opening to try them.

Chapter 5

Where to Invest

IT MAY SEEM CRASS TO THINK ABOUT INVESTMENTS WHEN THE world feels as if it's ending. In the climate-change era, billions of people will suffer—mostly poor people. Is that the right context for thinking about making money?

Yes, for one whopping big reason: Climate investing, at its core, supports companies that are trying to *mitigate* the climate problem. You're backing clean energy, electric cars, public transportation, sustainable crops. You're helping to *do* something about the problem.

"It's what I call 'No duh' investing," says Christine Harada, president of i(x) investments and former U.S. chief sustainability officer under President Obama. "You can make green while doing green things for the planet. Why wouldn't we do it?"

This chapter outlines the industries and companies that are likely to reward investment in a future where the climate is changing fast. But first: a crash course in environmental investing.

Intro to Sustainable Investing

The financial world teems with investment firms and funds that prioritize doing good in the world. They're responding to the 85% of American investors (and 95% of millennials) who've expressed an interest in sustainable investing, also called responsible, ethical, or principles-based investing.

These investments are white-hot. By mid-2020, total investments in them stood at over $40 trillion—a doubling in four years.

No wonder so many investment funds describe themselves as "sustainable" or "responsible" these days. But pinning down what they actually mean can be tricky. Furthermore, not many funds are designed to invest *exclusively* in climate issues; most consider other "make the world a better place" factors in the bargain.

You may, for example, hear some of these terms used interchangeably, even though they mean different things:

- **SRI** stands for "sustainable, responsible, and impact investing." These investors use "negative screens": They rule out investments in companies that violate their principles—say, tobacco, oil, and firearms companies.

- **ESG** stands for "environmental, social, and governance"—three factors that the stock pickers consider when choosing companies for investing. *Environmental* encompasses things like clean tech, carbon reduction, water use, pollution, green buildings. *Social* refers to workplace safety, hiring diversity, human rights, and benefits. *Corporate governance* incorporates executive pay, board diversity, anticorruption policies, and board independence.

  ESG traders *consider* sustainability factors, but don't flatly eliminate any companies on the basis of their E, S, or G qualities alone. You might be startled to learn that, for example, an ESG fund you own has investments in oil companies.

- **Impact investing** is the most direct idea of all. This time, you're investing in what the companies *do.* Their products and services are *directly* dedicated to clean energy or water access, for example.

The age-old wisdom is that sustainable investing may feel good, but doesn't make as much money as traditional stock picks. Turns out that hasn't been true for several years. For example, sustainable stocks have done *better* than the stock-market average every year since 2016. And during the pandemic, "the impact portfolios are very significantly outperforming the traditional ones," Brad Harrison of Tiedemann Advisors told the *New York Times*.

And now, the usual disclaimers: The stocks and funds described in this chapter are examples, intended to jump-start your research. Things

Where Not to Invest

Investing doesn't always entail betting *on* a company or industry; it can also mean betting *against* the ones you believe to be losers.

In the climate-change world, some of that's pretty obvious. Humanity's shift to clean, renewable power is well underway, and it's only begun; a massive global recoiling from disposable plastics is taking place, too. In time, both of those megatrends mean bad news for oil and gas companies.

Already, professional investors now lower a company's stock value if it owns oil reserves. The risk, according to the *Wall Street Journal*, is that "climate policies will curb future oil demand and leave some resources permanently underground and worthless."

If you'd invested $1,000 in clean-energy stocks at the beginning of 2017, you'd be up to $1,600 three years later. If you'd invested it in fossil-fuel companies instead? You'd have *lost* $120.

Another risky idea: investing in coastal real estate, where hurricanes and flooding are getting worse. Already, Miami homes near the water are losing value relative to higher-elevation homes.

Any industry that's 100% reliant on consumption of fossil fuels could also be in trouble—not only because the public will use less of their services in time, but also because the passage of a carbon tax would radically raise their costs. That includes airlines, cruise lines, and shipping lines, none of which can adopt electric propulsion anytime soon. All three of those industries are also susceptible to business disruptions whenever the weather gets crazy.

You also have to beware of companies whose *facilities* are vulnerable to violent-weather disruptions. Based on that principle, the market-intelligence firm Four Twenty Seven identified the top 15 corporations most vulnerable to climate disaster. The list includes a number of hard-drive and chip makers (Western Digital, Micron, Seagate, Applied Materials), which have major factories in flood-prone Asian regions like Thailand; Merck and Bristol-Myers Squibb, which manufacture their drugs in floody, stormy places like

Japan, Puerto Rico, and Florida; and power companies like ConEd, Dominion, NextEra, and PG&E, whose plants are on risky coasts.

Also in the vulnerable-location category: cruise lines like Norwegian and Royal Caribbean, whose headquarters, terminals, and travel agencies are in Miami, one of the most climate-doomed cities in America.

In time, more companies will begin to disclose their vulnerabilities to organizations like CDP.net, or to their shareholders. At that point, it will be easier for you to judge for yourself a stock's climate risk. Until then, let logic and Google be your guide.

may have changed since these pages were written. As in any kind of investing, do your homework before you invest. Past performance is no guarantee of future results. Invest for the long term, *especially* when it comes to climate ventures.

Above all, don't sue the book author if things go bad.

Stocks versus Funds

If you're an individual, nonpro trader, you won't be able to get in on exciting companies while they're still small. Some start-ups are looking only for venture capital, for example, or they're still privately held.

If you're inclined to buy stock in publicly traded companies, it's worth sniffing out how dedicated they are to climate-change mitigation.

For starters, you can check CDP (www.cdp.net), a not-for-profit charity that maintains a massive database of companies' carbon emissions and climate plans. (CDP is no longer short for the *Carbon* Disclosure Project, since it now asks companies to disclose details on their water use and deforestation, too.)

About 8,500 publicly traded companies now offer their climate data to the CDP. Increasingly, if companies are *not* reporting their climate plans, that's a bad omen for their success as a company. "At this point, if you're not reporting, it's a signal of poor quality," says Anant Sundaram, professor of business and climate change at Dartmouth's Tuck School of Business. "People will shake their heads and say, 'What is wrong with you?'"

You can also look over a company's annual report, its website, or its 10-K report: a mandatory, comprehensive report detailing a company's performance. If the company has anything to say about its exposure to climate problems, you'll find it in section 1A.

Mutual Funds

If you're a non-full-time, non-millionaire investor, though, there are solid reasons for investing in *funds*, like mutual funds and bond funds, instead of (or in addition to) individual stocks.

First, investing in a fund is far safer than buying stocks one at a time because you're investing in a portfolio of many companies simultaneously and blending their results. If one company's stock tanks, you won't be wiped out.

Of course, you're also less likely to get a huge windfall when one stock surges, but most people are willing to make that bargain in the name of safety.

Second, you don't need a lot of money to invest in a fund. You can pop over to your favorite trading website (E-Trade, Schwab, Fidelity, and so on) and buy as much or as little of a fund as you like.

Mutual funds bring you the wisdom and full-time dedication of managers who do all the research, the buying, and the selling. Their goal is to identify risks and opportunities before the rest of the world, in the name of making your money grow. They can also pressure individual companies to clean up their environmental acts.

Each fund posts its prospectus—a detailed description of its investment strategies—on its website. If the fund cares about things like environmental responsibility, you'll find it there.

Some impact funds invest in companies whose business is directly related to climate adaptation or mitigation. Examples: the Templeton Global Climate Change Fund, Aviva Investors Climate Transition European Equity Fund, Impax Environmental Markets, and Hartford Environmental Opportunities Fund.

ETFs

ETFs (exchange-traded funds) are increasingly attractive to many investors. In one way, they're like mutual funds: each share represents tiny slices of many individual stocks.

But ETFs generally cost far less in fees because they're based on stock-market indexes (the S&P 500 is an index, for example, and so is a gold-price-tracking index). You're not paying 1.1% of your entire holdings every year to a bunch of highly paid stock pickers, analysts, industry researchers, and so on. The average annual fee for ETFs is about half as much.

ETFs have favorable tax ramifications, too; you pay capital gains taxes only when you sell the entire investment.

Finally, you can buy and sell shares in ETFs at any time during the business day, as though they're shares in stocks. That way, you can pounce on breaking news, if the spirit moves you (or the fund manager). You can trade mutual funds, on the other hand, only at the end of each trading day.

Examples of climate-mitigation ETFs:

- **Etho Climate Leadership US ETF** has no holdings in fossil fuel, tobacco, weapons, or gambling companies. It's based on an index of stocks chosen on the basis of their companies' carbon footprints.
- **BlackRock iShares Global Clean Energy ETF** tracks an index of companies that produce energy from solar, wind, and other renewable sources.

Green Bonds

You, the lowly individual, are not the only person investing in a cleaner world. Since 2016, big nonprofit institutional investors (universities, foundations, pension funds, and nonprofit organizations) and commercial investors (banks, insurance companies, pension funds, mutual funds) alike have poured half a trillion dollars into *green bonds*.

A *bond* is like a loan that investors make to an institution—a city, government agency, country, or company. During the agreed-upon duration of the bond, the investor earns interest on that money—in many cases, tax-free. At the end of the bond period, the investors get the original amount back. They've earned income from the bond without risking their initial investment, as they might have in the stock market.

A *green* bond is one that a city, state, or company sells explicitly for the purpose of funding specified environmental projects. It might be energy-efficiency upgrades, pollution reduction, clean transportation, water management, or protecting ecosystems.

If you're a city, for example, and you need $500 million to build a new, clean rapid-transit line, you might offer a green bond to raise the money.

You, the city, get an instant influx of capital—with a lower interest rate than you'd have gotten from a bank loan, and with fewer restrictions. For the next five years (or whatever), your investors get income, maybe some tax breaks, the confidence that they'll get all their initial investment back, and the knowledge that they're helping the planet.

Green bonds are all the rage; investments in them have surged from $2.6 billion in 2012 to a projected $375 billion in 2020.

One key reason is that many huge industrial investors, like CalPERS (California's pension department), are *required* to invest sustainably.

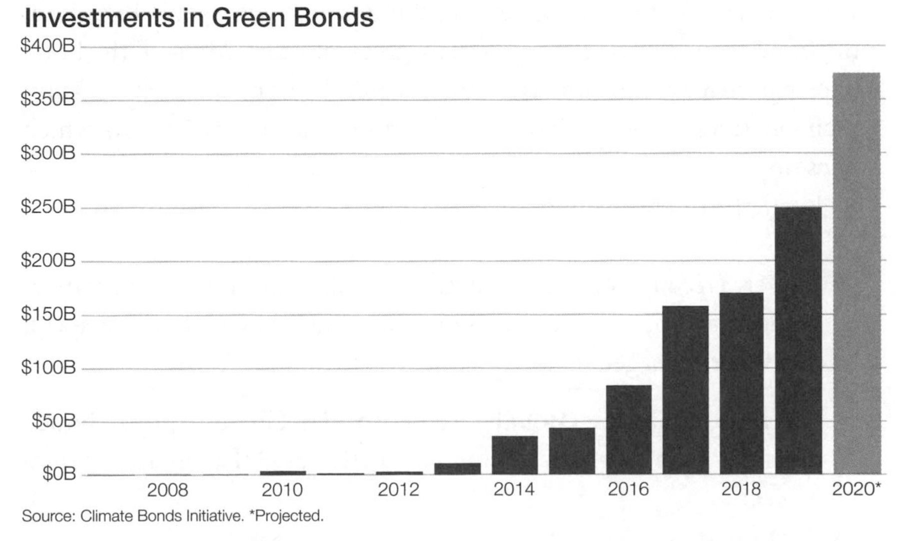

Figure 5-1. The investing world has been pouring money into green bonds.

Even bigger entities, like sovereign wealth funds, which invest a *country's* financial reserves, have similar mandates. At over $1 trillion, for example, Norway's sovereign fund is the world's largest. It does not invest in coal, gas, or oil companies.

All of these massive investment factories are constantly looking for environmentally responsible places to invest. That's the good news.

The bad news is the specter of *greenwashing*, which is a cynical and misleading attempt to make a product appear environmentally friendly in order to sell it.

"Just because it's green doesn't mean it's good. You have to look both ways," says Patrick Drum, portfolio manager of the Saturna Sustainable Bond fund.

He's full of examples of green bonds whose green bona fides were "a little nefarious." He mentions, for example, the 13 Chinese green bonds in 2019 that funded new *coal* plants, or the Massachusetts green bond that was intended to pay for a new parking garage. Yes, sure, the garage's design included charging stations for electric cars, but the net result of any new garage is to encourage *more driving*. "They're going to throw a few plugs up and call that a green bond? Those things become questionable," says Drum.

In general, individuals can't buy individual green bonds directly. But you can invest in them via green-bond *funds*. That way, you'll know that the bonds you're buying are genuinely green at heart. Many of the bond offerings' managers voluntarily hire independent organizations to audit their offerings—a good sign—and the fund manager will know which ones do.

Examples of green-bond funds:

- **Calvert Green Bond Fund** buys bonds that fund projects in renewable energy, energy efficiency, green buildings, low-carbon transport, and water management.
- **BlackRock iShares Global Green Bond ETF** buys green bonds primarily from overseas, giving you a slice of global environmental projects.

- **VanEck Vectors Green Bond ETF (GRNB)** buys green bonds both in the United States and abroad, all of which have been certified by the Climate Bonds Initiative, a non-for-profit investor group that certifies bonds as being genuinely green.

You're not going to double your money in six months, that's for sure; most bonds, especially these bonds, are slow-and-steady, supersafe investments. But you'll know that your money is pushing back against the earth's destruction.

The Climate-Change Sectors

For 100 years, economists' calculations considered natural resources to be infinite. They ignored "externalities" like pollution.

"If I'm a copper miner in the early 1900s, I go into a mine in Arizona and I pull out a bunch of copper. I extract it with poisonous chemicals, and I dump the chemicals. And it doesn't matter, because I'm in the Arizona desert. As long as population is low, those externalities don't matter. They can be ignored," says Erik Kobayashi-Solomon. He's the founder of IOI Capital, a private partnership that invests in companies focusing on climate-change mitigation and adaptation.

But now the rules have changed. *Trillions* of dollars will soon be spent decarbonizing the economy and overhauling our energy grids. And wherever that much money flows, there are huge investment opportunities. "Someone's going to make a lot of money investing in efforts to address climate change," says Lucas White, portfolio manager for the GMO Climate Change Fund. "Someone's going to nail it and do very, very well."

That's the premise of climate-change investing: putting your money into companies that will profit from the shift. In general, investment pros assume that those corporations will occupy certain predictable sectors, like clean energy, agriculture, electric vehicles, water access, and pollution-reduction technology.

Here, for your climate-investing pleasure, is a roundup of the major categories.

Clean Energy

Solar, wind, hydro, geothermal, and tidal power are almost overwhelming in their wonderfulness. Imagine: free power. Electricity that consumes no fuel, ever—and therefore ends all wars with countries where we have to dig up that fuel. And no emissions, ever. Build it once, and enjoy the power for decades, yours free.

Decarbonizing our power is a big, big deal for the planet. Power production is responsible for 28.4% of U.S. greenhouse-gas emissions.

The systems we've developed to get power have become massive, sprawling, entrenched, complex, and dirty. That's why overhauling the whole thing will involve such a massive amount of money changing hands. The trick is to be on the winning end of those transactions.

Solar Power

Your first thought upon hearing the term *climate investing* might be: solar panels. Surely they're the future of energy on this planet!

Well, yes, they are. But that doesn't make them good investments.

Solar panels are commodities. The barriers to entry are low, meaning that anyone with some start-up cash can begin manufacturing them, and many companies have. Worse, says Lucas White, there's no way to differentiate your product from everybody else's. "Apple doesn't have a cool white solar panel. They're all these boring, dopey gray panels. Your entire choice is based on cost per unit of electricity produced."

The lowest price wins, and Chinese manufacturers have been dumping panels on the market at a loss.

That's why the price of solar has plummeted 60% in ten years, much faster than experts were predicting. That's great news.

But that makes solar panels a tough investment. "It's a race to the bottom," says Vivek Tanneeru, portfolio manager of Matthews Asia, the largest Asia-only investment specialist in the United States. "Usually, the lowest bidder gets the order, and they're barely making profits. It's just a terrible place to be."

Panels aren't the only part of the solar story, however. Solar more generally—panel makers *and* installers *and* utility companies—can be a reasonable investment. "You won't get crazy 25% returns, but it's going to be slow and steady," says Christine Harada. "And it's countercyclical, meaning that it will withstand any kind of recession."

She points out that 38 states (and Washington, DC) have now established renewable-energy production goals; New York, Hawaii, and California, for example, plan to reach 50% renewable energy by 2030. "In those states, solar's taken off. It's not going to stop for the next five or ten years."

The biggest companies involved in solar include panel makers like SunPower (SPWR) or First Solar (FSLR), panel installers like Sunrun (RUN), and solar-leaning utilities like NextEra (NEE) and AES (AES).

Or, for safety in numbers, you could buy into an ETF that invests in solar, like the Invesco Solar ETF (TAN), which owns shares in about 24 solar companies.

Wind Power

Wind power is growing about 9% a year. In the United States, wind power now supplies about 7.3% of all electricity; in six states, it provides over 20%.

But wind, too, faces challenges as an investment. One of them is that the actual turbines are often built hundreds of miles away from the cities that will use the power they generate. Yes, we can expand the electrical grid to reach the turbines—but that takes time, and as with many other huge infrastructure projects, those projects are often delayed.

Still interested? The industry is dominated by three giant turbine companies: Vestas Wind Systems (VWDRY), Siemens Gamesa (SGRE), and GE Renewable Energy, which is a subsidiary of General Electric. (You can't invest purely in GE's wind division.)

Morgan Stanley analysts recommend companies like NextEra Energy (NEE), a massive energy company that runs electric utilities all over America. It's the world's largest generator of wind and solar power, although some of its plants burn fossil fuels. They also like TPI Composites (TPIC), the biggest maker of wind-turbine blades in the United States.

You could also limit your risk by buying into an ETF, like the First Trust Global Wind Energy ETF (FAN). It owns shares in the biggest wind-turbine companies and wind-turbine part makers.

Geothermal, Hydro, and Tidal Power

Solar and wind aren't the only ways to generate clean power.

- **Hydroelectric power** comes from running or falling water, which is why you often see power stations next to big dams. Hydro provides about 7% of all power in the United States, and 20% of China's.
- **Geothermal power** exploits the hot innards of the planet, turning water into steam, which drives a turbine. Geothermal is a fairly small player; by 2050, if all goes well, geothermal *may* produce 5% of the world's power.
- **Tidal power** uses the massive forces of the incoming and outgoing tides to generate power. Nobody's generating much meaningful juice from these systems yet, but enthusiasts point out that the technology is ready for broader experiments. One great feature of tidal power: Unlike wind and solar, it's predictable. We know *exactly* when the tide is going to go in or out.
- **Wave power** is another superniche player, attractive because of its unlimitedness and lack of emissions, but still mostly just a cool idea. Various proposed projects would turn the motion of the waves into harnessable energy.

All of these types are clean and renewable, but most of them are also limited geographically. They're not generating a lot of tidal power in Oklahoma.

To invest in these alternative sources, your best bet may be diversified clean-energy funds. For example, the Invesco WilderHill Clean Energy ETF is based on a clean-energy index of 40 stocks—solar, wind, hydrogen fuel cell, and others. iShares Global Clean Energy ETF takes an international approach, with investments in 30 solar, wind, and, especially, public electricity companies around the globe.

Utility Companies

As the Great Decarbonization Wave gets underway, power companies stand to profit in two ways.

First, as the price of clean, renewable power continues to fall, it will cost them less to produce power. Already, the cost of wind power has dropped by 69% since 2009; the price of solar has dropped 88%. Their prices are close to the price of coal and gas power—lower, in some places—and their prices aren't finished falling.

In some places, it actually costs more to keep operating a coal plant than to build a whole new wind farm.

For solar and wind power, a utility doesn't have to buy any raw materials to burn. There's no legally required pollution-control equipment to buy and maintain, either, such as catalytic converters and scrubbing units. So the power company's costs go down, but its income stays the same. Translation: more profits. (You don't really think they'll *lower* your electricity rates, do you?)

Second, there's the threat of regulation. Most experts believe that the United States will eventually institute a carbon tax, as described in Chapter 17. The day that law goes into effect, utilities that are still burning fossil fuels will be in deep financial trouble. The ones who've decarbonized will suddenly look like geniuses.

It's already happening. According to a Morgan Stanley report, North American utilities who've taken steps to decarbonize have already improved their earnings per share by 6% relative to their peers.

All ten of the ten largest electric utilities have set emissions-reduction goals; most have committed to becoming net-zero (producing no carbon dioxide at all) by 2050, in keeping with the Paris Agreement's goals.

If you were to invest in utility companies, therefore, you'd want to choose the ones that will be the most profitable as time goes by. Some good options include:

- **Xcel Energy (XEL)** committed early to delivering 100% carbon-free energy by 2050. It has rapidly built up its access to wind power, and it plans to retire 23 coal plants by 2027. Already, Xcel has decreased its carbon output by 38% since 2005—while doubling its power output.

- **NextEra Energy (NEE)**, which calls itself "the world's largest producer of wind and solar energy," has set a goal to reduce its carbon emissions by 67% by 2025 (from 2005). If NextEra were a country, it would rank number eight among all nations in wind-power capacity. Its Florida utility is installing the world's largest grid-storage battery project.

- **AES Corp. (AES)** describes its mission as becoming "the world's leading sustainable power company." It has set, as its 2030 goal, a 70% reduction in emissions (from 2016).
- **California utilities.** It's too bad that PG&E, California's largest utility, was driven into bankruptcy by the state's recent devastating wildfires, because it was working hard toward cleaner energy.

  But Southern California Edison (EIX) is a healthy alternative. "They're a *little* dinosaurish, but still forward leaning when it comes to deploying solar," says Christine Harada.

If a mutual fund or ETF is more your style, your best bet is one of the clean-energy funds described above.

Grid, Power Storage, Generators

There's a massive fly in the solar + wind ointment: intermittence. When the sun isn't shining or the wind's not blowing, there's no power. That's a profound problem, one that's preventing these free, unlimited sources from supplying all of our power. There's only one obvious way to fix the intermittence problem: batteries.

We're talking about massive, heavy, industrial storage systems that can store the power that the sun generates during the day, and then deliver it to buildings at night. Or store the wind energy when it's blowing and feed it back to us when it's not.

Nobody's sure what form these grid-storage batteries will take. All kinds of systems are in limited or experimental use. There are *pumped-storage* systems (sunlight powers pumps that feed water uphill; the water flows down at night); rail-energy storage (sunlight powers heavy rail cars uphill; they roll down at night to generate power); flywheel energy storage (the sunlight powers heavy, frictionless flywheels to spin up at incredible speeds—then at night, the spinning wheels power a generator); flow batteries (gigantic batteries whose chemical components are stored in huge tanks); and even massive banks of regular lithium-ion rechargeable batteries.

But talk about a market poised for growth: Grid battery storage is expected to grow 1,700% by 2030. "Literally trillions of dollars are going to flow into modernizing our electric grids worldwide," says GMO Climate's Lucas White. That flood of money will begin with $620 billion in investments by 2040.

Energy efficiency is a cousin to these industries, and it's exactly as vague and broad a category as it sounds. But the idea is compelling: If your home or company can serve its purpose with, say, 40% less energy, that's a more effective way of reducing greenhouse gases than adding, say, 10% more renewable power to the mix.

Efficiency enhancements can take all kinds of forms: LED lighting, more efficient appliances, better building materials, smarter architecture, shorter resource chains, and so on. Hundreds of companies toil in the folds of these sectors.

Don't forget about home generators as an investment opportunity, either (Chapter 3). As extreme weather gets ever more extreme, more people will lose power—then rush out to get a generator so they'll be prepared the next time. Generac, which holds 70% of the U.S. generator market, has been experiencing surging sales, along the lines of 45% a year. It's no shock that sales skyrocket before and after every hurricane; Briggs & Stratton, another manufacturer, saw a double-digit sales increase for its standby generators in the year following Hurricane Sandy.

And in 2019, when Pacific Gas & Electric deliberately cut off power to 570,000 California customers to avoid a dry-season wildfire, calls to some generator dealers increased 1,400%.

You can hit a lot of these corporate categories in one fell swoop with shares in funds like these:

- **Schroder ISF Global Energy Transition (SCHGLAA:LX)** invests in grid storage, along with renewable power production, renewable energy equipment, transmission and distribution, smart-grid technologies, and electric-car charging. No fossil fuels or nuclear.
- **First Trust NASDAQ Clean Edge Green Energy Index (QCLN)** is tied to an index of 40 clean-energy technology stocks: solar panels, biofuels, advanced batteries, and so on.

- **Fidelity Select Environment & Alternative Energy Portfolio (FSLEX)** invests in renewable energy, energy efficiency, pollution control, water infrastructure, waste and recycling technologies, and so on.

Efficiency

How would you like a company to come to your building, inspect its systems top to bottom, and then give it a complete energy makeover—for *free*? That's right: They'll design, build, buy the equipment for radically improved efficiency and cost savings—and pay for it all themselves.

In exchange, all they ask is a percentage of the money you *save* on energy thereafter—for between 7 and 20 years. The more money their upgrades save you, the happier they are. They have an incentive to do a spectacular job.

They're called Energy Savings Companies, or ESCOs. Since 1990, these firms have performed energy retrofits on government buildings, schools, hospitals, universities, and commercial buildings, saving $55 billion in energy costs and keeping 450 million tons of carbon dioxide out of the air.

The biggest five ESCOs are Schneider Electric (SBGSY), Siemens (SIEGY), Ameresco (AMRC), Noresco (part of United Technologies, UTX), and Trane (part of American Standard, ASD).

Water

The shortage of fresh water, as you know by now, is a global crisis. Climate change is contributing to droughts, depleted aquifers, shrinking snowpack and glaciers, more regions turning to desert—while the growing and upwardly mobile population wants more fresh water than ever before. Today, about 4 billion people experience severe water shortages for at least one month a year.

The days of water feeling like a free and unlimited resource are ending. "Water is increasingly likely to be recognized as the valuable finite resource that it is," says Lucas White.

As an investor, you can consider the growing problem in either of two ways. First, you can recognize the business categories that will be in trouble as water becomes scarcer and more expensive:

- **Agriculture** gulps down 69% of all the fresh water we consume on the earth. All that irrigation, all those cows.
- **Oil and gas.** Extracting oil and gas takes massive amounts of water—especially in fracking—and produces staggering volumes of *contaminated* water. The clean water is increasingly expensive to acquire, and the dirty water is increasingly expensive to dispose of.
- **The power grid.** Getting electricity to your plugs is also a massively water-thirsty operation. The power system needs water to power the turbines, to cool the equipment, and to clean the emissions. About half of all power plants are in regions that already experience water shortages.
- **Electronics.** Yes, the chip-making industry, without which you would not have a smartphone, car, microwave, or TV, requires vast amounts of ultrapure water to rinse the silicon wafers that get etched into circuits. It works out to 3,200 gallons of water per iPhone, or 4 million gallons a day *per factory.*

The other way to invest is to look at companies that focus on efficient use of water resources, water recycling, purification, treatment, desalination, and so on.

If you're shopping for individual stocks, you could look at American Water Works (AWK) and Aqua America, the two largest U.S. water utilities. ITT Industries makes purification systems, among many other forms of high-tech equipment.

Or diversify with a mutual fund like the Calvert Global Water Fund (CFWAX) or the AllianzGI Global Water Fund (AWTAX).

ETFs that traffic in water stocks include Invesco Global Water Portfolio ETF (PIO), First Trust ISE Water Index Fund (FIW), and iShares Dow Jones US Utilities Index ETF (IDU).

Farming

Farming has always been a difficult line of work, and climate change is making it a lot harder. But for investors, agricultural companies are attractive for a couple of reasons. First, there's no such thing as a slump in demand for *food*; people always need to eat. And by 2050, there will be a lot *more* people eating—9.7 billion. That's what you call a growth industry.

Second, the changing climate is driving billions of dollars of innovation in every realm of food production. For example:

- **Smart irrigation.** These days, irrigation is a lot more sophisticated than a bunch of pipes spraying water onto the crops. The "smart" systems monitor the soil, the weather, evaporation rates, and how much water the plants are actually consuming and adjust automatically, saving water and money.

  Many of the players (Hortau, HydroPoint, Arable) are privately held. But you could always buy shares in Valmont (VMI) or India's Jain Irrigation (JISLJALEQS), both of whom specialize in high-tech irrigation.

- **Drought-resistant seeds.** The crops we've been growing for decades don't grow well in the hotter, wetter, crazier new weather era. If the temperature rises 7°F—which, at the rate we're going, is what we'll have by 2075—U.S. corn production will drop by *half.*

  The great hope is that Big Ag can come up with more resilient breeds of seeds. That would include seed developers like Monsanto (MON) and Syngenta (SYT).

- **Fertilizer.** The invention of man-made fertilizer in the early 1900s was a blessing, because it increased agricultural production enough to feed 2 billion more people. But it's also been a curse, because making fertilizer involves massive burning of fossil fuels.

  The world needs to figure out cleaner ways to make the stuff. In the meantime, the world will need *more* fertilizer as the population grows and the growing regions shift. That's an argument for fertilizer companies like CF Industries Holdings (CF), Mosaic (MOS), and Nutrien (NTR), which resulted when two biggies (Potash and Agrium) merged in 2018.

- **Pest resistance.** Climate change is bringing more bugs to more places. Farmers are desperately seeking ways to protect their crops from pests that will decimate their production. Companies like Monsanto (MON) and Ceres (CERE) are tackling the problem with biotechnology, developing both pest-resistant breeds of crops and pesticides.

- **Equipment.** In the northern hemisphere, the "breadbasket of America" growing regions are getting hotter. To adapt, farms will either have to start growing different crops or move north. In both cases, there'll be a lot more people buying new farm equipment.

  That bodes well for companies like Deere & Company (DE), AGCO (AGCO), and Navistar (NAV, formerly International Harvester).

- **Precision agriculture** is a hot new area, expected to reach $240 billion in sales by 2050. The idea is to use sensors on the ground, on the equipment, and in the air to detect weather, soil conditions, pest effects, moisture levels, and microclimates. The farmers, or even their automated equipment, can then plant, water, or fertilize accordingly. GPS-driven tractors and drones (for sensing and imaging field areas) fall under this umbrella.

  Example companies: Trimble Navigation (TRMB), AGCO (AGCO), AgEagle Aerial (UAVS), and, inevitably, Deere & Company (DE).

- **Land.** Oh, yeah, there's one more thing we're going to need a lot of: farmland. You can, in fact, invest in companies and funds that manage the farms themselves, which make money from the rent that farmers pay. The price of an acre of Iowa farmland jumped 400% from 2001 to 2018.

  You may not want to buy the farm. Buying shares in a farm *real-estate trust* carries a lot less risk because you're owning slices of farms in many different places. Farmland REITs (real-estate investment trusts, which buy real estate that generates income) include Farmland Partners (FPI) and Gladstone Land (LAND).

As always, you may not want to invest in individual stocks; it's safer to buy shares in a mutual fund or an ETF. Promising options include Invesco DB Agriculture (DBA) and VanEck Vectors Agribusiness ETF (whose stock ticker, hilariously, is MOO). They invest in seed companies, food companies, fertilizer companies, and so on.

For agribusiness mutual funds, consider Fidelity Global Commodity Stock Fund (FFGCX) and CI Global Infrastructure, Timber, and Agribusiness (INNAX).

Food Tech

Everybody's got to eat, every day. As the population balloons to 9.7 billion in 2050, somebody will have to figure out how to feed them.

For investors, the usual suspects are the huge agribusiness companies, but also Archer Daniels Midland (ADM), the behemoth, and Bunge (BG). They buy crops from farmers, turn it into packaged food, and get it to the grocery stores.

But it's the food *tech* developments that are getting investors' interest. These are the two everybody's talking about:

- **Animal-free meat.** Already, the burgers from plant-based meat companies like Beyond Meat (BYND) and Impossible Burger (not yet public) are crazily popular, utterly delicious, and 100% cow-free. Plant-based products already make up 5% of all U.S. meat purchases, and that's expected to triple by 2030. That would make meatless meat a $140 billion a year industry in ten years.

  Then there are the companies making *real* meat without cows—by growing cow muscle tissue in an incubator. They include Memphis Meats and Mosa Meat, neither of which has gone public.

  Other companies are developing animal-free fish, eggs, chicken, and dairy—a field that's expected to grow 28% a year.

  Be careful, though: You can't create a new product that successful without attracting the attention of Big Meat. Already, Nestlé, Tyson, Conagra, Kellogg's, Hormel, and Kroger have all launched, or are preparing to launch, their own no-animal meats. The category is going to be enormous, yes—but it will also be teeming with competitors.

- **Vertical farms.** In these buildings, farmers (or robots) grow crops in stacked layers, indoors, under perfect temperature, light, and humidity conditions. The advantages: year-round growing, no weather surprises, no weeds, no bugs, no emissions from plowing or planting machinery, no herbicides or pesticides, minimal fertilizer runoff, perfectly controlled growth, very little land use, no wasted water or nutrients, and a short distance from "farm" to customers' tables, especially in cities.

  All of this is an enticing prospect—both for anyone who eats food and for investors, who expect vertical farming to be a $3 billion industry by 2024.

Figure 5-2. Vertical farms don't worry about weather, because they're indoors.

Dozens of these farming companies are already in business; you can already buy their crops in grocery stores and restaurants. At this point, it's mostly lettuces, herbs, and various baby greens, all unbelievably pure, fresh, and local.

AeroFarms, for example, doesn't even use soil; it sprays a nutrient mist on the suspended roots, requiring 95% less water than traditional farming. Whole Foods carries AeroFarms baby greens under the Dream Greens label.

Other U.S. entrants include Plenty (received $200 million in funding from SoftBank), Bowery Farming ($90 million from Google), Green Spirit Farms, Babylon Micro-Farms, Iron Ox ("grown by robots with love").

All of them, alas, are privately held; you can't buy stock in them except through rich-people's funds.

Transportation

The single biggest source of greenhouse gases (29%) is machines moving us around the face of the earth. Unless we can decarbonize transportation, we won't decarbonize enough to hold off climate change.

That's why so many companies are racing to electrify our wheels—and so many investment dollars are pouring into them.

Electric Cars

The global decarbonization effort won't succeed until we can find a way to get around without burning fossil fuels on every trip. The obvious solution: electric cars, which produce zero emissions while you drive.

In the last five years, the prices of EVs have fallen, the recharging time has fallen, and their driving range has skyrocketed. They're ready for prime time; analysts expect sales to grow by 24% a year, reaching 28% of all car sales by 2030.

With varying degrees of credibility, 18 countries have already announced plans to begin *banning* gas cars by 2040, including Canada, Singapore, Germany, China, France, and the United Kingdom. All of this adds up to a tsunami of EV research, development, and sales.

The Electric-Car Conversation

Electric cars are a blast to drive because they offer *insane* acceleration. They're quiet. The electricity they use costs about a third of what gasoline does.

Electric vehicles (EVs) also cost next to nothing in service and maintenance. An EV has a *motor* (the thing that turns the wheels) but no *engine* (the thing that burns fuel to generate power). Therefore, it has no transmission, spark plugs, fan belts, air filters, timing belts, or cylinder heads. It never needs emissions checks, oil changes, or tune-ups. Your brake pads and rotors go years without needing replacement, too, since just lifting your foot from the accelerator slows the car down (by recharging the batteries).

The drivetrain of a gas car has over 2,000 moving parts. An EV's drivetrain has 20. You can guess which one is more reliable—and why service departments generally *hate* electric cars. Car dealerships make twice as much profit from servicing a car as they do from selling it.

And, of course, an EV produces zero emissions as it drives.

Now, change is always scary, especially technological change. It's no surprise that electric cars have had to drive through clouds of bad information and mythology. For example:

"There's no emissions advantage; the electricity still has to come from somewhere." The electricity to charge an EV usually comes from the electrical grid. And true enough: In most states, some of that power comes from fossil fuels.

Even so, an EV is still far better for emissions, for two reasons. First, the proportion of renewable, emissions-free energy used by today's power plants is steadily rising. The more clean energy the world produces, the better EVs look.

Second, a municipal power plant can generate electricity far more efficiently than millions of individual engines built into cars can, thanks to a little thing called economies of scale.

"There's no emissions advantage, because EVs produce more gases in manufacturing." Lithium batteries do require a lot of energy to manufacture. All told, the *manufacture* of a typical EV produces 15% to 68% more greenhouse gases than a gas-engine car, depending on the car's range and where it's built. (Unless, of course, the factory that makes the batteries is powered by renewable energy, as Tesla's factories are.)

But you have to consider the entire life cycle of a car. Once it's on the road, an electric car produces far fewer emissions than a gas car. After about 1.5 years, an electric car has caught up with the gas car's life-cycle emissions and begins to surpass it. Over the vehicle's ten-year life cycle, from manufacturing to disposal, a modern EV produces about *half* the emissions of a gas car.

"There were five million electric cars on the roads at the end of 2018," says Matthews Asia portfolio manager Vivek Tanneeru. "By 2030, there might be seventy-five to one hundred million. That's an enormous opportunity."

Your first thought of electric-car investment might be Tesla. After all, 70% of all electric cars sold in the United States are Teslas. In 2019 and 2020, the Tesla Model 3 outsold all other electric vehicles *combined.* Its cars are fast, stylish, and adored by their owners.

But the rest of the industry isn't standing still. Worldwide, the Nissan Leaf is the biggest selling EV. Toyota, Volkswagen, Hyundai, GM, Ford, Nissan, Honda, Fiat Chrysler, Daimler, Renault, Mitsubishi, Audi, Mini, Subaru, Kia, Mercedes, BMW, Volvo, Porsche, Jaguar—virtually every carmaker on earth is now making or will soon make lines of electric vehicles.

So you could invest in individual car companies. But GMO Climate's Lucas White points out that you can also invest in the *parts* that all of the car companies will have to use.

How can you get in on this revolution?

- **Batteries.** One hundred percent of electric-car batteries are made by Asian companies. (That even includes the batteries that Tesla makes in its vast Nevada battery factory, the Gigafactory. Panasonic manufactures the battery cells in one side of the building, then passes them to the Tesla side for assembly.)

It's incredibly difficult to start up a battery-making outfit. European companies like Bosch and Dyson spent billions trying; both ultimately gave up. "That's why I think that the incumbents have a good chance of keeping their lead, over the coming five to ten years at least," says Vivek Tanneeru.

All current electric cars use lithium-ion batteries, which are dominated by four Asian manufacturers: Panasonic (PCRFY), Contemporary Amperex Technology (300750), Samsung SDI (006400), and LG Chemical (051910). (Asian stock symbols are generally numeric instead of alphabetic.)

- **Lithium.** The lithium-ion batteries in the electric cars use the same chemistry found in laptop and phone batteries. And lithium comes from mines. And those mines are run by mining companies like Albemarle (ALB) and Chile's Sociedad Química y Minera de Chile (SQM).
- **Components.** Somebody's got to make the charging ports, high-voltage connectors, and electrical shielding required in an EV, no matter what the brand. Those are some of the specialties of Aptiv (APTV) and Valeo (VLEEY).

If you think it might be simpler and safer to invest in a *fund* that benefits from the rise of EVs—good call!—you could consider the Global X Autonomous & Electric Vehicles ETF (DRIV) or the KraneShares Electric Vehicles & Future Mobility ETF (KARS).

Mass Transit

If you're an American, you might be completely oblivious to what's happening in public-transit systems. That's because, while not many new ones are popping up on our own soil, they're blossoming all over the rest of the world. You can now find metro rail systems in 178 cities around the world—and *42%* of them have been built since 2000.

Two hundred more metro lines are planned for construction in the next five years, in places like Saudi Arabia, New Zealand, Colombia, Nigeria, Singapore, and Indonesia.

Seven of the ten busiest systems are in Asia. And because Asian governments are dead set on reducing emissions, Tanneeru says, the companies who make, own, and operate public-transit systems are all good bets for investment.

He singles out, for example, Hong Kong's MTR, which he says is the best operator of mass-transit systems in the world. What makes it a good investment is its staggering growth opportunities—*outside* Hong Kong. MTR now operates rail services in huge Chinese megacities (Beijing, Shenzhen, and Guangzhou), as well as Stockholm, London, Melbourne, and Sydney.

Most interesting of all, the Hong Kong government has given MTR the rights to develop the properties around its stations: grocery stores, restaurants, cafés, drugstores, and even housing.

So where should you invest? Plenty of mutual funds and ETFs invest in transportation—but most include stocks from trucking, rail cargo, airlines, and other fossil-fuel-burning companies. You'd be hard-pressed to find any that traffic exclusively in mass-transit projects. For that, you'd be better off buying into a general clean-energy fund.

Semiconductors

There's a recurring theme in the descriptions of up-and-coming climate-investing areas: technology. Inside every car, every wind turbine, every solar panel, every nuclear plant, every mass-transit system, and every precision-agriculture machine, there's software—and semiconductors (chips and circuit boards).

Some semiconductor companies, like Intel and Texas Instruments, manufacture their own chips. But many others only *design* their chips—then outsource to Asia for manufacturing.

And not just anywhere in Asia. Taiwan Semiconductor Manufacturing Company (TSM) makes 50% of the world's semiconductors, including chips for AMD and Qualcomm, and the chips in every iPhone.

Copper, incidentally, is also a key component of such climate-change components as wind turbines, solar projects, and EV charging networks; an electric car uses as much as four times as much copper as a gas car. That helps explain why 10% of Lucas White's GMO Climate fund is invested in copper-mining firms, like Freeport-McMoRan (FCX).

Insurance

If there's any sector rocked by climate change, it's insurance. The insurance industry profits when disasters don't happen. So what are the insurance companies supposed to do in the new climate, where disasters strike more often?

They can raise rates sky-high. They can deny coverage to people in high-risk areas. They can exit disaster-prone states altogether, as described in the next chapter.

Furthermore, insurance companies make money in two different ways. They collect premiums from customers, of course, but they may only break even on the premiums.

The second way they make money: by investing it. "Your money doesn't just sit there. They make investments. They buy buildings," says Amy Bach, executive director of United Policyholders, a consumer advocacy group. "They make your money grow."

Insurance companies' investments are regulated—limited to safe, low-yield investments. The government doesn't want them caught without money to pay claims when a disaster strikes.

But Christine Harada says that overall, insurance stocks are great to own, especially if you hold them for the long haul. They're "low yield, but stable." Had you invested in, for example, Progressive Corporation (PGR) in 2004, for example, you would have averaged an enviable 13% annual return.

You can buy stock in individual insurance companies, like the Hartford Financial Services Group (HIG), Chubb (CB), or MetLife (MET). Some *re*insurance companies are worth investigating, too, like Greenlight Capital Re (GLRE) or RenaissanceRe (RNR). These are insurance companies *for* insurance companies; they come forward with money when some catastrophe threatens to wipe out a regular insurance company.

Or you can buy shares of an insurance ETF, like the SPDR S&P Insurance ETF (KIE) or the iShares U.S. Insurance ETF (IAK).

Medicine

In the warming world, more people will get sick. That's bad news for them, but good news for drug companies—and the people who invest in them.

Pharma giant Eli Lilly, for example, noted in its disclosure to the CDP that the growing numbers of mosquitoes in more regions (see Chapter 14) mean more disease. "This may then increase demand for certain medicines we produce," it notes. Potential upside from these diseases alone: $30 million.

The company also notes that climate change means more diabetes patients—because in hotter weather, people don't get as much physical activity, especially when traditional food sources are getting interrupted. These risks, the company notes, "may drive an increased demand for diabetes education and awareness and for our diabetes products."

Of course, these companies make all kinds of drugs that are affected by all kinds of factors. But clearly, as a species, exploding in population but living on a shrinking planet with constrained food and water, we're not going to be getting *healthier.*

It's easy to invest in these companies, if you're sold on the concept:

- **Individual stocks.** They're all out there, all publicly traded: Johnson & Johnson (JNJ), Roche (ROG), Pfizer (PFE), Novartis (NVS), Eli Lilly (LLY), and so on.

- **Funds.** To diversify your investment, mutual funds are a natural—either pharma-only mutual funds like Fidelity Select Pharmaceuticals Portfolio (FPHAX), or broader health-care funds like T. Rowe Price Health Sciences Fund (PRHSX).

- **ETFs** include KraneShares MSCI All China Health Care Index ETF (KURE), Invesco Dynamic Pharmaceuticals ETF (PJP), and iShares U.S. Pharmaceuticals ETF (IHE).

The Carbon-Tax Promise

A vast population of experts, economists, scientists, executives, government officials, and oil-company executives are weighing their own investments with an eye on a huge what-if: a carbon tax.

Chapter 17 describes this concept in more detail, but the idea is that the United States would charge a tax on fossil fuels—as they're extracted, imported, or burned. Overnight, there'd be a huge, society-wide incentive to embrace, buy, and invent cleaner power sources. And overnight, the value of every publicly traded company would radically change.

"The cost implications for many of these firms are massive—massive!" says Tuck Business School's Anant Sundaram. Already, he says, the risk managers at every major corporation keep a set of spreadsheets that forecast how the company would do in a carbon-taxed parallel universe. "'What would the economics would look like if we had fifty- or hundred-dollar-per-ton price of CO_2?' It's all part of what they call resilience strategy," he says.

The great thing about investing greenly today is that you'll *really* hit pay dirt if a carbon-pricing program comes to pass. Stocks of the biggest carbon-producing companies, like utilities and airlines, would tank. Conversely, a carbon tax would do amazing things for stocks and funds that are green now. "If the world freaks out about climate change the way it should," says GMO Climate's Lucas White, "you'll get *incredibly* good returns."

The Big Picture

Pouring money into stocks and funds isn't the only way you can use climate investing to your advantage—and the planet's.

Each company whose stock you hold hosts an annual meeting; at each meeting, shareholders vote on a bunch of resolutions. More and more often, climate-related issues come up. For goodness' sake, vote! You own a tiny piece of the company; use your power.

At GreenAmerica.org, you can view a list of upcoming shareholder resolutions from various companies. Here's a short link to the actual page: http://j.mp/331zRGg. Shareholder resolutions on environmental issues don't usually pass, but if they get even 20% approval, they get the company management's attention.

Your vote counts even if you've bought into a mutual fund. You can find out its managers' votes on various corporate resolutions (like environmental ones) in the fund's so-called N-PX report, which you can find at www.sec.gov/edgar/searchedgar/n-px.htm. Search it for keywords like *carbon* or *GHG*, or the name of the invested company, to see how your fund manager voted on those resolutions.

And if they're not voting the way you'd like, well, call the fund's 800 number, or click the Contact Us link on the website, and complain. They care about public opinion just as much as anyone else. That all goes for your IRA, 401(k), or retirement plan, too.

Keep in mind, too, that *institutions* all around you are also making investments: universities, nonprofits, churches, pension funds, even your local government. You're certainly a member of some of those outfits.

As such, you're entitled to have a voice, and to ask a few questions—like, "Do you guys consider climate when you decide how to invest?" Or, "Hey, how do you vote on climate issues when shareholder votes come up?"

Finally, remember that investing doesn't always mean handing over dollars. "Everything that you do is a form of investment," says Erik Kobayashi-Solomon.

Installing solar panels on your house or your commercial property is a climate investment. Writing to, calling, or visiting your state or congressional lawmakers is a climate investment. Improving your home's energy efficiency is a climate investment.

Whether it's time, effort, or money, all of it is spending something now—in the expectation that you'll be handsomely rewarded later.

Chapter 6

How to Insure

THERE'S A GOOD CHANCE THAT YOU HAVE INSURANCE. BUT there's also a good chance that you don't truly know what's in your policy.

That's understandable, and common. The policies are pages long, dense with terminology and legalese. Most people signing up just want to pay the lowest amount possible and get out of there.

But climate change is taking insurance out of Hypothetical Land and into your everyday concerns. In 2018, wildfire insurance claims hit their highest rate in recorded history; 2020 is likely to break the record again. The number of flood-insurance claims has more than quintupled since 2013—and the *size* of those claims has jumped 340%.

Suddenly, not having the right insurance is becoming a life-changing matter. Often, people realize that only when it's too late: after the disaster.

"When you discover that you're grossly underinsured, it's like having the rug pulled out from under you *again*," says Linda Masterson, whose house burned to the ground in 2011, prompting her to write *Surviving Wildfire: Get Prepared, Stay Alive, Rebuild Your Life*. "It's a horrifying 'Oh my God' moment to everyone."

We chronically underestimate our extreme-weather risk. Half of all Americans live on the coasts, but only 18% of people who live in flood zones have flood insurance. There's always a spike in people signing up for flood insurance after a major hurricane or flood—but then, when a couple of years go by without another storm, they let it expire.

If you got a thousand insurance experts into a hotel ballroom and forced them to agree on a single piece of advice, they would say, "Know thy policy." Understand what you have, meet with your agent, take photos.

"You could get yourself much better prepared in a day," says Masterson. "Is a day worth having 25, 35 percent more assets to start over with? A *day*."

Amy Bach, founder of United Policyholders, a consumer advocacy group, concedes that studying your insurance policy is nobody's idea of fun. And it's certainly not the message you hear from the insurance companies.

"Fifteen minutes will save you 15 percent, right?" she says. "But fifteen minutes is not enough time to put financial protection in place for what's probably your biggest asset."

Understanding Your Insurance

The first document to dig up is the declarations page. It's the first page, the most important page of your policy, always clearly headlined "Declarations Page." Here's where you find out the dates you're covered, the amounts, and other essentials.

Then make an appointment with whoever sold you the insurance, especially if you haven't revisited your policy in a while; things change. That broker has made money from you and owes you the time to explain your policy.

Here's what you need to find out.

Amount of Coverage

This is the big one. Two-thirds of American homes are underinsured.

Over half of people who lose their homes to wildfire, for example, discover that they're underinsured by at least 25%. That would mean that if you lost a $250,000 home, insurance would cover only $187,500 of it.

Maybe you never had enough to begin with. You were trying to keep your premiums to a minimum, so you accepted a basic insurance plan, recommended by your mortgage bank, right out of the gate.

Or maybe you've made upgrades to your place since taking out the policy—but you never told the insurance company. That's a risky spot to be in, thanks to the mathematical fine print described in the box on page 196. You can really get hosed if you fail to report home improvements.

There's a tiny glimmer of happier news hiding in this advice: If the

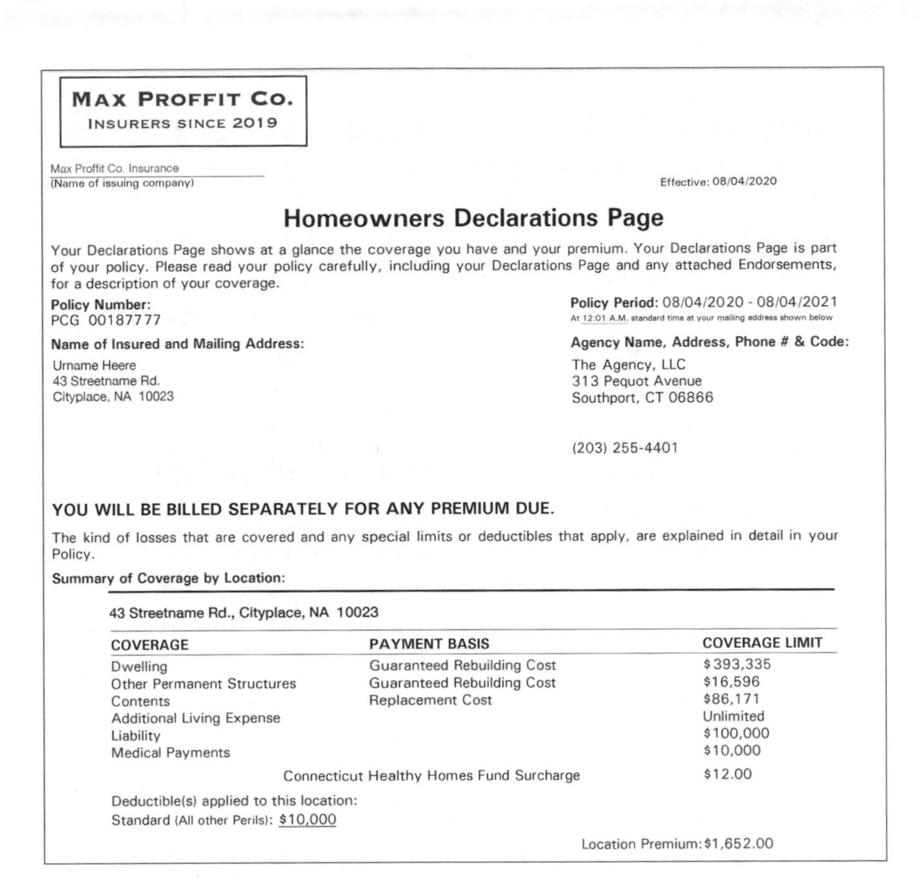

MAX PROFFIT CO.
INSURERS SINCE 2019

Max Proffit Co. Insurance
(Name of issuing company)

Effective: 08/04/2020

Homeowners Declarations Page

Your Declarations Page shows at a glance the coverage you have and your premium. Your Declarations Page is part of your policy. Please read your policy carefully, including your Declarations Page and any attached Endorsements, for a description of your coverage.

Policy Number:
PCG 00187777

Policy Period: 08/04/2020 - 08/04/2021
At 12:01 A.M. standard time at your mailing address shown below

Name of Insured and Mailing Address:
Urname Heere
43 Streetname Rd.
Cityplace, NA 10023

Agency Name, Address, Phone # & Code:
The Agency, LLC
313 Pequot Avenue
Southport, CT 06866

(203) 255-4401

YOU WILL BE BILLED SEPARATELY FOR ANY PREMIUM DUE.

The kind of losses that are covered and any special limits or deductibles that apply, are explained in detail in your Policy.

Summary of Coverage by Location:

43 Streetname Rd., Cityplace, NA 10023

COVERAGE	PAYMENT BASIS	COVERAGE LIMIT
Dwelling	Guaranteed Rebuilding Cost	$393,335
Other Permanent Structures	Guaranteed Rebuilding Cost	$16,596
Contents	Replacement Cost	$86,171
Additional Living Expense		Unlimited
Liability		$100,000
Medical Payments		$10,000
	Connecticut Healthy Homes Fund Surcharge	$12.00

Deductible(s) applied to this location:
Standard (All other Perils): $10,000

Location Premium: $1,652.00

Figure 6-1. A typical declarations page identifies your coverage and deductible.

renovations improved the *safety* of your home, the insurance company is likely to *decrease* your premiums, sometimes substantially. Keep that in mind if you install stronger doors, put in smoke detectors or sprinklers, renovate your plumbing, add a security system, install storm shutters or shatterproof glass, and so on. A new roof alone can easily lop 20% off your insurance price.

Of course, maybe you *know* that you're underinsured. Maybe you're trying to save money on the premiums.

Linda Masterson strongly urges you to resist that temptation. "I can only tell you from talking to hundreds of survivors that nobody feels that way after their home burns down," she says. "I've never talked with anyone who said, 'Gee, I wish we'd had less insurance.' "

How You Might Get Shafted When You Upgrade Your Home

Suppose you've got $300,000 of insurance on a home that, because of improvements you've made, is now worth $500,000. Maybe you refinished the basement or the attic or remodeled the kitchen—or all of the above. You had a good year.

Suppose, furthermore, that you didn't let the insurance company know that your home is now worth more. You suspected that the cost of your insurance would go up (and you'd have been right).

Okay then: Now a fire causes $200,000 in damage. You might say, "Well, thank goodness I have an insurance policy worth $300,000! That's more than enough to cover the $200K in damage!"

But you'd be wrong. All you'd get is $150,000. Less, actually; your deductible would make that number even lower.

Here's the reason. Insurance companies require you to carry insurance worth at least 80% of your home's replacement cost. If you don't, they're allowed to lower their payouts.

The reduction they're permitted to take matches the ratio of your actual coverage ($300,000 in this example) and the 80%-of-actual-value figure ($400,000).

In this situation, $300,000/$400,000 = 75%. That's why they'll pay you only 75% of the $200,000 fire claim, which comes out to $150,000.

And you know what they'll say? "You should have read your policy!"

Changing Insurance Policies

It's no longer enough to buy insurance and forget about it. Your insurer may quietly *change* your coverage as it perceives your risk to be changing.

The trouble is that insurance companies may calculate your risk—and therefore your rate—based on your zip code. Then, if a few people in your zip code file claims, the insurance company may raise everyone's rates—or drop your policies entirely. You're under no obligation to take that lying down, especially if you feel that *your* property isn't as risky as the others in your area.

"You have to be an advocate," says Jason Thistlethwaite, who studies insurance and climate change at the University of Waterloo. "You have to bother your insurance company and aggressively shop around. Play them off against each other."

You can get quotes from various companies at a comparison site like insurancequotes.com or policygenius.com. Just be aware that once you register at these sites, you'll be on the receiving end of a *lot* of attention from insurance salespeople thereafter—calls and emails. That'd be an argument for supplying a rarely used, backup email address and maybe a fake phone number. You just want the quotes.

You can also investigate various insurance companies at insure.com—to find out if they're under investigation, for example, or to see how financially sound they are.

You should also ask the insurance company what it's *doing* about climate change. "If they're at least doing something, that's a signal that there's some better management there," Thistlethwaite says.

Lowering Your Rates

Suppose that, convinced by these pages that it's not worth being underinsured, you reluctantly upgrade your coverage. Now you're paying more for insurance.

You can lower your rates in any of four ways:

- **Make improvements.** Insurance companies *hate* paying out money. Their favorite customers are the quiet ones who pay their premiums and ask little in return.

  To make that point clear, the insurers usually offer you cheaper policies if you make improvements that reduce their risk (and yours).

  For example, flood insurance is less expensive if you've elevated your living space. In wildfire country, insurance costs less if you've cleared your yard as described in Chapter 3. In tornado states, they'll cut you a break if you retrofit your home to resist wind damage, also described in Chapter 3.

You don't have to make big, expensive renovations, either. Some of the features that matter to insurance bean counters are smoke detectors, fire extinguishers, sprinkler systems, burglar and fire alarms monitored by a company, dead-bolt locks, fire-safe window grates, fire-rated roof materials, wind-resistant shutters, water-storage tanks, modernization of plumbing, heating, and electrical systems, and designing your yard against wildfire as described in Chapter 3. If you do make these changes, be sure to contact your insurer and let its agents know, so you can start enjoying the lower premium payments.

- **Agree to a higher deductible.** The deductible is an amount that *you* have to pay before insurance contributes anything. You can find out your current deductible by checking your policy's declarations page.

  Suppose, for example, that your deductible is $5,000. Now you have a flood that requires $30,000 in cleanup. In that case, *you* have to pay the first $5,000; the insurance pays the rest.

  If there's a small fire that causes only $3,000 in damage, you have to pay for *all* of it, because that's lower than $5,000.

  In any case, the bigger your deductible amount, the lower the cost of your insurance premiums.

- **Work on your credit score.** The insurance companies have determined that people with low credit scores tend to file more claims. You know what that means: higher rates for those customers.

- **Bundle your insurances.** The big insurers routinely cut you a deal when you buy more than one kind of insurance from them: medical, auto, homeowner's, and so on. If you currently have insurance from multiple companies, it's worth a phone call or website visit to find out how much you could save by combining them. It's usually 5% to 15%.

Finally, insurance companies also generally offer lower rates if you're a first-time homeowner, a nonsmoker, part of a couple, 55 or older, retired, or a longtime customer, and if you have a history of *not* making claims.

Flood Insurance

If you own a home, you probably have homeowner's insurance. But most people are astonished to learn that *it doesn't cover flooding.* Flood protection is almost always a separate purchase, usually from a separate company.

The first question, then, is, Do you *need* flood insurance?

Here's one way to tell, courtesy of former FEMA executive Roy Wright: "Pull out your driver's license. If it says 'Florida' at the top, you need flood insurance."

He's kidding. The actual answer is, if you live near the coast in *any* state, then you need flood insurance.

Even if you're *not* on the shore or in Florida, you're at greater chance of flooding with every passing climate-change year. According to FEMA, by 2100, the flood-vulnerable areas of the United States will grow by 45%, the number of homes its National Flood Insurance Program covers will more than double, and the cost of each payout will increase by as much as 90%.

All kinds of events can create flooding: coastal storms, heavy rainfall and runoff, freak tides, sea-level rise, and so on. As a result, floods are increasingly striking inland towns that nobody had ever considered susceptible. Consider:

- Every year, 20% of flood-insurance claims come from people who don't live in flood zones.
- In New York City, over 80% of homes flooded by Hurricane Sandy had no insurance at the time.
- Nashville is in the center of a landlocked state; it's 400 miles inland from the nearest ocean. In 2010, a single rainstorm caused $2 billion in flood damage, with 11,000 homes and businesses damaged and 11 people killed.
- About 75% of homes that suffered losses in Hurricane Harvey were outside of the government-measured floodplains.

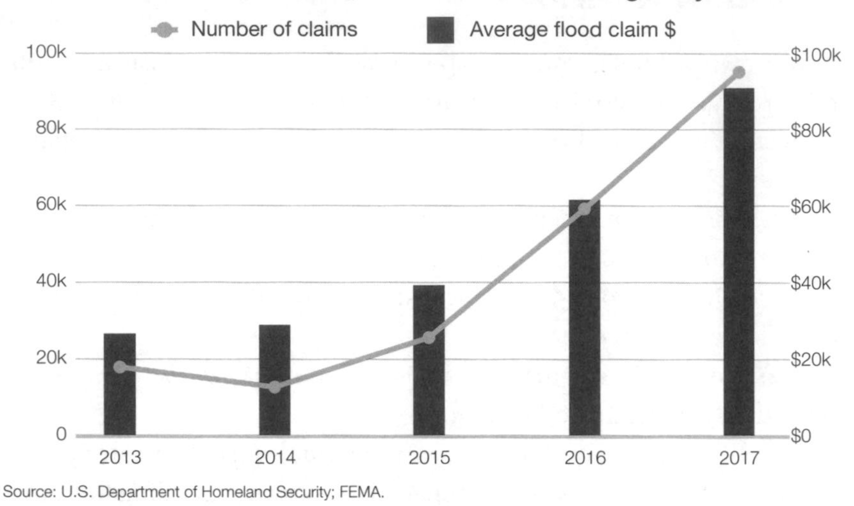

Figure 6-2. Government flood-insurance claims are skyrocketing, both in number (the bars) and dollar amounts paid (the sloping line).

That included the houses in Nederland, Texas, which is not on the coast; it's 19 miles inland from the Gulf of Mexico. In two days, this Houston suburb got over *five feet* of rain.

It doesn't take Noah's flood to make your life miserable, either. Even one inch of water in your building costs about $27,000 in cleanup and repair.

All right, suppose that you're persuaded; you've decided to look into flood insurance. You might make http://nationwide.floodrisktool.com/Map.aspx your first stop. You can call up overlays that show you the floodplain, the flood history, and tracks of past hurricanes for any address.

The government maintains its own maps, too. They show the data they'll use to calculate your home's risk if you apply for National Flood Insurance. In this case, though, the information has carefully been translated into Govtspeak, and you may not make much progress interpreting it without the crash course in the sidebar on page 201.

Figure 6-3. At Nationwide's flood mapping site, you can type in your address and view a current map of your home's flood risk.

How to Use FEMA's Flood-Insurance Maps

FEMA maintains a set of detailed U.S. floodplain maps, illustrated in Figure 6-4. Cities and towns use these maps to write their building codes and floodplain regulations. *You* can use them to see which areas have at least a 1% chance of flooding each year—and therefore if you're required to have flood insurance.

To view these maps, visit http://fema.gov/national-flood-hazard-layer-nfhl. Click "NFHL Viewer," which stands for National Flood Hazard Layer Viewer. (As you know, EGLA—"every government likes acronyms.") You can plug in any address or double-click repeatedly to zoom into the map. Wait 15 seconds and, presto: You're seeing exactly how flood-risky the government thinks your property is—or you will, once you understand the wonky codes and colors.

If you zoom in enough, areas of red and blue tinting appear. These are the Special Flood Hazard Areas (SFHA)—the places where flood insurance is mandatory if you want a government-backed mortgage. Zoom in still more, and you see zone labels:

Figure 6-4. FEMA's own flood-risk maps are available to the public.

Zone V and VE are the highest-risk areas, where flood insurance is mandatory. These are coastal areas, subject to rising water, storm surges, and *waves* over three feet high, which can cause structural damage and require stricter building codes. (You can remember it this way: "*V* for *velocity* of the waves.")

Within the VE zones, the elevation notation ("EL 11 Feet") shows how *high* the waters might rise in that 1%-chance-a-year flood.

The E areas—VE (and AE, described next)—get blue tinting. They represent areas where FEMA has performed detailed studies and calculated those "EL" measurements.

Zone A & AE indicate flood risk *without* waves, either coastal *or* inland. You still need flood insurance, but the building codes might not be as strict.

Zone AO designates areas of shallow flooding, usually caused by ponding—for example, when water has "overtopped" a barrier and collected on the other side.

Zone X means "Good news! You're not required to have flood insurance!" Your place can still flood, of course—just about *any place* can flood—but you're not in the 1%-chance flood zone. The orange-tinted X areas have moderate flood risk (between 0.2% and 1% chance each year); untinted X areas have even lower risk.

Yellow writing ("450290689G, eff. 12/21/2017") identifies each flood map panel and when the community adopted it. That's important when there's an argument about which building codes affect you. You may also spot slanted lines or dashed lines, labeled CBRS or "Otherwise Protected Area." These are legally protected nature areas, where nobody's allowed to build.

These FEMA maps are not the same thing as the Surging Seas future-predicting flood maps at ss2.climatecentral.org. The FEMA flood maps don't indicate *historical* flooding, they don't predict *future* flooding or sea-level rise, and they don't depict all *possible* flooding. They're a snapshot of flood risk right now, for the purposes of floodplain management, making smart building decisions, writing building codes, and calculating insurance.

If you decide to proceed, your next stop is an independent insurance agent, which sells both government and private flood insurance.

And now, an introduction to each.

National Flood Insurance

In the 1950s and 1960s, most insurance companies got out of the flood-insurance game. They couldn't build a business on something as unpredictable and expensive as flooding.

In 1968, Congress rode to the rescue with its *own* flood-insurance agency: the National Flood Insurance Program (NFIP).

NFIP was never intended to be a straight-up insurance program. It's "a hybrid of insurance and a social program," according to Chris Heidrick, chairman of the flooding task force of the Independent Insurance Agents & Brokers of America.

NFIP's first mission was to offer affordable insurance to anyone who lives near water. Its second mission, and the real genius of the concept, was that NFIP would steer people away from disasters in the first place—by offering insurance only in towns that developed their coasts responsibly: all new construction must be elevated, away from the floodplain, and so on.

When not many people signed up for NFIP, Congress, in 1973, turned flood insurance into an offer that citizens couldn't refuse: If you live in the flood zone, you can't get a federally backed mortgage *unless* you have the insurance!

Today, NFIP provides 95% of all the flood insurance in the United States. It covers 5.3 million homes, with policies covering $1.3 trillion.

Life in the 1% Zone

What you pay for NFIP insurance depends on where you live, and what you live in.

Over the decades, the NFIP has developed a complex system of maps. The central feature is the "100-year" floodplain, which is a point of chronic confusion for the public. It's a wildly misleading term. What it's supposed to mean is that in any given year, there's a 1% chance of flood. In fact, though, when you get down to mathematical brass tacks, there's actually a *63.4% chance* of a "100-year flood" in any 100-year period. You can read the mathematical reasoning in the "100-year flood" entry of Wikipedia, if you're handy with binomial distribution. For now, it just means "a good chance of flooding from time to time."

FEMA, which runs the NFIP, uses the 1%-chance flood maps to determine the likely flood zones. It doesn't call them "flood zones," though; FEMA ran that phrase through an English/Government dictionary and came up with the term Special Flood Hazard Areas (SFHAs).

In a Special Flood Hazard Area, you have the greatest flooding risk. In these areas, you can't get a federally backed mortgage unless you have flood insurance. And your home fetches a lower price than identical homes *outside* the flood zone.

How NFIP Insurance Works

You can't buy FEMA's flood insurance from FEMA directly—only through insurance agents. If necessary, you can find one at FloodSmart.gov/purchasefloodinsurance. The prices are identical no matter which agent you go to.

And don't think you can wait to get insurance until you hear news of a megastorm approaching. Once a storm watch is in effect, flood-insurance companies no longer accept new customers, just to shut down slicksters like you. And besides: It takes 30 days for your coverage to kick in after you sign up.

Depending on the value and size of the house, how much it's been elevated or floodproofed, and its distance from the hazard area, the cost for NFIP coverage ranges from $119 to $4,500 a year.

It covers up to $250,000 in damage to your home, and $100,000 for everything inside it. For nonresidential buildings, NFIP covers $500,000. Check the fine print to see what things are *not* covered, like anything in your basement and anything *outside the building*, like cars, landscaping, decks, patios, fences, hot tubs, and swimming pools.

Note, too, that you're reimbursed for the present cash value of what was destroyed—what you'd get if you sold it used—not what it would cost to replace it. And of course, there's a deductible: $1,000 if your policy covers less than $100,000 in building damage, $1,250 for more than that. You can also choose a higher deductible—all the way up to $10,000—in exchange for lower premiums.

Private Is Coming Back

In the last few years, something amazing has happened: For the first time in decades, a few private insurers have started offering flood insurance again.

They're not the big boys like Nationwide and Allstate; they're smaller companies your broker knows about.

You might not think that would make any sense. In this era of mega-hurricanes causing billions of dollars of damage a year, why would any company be nuts enough to start offering to insure homes against flooding?

Three things have changed in the insurance industry:

- ♦ **Rising NFIP rates.** In 2014, Congress instructed NFIP to begin standardizing its rates, allowing them to rise, slowly, to more realistic market levels. Private insurance companies are saying, in effect, "Well, if the Feds are gonna quit undercutting us, maybe we can actually play this game once again."

- ♦ **Better simulations.** For decades, insurance companies based their predictions of future disasters on historical weather patterns. "That's like looking in your rearview mirror to understand what's coming down the road ahead of you," says the University of Waterloo's Jason Thistlethwaite.

  Today, it's all about software models. Every major insurer employs a team to write computer simulations. The goal is to calculate the effect of the warming climate on natural disasters that will cost the companies money.

  "Insurance companies take a given storm, like Katrina, and they'll run that scenario one hundred thousand times with minor tweaks on the variables," says Chris Heidrick. Then, to calculate the average expected losses, they average those simulations together.

  Nobody can perfectly predict the future. But these days, the insurance companies are squeezing many more drops of useful analysis from the data they have.

- ♦ **Catastrophe bonds.** These days, insurance companies feel a lot less stressed about the next big disaster, thanks to a new kind of investment they call cat bonds.

  Here's how they work: The insurers offer special bonds to huge institutional investors, like hedge funds and pension funds. The investors receive nice interest for, say, three years. If the world is calm during that period—what a deal! The investors get great interest for three years, and at the end, they get *all their money back.*

  But if there's a natural disaster in that time, the insurance company *keeps* the investors' money to cover their losses from the disaster payouts. The investors lose their bet, and their money.

  Catastrophe bonds have become massively popular with big investment organizations, who are always under pressure to find investments with good returns; they've been pouring about $10 billion a year into them.

Together, these three factors make insurance companies less nervous about being washed away with the next disaster—and with all of that investor money in hand, they're looking for ways to expand their businesses.

How Private Insurance Works

There are plenty of reasons you might want to look into private insurance instead of NFIP coverage.

For one thing, NFIP isn't available everywhere. You can get it only in the 22,000 communities that have agreed to enforce the government's guidelines for flood resistance.

Private insurance is also a lot easier to get than NFIP coverage; often, you can get quotes right online. Your rate is likely to be based on far more recent and accurate flood-risk data than NFIP's old flood maps are.

You'll also need private insurance if you hope to cover your house and belongings for more than the NFIP's $350,000 limit.

Unfortunately, private insurance costs more than an equivalent NFIP policy, especially in flood-prone areas. For a risky house in Miami Beach, for example, private flood insurance that covers the building for $300,000 and its contents for $100,000 will cost you a staggering $20,000 a year. For an identical house on higher ground, you'd pay only $3,700.

And by the way: Every five years, you should check those online flood maps to see if anything has changed. New construction, changes in floodplains, or changes in how your house's flood risk is calculated can affect what you're paying, and for how much coverage. That number can go up or down.

NFIP's Future

NFIP is not, ahem, the government's finest creation. Its critics point out:

- The NFIP no longer pays for itself. Thanks to the recent rash of hurricanes like Katrina and Sandy, the program is $25 billion in debt.

- NFIP's low, low rates—as low as one-third what they should be—encourage people to build homes where they shouldn't, in areas that tend to flood.

- The government doesn't raise your rates after you've collected insurance money for a flood, the way car insurers do after you've had an accident. One home in Texas has been flooded 22 times since 1979—and the government paid to rebuild it every time, for a cumulative payout of $2.5 million. Those homeowners' premiums never go up, and their policies never go away.

- Some of the NFIP's maps are still based on old data, bad data, or ancient methods. And the climate is changing faster than NFIP's mappers can keep up.

- The NFIP makes no attempt to calculate *future* flood risk. And it doesn't account for erosion, sea-level rise, or changing rain patterns in its mapping. That's a problem because if your house is just outside an NFIP Special Flood Hazard Area now, it probably won't be outside that zone in a couple of decades.

  (In 2015, President Obama signed an executive order directing the NFIP to begin taking future flood risk into account. In 2017, President Trump signed an executive order canceling it.)

- Just because people have to sign up for NFIP when they get their mortgages doesn't mean they *keep* it. Mortgages usually have 15- or 30-year terms; people generally keep their NFIP flood-insurance policies for two or four years. After that, they're sitting ducks for floods.

Fortunately, reform is in the air. A NFIP modernization program called Risk Rating 2.0 will drastically overhaul those 1970s-era flood-risk calculations. Your rates won't be based on some notion of "1% chance of a flood each year." Instead, it will be based on your *specific* house: its value, distance from water, terrain, flood-resistant features, and so on.

The new rating system, scheduled to take effect in late 2021, won't just raise rates for the riskiest homes; it will also *lower* rates for homes that were unfairly swept into broad-stroke "100-year" areas. Half of NFIP's customers will be thrilled to discover when their rates suddenly drop; the other half will be furious that their rates suddenly shoot up. Good luck with that one, Congress.

Homeowner's Insurance

Why do they even call it homeowner's insurance? This kind of insurance doesn't cover things that happen to *you,* the homeowner. It covers the *home.*

In any case, no state requires home insurance. But if you want a mortgage, the bank will probably make you get it. Once you've paid off your mortgage, nobody cares if you have home insurance, although having it is not a terrible idea. Your home is probably the most valuable thing you own.

The insurance business is fantastically complicated. In terms of choice, shopping for a policy feels more like you're entering a Home Depot than, say, a Subway.

For starters, there are eight kinds (or "forms") of homeowner's insurance, labeled HO1 through HO8. (*HO* presumably stands for "home owner's.")

HO1 and HO2 are cheap, but they also cover the fewest kinds of disaster; in fact, they're called "named peril" policies because they cover *only* a short list of pre-identified disasters (perils): fire, lightning, theft, vandalism, volcanoes, and so on.

HO4 is renter's insurance; HO6 is for condos or co-ops; HO7 is for mobile homes; HO8 is cheap, basic insurance for old homes (over 40 years old). It's for people who don't want to upgrade their homes to meet current building codes.

By far the most common forms, though, are the HO3 and HO5 policies. They're "open peril" policy, which means that they cover *anything* bad that happens, except for a short list of pre-identified disasters that they *don't* cover.

Those not-covered things are earth movement (earthquakes, landslides, and sinkholes); ordinance of law (translation: "The cost of meeting new building codes that have cropped up since your house was originally built"); water damage from flooding, tides, groundwater seeping up, or sewers backing up; power failure; neglect; war; nuclear hazard; intentional loss; government action; collapse; theft while under construction; vandalism if you haven't been around for two months; mold, fungus, or wet rot; wear and tear; mechanical breakdown; smog, rust, and corrosion; smoke from agricultural and industrial operations; pollutants leaking or seeping; the building's settling, shrinking, bulging, or expanding; birds, bugs, and rodents; and animals you own.

An HO3 or HO5 policy covers what insurance people call the Six Basic Protections, which give you a nice financial cushion in times of awfulness:

- **Dwelling** means your house, the structure itself. It covers, as Linda Masterson puts it, "everything in your house that wouldn't fall out if you turned it upside down."
- **Other Permanent Structures** on your property includes things like your garage or a detached barn.
- **Contents** refers to all your belongings, inside and outside the house—even when you're traveling.
- **Additional Living Expenses** means reimbursement while your home is being repaired. Usually, you're covered for a year or two, and they'll pay for the same size and niceness of living space that you're used to.
- **Liability** covers you for legal fees and medical bills if your negligence is responsible for someone getting hurt on your property. The usual examples are a delivery person tripping on a cracked stone and suing you, or your dog biting a neighbor.
- **Medical payment** pays the doctor and hospital bills for guests who get hurt in your house, no matter whose fault it is.

(The difference between an HO5 policy and an HO3: The HO5 is an open-perils policy for both your home *and* your belongings. The HO3 limits coverage of your belongings to a list of 16 named perils. However, HO5 is available only to newer homes—and it costs more.)

There's a lot of fine print in these policies, with deductibles and limits to each category. In particular, each one may offer either the *replacement* cost (enough money to buy *new* copies of what you lost) or the *actual cash value* of everything you lost—that is, what it might have fetched as a used item. If your two-year-old iPhone gets destroyed in a fire, you might get reimbursed only $450—what it would fetch on eBay today—even if you paid $1,100 for it when it was new.

Obviously, you want replacement-cost coverage for as many things as possible. Replacement-cost coverage costs more.

Homeowner's Insurance and Flood

In an era where flooding is becoming more common just about everywhere, it's worth noticing some items that *none* of these policies cover.

Your homeowner's insurance covers damage from water that pours in through your roof, or from rain that whips in through a broken window.

But it doesn't touch any damage that the insurance company's detectives determine came from water from the *ground*—that is, flooding.

Nor will homeowner's insurance help you with anything *related* to flooding, like mold, wet rot, pollutants from floodwaters, or flood damage to your cars. If you want flood protection, you have to buy it separately.

Homeowner's Insurance and Hurricanes

A typical policy covers damage from wind, hurricanes, and hail. Trees that fall, windows that break, shingles that fly away: all covered.

But insurance companies aren't stupid. If you live on the coast, they'll probably charge you for a separate windstorm policy just to get that same amount of coverage.

In the 19 floodiest states, you'll also be socked with *hurricane deductibles*. That means you, Alabama, Connecticut, Delaware, Florida, Georgia, Hawaii, Louisiana, Maine, Maryland, Massachusetts, Mississippi, New Jersey, New York, North Carolina, Pennsylvania, Rhode Island, South Carolina, Texas, and Virginia, plus Washington, DC.

As you know, a traditional deductible is a fixed amount—say, $1,000—that you have to pay for losses yourself before the insurance kicks in. A hurricane deductible is worse because it's a *percentage* of the losses. If the hurricane does a massive amount of damage, the amount you pay goes up proportionally. The percentage is usually between 2% and 5%; in Florida, it can go as high as 10%.

That can hurt. If you have $350,000 of insurance on your house, and it has a 5% hurricane deductible, then you get to pay the first $17,500 in repair costs.

Some policies require that you have to start fresh paying this deductible with *every storm*. In others, you can tally them up over a year—two storms in the same season can count toward the same deductible.

Here's another way the insurance companies aren't stupid: Once the government announces a hurricane watch, you can't make any changes to your policy. You won't get away with beefing up your policy with more coverage as the storm approaches.

The point, therefore, is to research this question *now*, before the next big storm comes crashing in from the Atlantic.

Homeowner's Insurance and Wildfire

In most states, fire damage is covered by the good old HO3 policy. The building itself, your belongings inside it, and even trees and bushes on the property are protected against loss and damage from fire, smoke, and ash. Even theft is covered, in case of some postfire looting.

And if you have to live somewhere else while your place is being cleaned and rebuilt, your insurance covers your living expenses, too.

In one kind of place, though, the news isn't so good: wildfire country.

In 2017 and 2018, insurance companies wound up paying out $24 billion to the owners of burned homes—and they've vowed never to be caught again. They're jacking up rates or dumping customers entirely. If you're among the dumped, you have no choice but to look for a replacement policy that costs double or triple what you used to pay.

The upshot is that in wildfire country, you may get a special insurance-company treat: Wildfire damage isn't covered.

Check the "exclusions" section of your insurance policy to find out if you're in that category. Especially if you live in California, Alaska, Arizona, Colorado, Idaho, Montana, New Mexico, Nevada, Oregon, Utah,

Washington, or Wyoming. Even more especially if you live near what fire experts call the WUI—the wildland/urban interface—where forest comes right up against civilization.

If you wind up having to file a claim, the usual guidelines apply, as described at this end of this chapter.

But if you experience bad fire damage, it may be worth hiring a qualified inspector to document the damage, especially if the insurance company won't. Use a company that works with Certified Industrial Hygienists, experts who swab various surfaces in your home for the presence of smoke, soot, and carbon and sample the air to test for unhealthy gases seeping out. They also recommend how to clean and repair the place.

You'll pay about $3,000 for all of this testing and report writing, and you'll submit the results to your insurance company.

After the Disaster

Standing on the street, looking at the smoking/dripping/roofless mess that was once your home, you're likely to experience tremendous feelings of loss, frustration, and helplessness. But somehow, you'll have to summon enough additional energy to confront Job Number One: insurance paperwork.

No matter what kind of catastrophe has befallen you, the degree to which the insurance company will soften the financial blow depends on the steps you take now.

If repairing the damage will cost less than your deductible, well, obviously, there's no point in filing a claim. Even if the repair costs a little *more* than the deductible, consider paying the whole thing yourself instead of filing a claim.

The reason: Each time you file a claim, most insurance companies automatically raise your rates. On average, you can expect your rates to jump up 8% after a weather-related claim, 16% after a water-damage claim, and 33% after a fire claim.

All the insurance companies, furthermore, share their records with one another. If your old company drops you, you may have trouble getting new insurance.

Start the Claim

If you do decide to make a claim, *call the insurance company right away*. After an extreme-weather event, their phones are ringing off the hook with other victims. You want to be at the front of the line.

They'll assign a case number and an insurance adjuster to your situation. The adjuster's job is to estimate the costs for repairs and how much you'll get for them. The insurance company has specialists in natural-disaster claims and will send an inspector over to look at the damage right away—especially if your home is a total loss.

If the agents ask for photos or paperwork, get it to them as soon as you can. Every policy has a deadline for filing a claim after a disaster, typically a year, but don't wait!

Working through an insurance claim is a complicated, confusing, confounding process, often said to be more unpleasant than the disaster itself. Keep asking questions.

Keep a diary of your conversations and emails, including with the adjuster who's handling your case and the builders working on your house. After an important conversation, email a recap to your adjuster. "Thanks for our chat today, Chris," you might say. "Just want to make sure I understand what we covered." Not only are you creating a paper trail, but if you misunderstood something, your adjuster has a chance to set you straight.

If your house is mostly or totally destroyed, the process may take so long that a second insurance adjuster takes over for the first one. That's yet *another* reason to have copious notes on the conversations you've had before.

The bottom line: "He with the most notes wins," says Amy Bach of United Policyholders.

More insurance-hell tips:

- Take pictures and videos of the damage as soon as possible. Once things start getting cleared away, once the floodwaters recede, it will be harder to establish what you saw. And if the insurance company winds up lowballing their offer to reimburse you, visual records like this come in handy.

- Protect the place from *further* damage. Tape up broken windows to prevent more water intrusion. Tarp up holes in your roof. If you don't, the insurance company isn't likely to pay for any *more* damage that happens as a result. Keep the receipts for anything you spend on such temporary repairs. Your insurance covers all that.

- Same thing with living expenses while you're out of the house: Keep your receipts. You'll get reimbursement. That's even true for your pets' temporary housing, if you've had to pay to keep them at, for example, a kennel.

- You're supposed to get a starter check immediately to help you pay for your temporary living expenses. If you don't get it, ask for it.

- Make a list of your belongings. Compare it to the list you made *before* the disaster, which you already have on hand, because you've read Chapter 8 of this book. Round up receipts that indicate the value of everything that's been damaged or destroyed. Put together before/after photos, if you can.

- Don't throw away damaged possessions until you know what your insurance company intends to give you in the settlement (and you're satisfied with it).

- Get some estimates from a few contractors—detailed bids from licensed contractors. That's time you're saving the insurer, so the process will go faster.

- If your insurance adjuster isn't helpful in explaining the twisty prose of your policy, or isn't delivering on promises, you can ask to speak to a claims supervisor—the adjuster's boss. And if *that's* no help, you can contact your state's insurance commissioner's office. They exist to help you out.

Disaster Scammers

Nature's violence isn't the only upsetting force you have to contend with after an extreme-weather event; there are also the *people.*

"After a disaster, scam artists stay up nights thinking of ways they can defraud innocent individuals already hurting because of a storm or hurricane or other disaster," writes FEMA. Yes, there are actually disaster chasers, crisscrossing the country looking for people to rip off.

These phony contractors or adjusters may show up on your doorstep, offering to "help." They may offer you discounts on their work or a "government check" to tide you over (it will bounce). They're well aware that you may not be thinking entirely straight.

FEMA and the Insurance Information Institute urge you to be careful in the aftermath as you begin repairs:

- Ask for name, picture ID, office address and phone, and license. Ask for references (and check them out). Ask if you'll get a written contract. All of this is entirely reasonable, and a legitimate contractor has no problem answering.

- Get more than one estimate from contractors.

- In a common post-disaster scheme, someone pretending to be a contractor talks you into paying a big deposit before the work starts. Work may even start—but then the "contractor" disappears for good.

  Investigate the contractors' reputations (at, for example, AngiesList.com). Get references, and contact those references. Don't give anyone a deposit until you've done this research.

- Some contractors encourage you to spend money on temporary repairs, like taping up windows or covering holes in the roof with a tarp. Problem is, your final insurance settlement includes a certain amount for those and a certain amount for the permanent repairs. If you've spent a lot of money on temporary fixes, you may not have enough left over for the permanent ones.

  Most of these things you can do yourself, anyway. Keep your receipts.

- If you discover that somebody's fishy, report them by calling FEMA's fraud hotline: 800-323-8603.

And what if you're not satisfied with the insurer's offer, or your claim is denied entirely?

In that case, you're not powerless. You're perfectly welcome to negotiate with them; make your case.

If they've still got their minds made up to shaft you, you can make a complaint through your state's insurance commissioner (find that office through Google—*Virginia insurance department*, for example). Or hire a public adjuster. Or get a lawyer.

Unfortunately, those roads can be expensive and slow.

The Public Adjuster

There's one final way to protect yourself from the insurance company itself.

If the damage is pretty bad, consider hiring a licensed public adjuster. That's a professional who works for you, not the insurance company. A public adjuster's job is to help you file a claim and to help you get a better settlement from the insurance companies.

You can find one through National Association of Public Insurance Adjusters (www.napia.com). If you find one with the CCPA designation, that means he or she has earned an advanced certification; SPPA means they've had at least ten years of experience.

The upside of public adjusters is that they have the technical ability to assess the damage (or to hire toxicologists or engineers who do), and they understand the labyrinths of prose in your policy.

The downside is that they collect a percentage of your award (10%, for example).

But there's a lot of upside. "If you get a good one, they are worth every dime you pay them," says Amy Bach. "Not only will they do all the paperwork to meticulously document your loss, and value it, but they also know how to be tough with an insurance adjuster and how to negotiate, which a lot of people don't know how to do."

A study of Florida claims found that hurricane-damage payouts using a public adjuster ranged from 20% to *750%* higher than you'd have gotten on your own.

No wonder, then, that insurance companies aren't big fans of public adjusters. They may even try to talk you out of using one.

Be careful: The person who works for you is called a *public* adjuster (or, outside the United States, insurance loss adjusters). You may also encounter a "company adjuster" or "independent adjuster." *They* work for the insurance company.

What's Happening to Insurance

If you had to guess what general effect climate change and related weather disasters are having on insurance rates, well, you'd be right: In the last ten years, homeowner's insurance rates have shot up, on average, 50%.

But here's the problem: Insurance companies are for-profit entities. If they pay out more than they take in, they go out of business. (The 2017 and 2018 California wildfifires alone wiped out 25 years' worth of the industry's profits.) Maybe you think of insurance companies as huge, money-grubbing corporate monoliths—and some of them don't do much to change our minds. But a bankrupt insurance company doesn't do anybody any good.

When insurers have to pay for more losses, thanks to more damaging disasters, they have only one lever to pull: charge more. But if the price gets so high that nobody can afford the insurance, the insurance companies give up. They stop selling insurance and start dumping existing customers (refusing to renew their policies).

"They're not going to have any interest in selling coverage to areas they know where those coverages aren't profitable," says Jason Thistlethwaite. "That'd be like selling insurance to a drunk driver. It's just not a smart business decision."

In 2018, for example, Florida insurance companies canceled 87,000 policies and raised everybody else's rates between 7% and 12%.

In California, it's hard to know how many homeowner's policies were dropped by insurers after each wildfire, because no law requires insurance companies to report them. But in the last four years, 340,000 homeowners reported that they've been dumped, prompting the state to impose a one-year ban on dumping customers who live near recent wildfires.

"That number is going to triple, at least," says Amy Bach.

What are you supposed to do? If the insurance company closes shop, drops you, or jacks its prices, your options are limited:

- **Sign up for "surplus lines" coverage.** Surplus insurance is offered from smaller companies you haven't heard of. It's designed to step in when regular insurance companies consider their risk too high. It may cost double or triple your original rates, though.
- **Look for government insurance.** In many cases, when private insurers pull out, government agencies step in to offer basic or last-ditch insurance options.

  Florida, for example, created the state-run Citizens Property Insurance. California developed the FAIR Plan property insurance. In Texas, there's the Texas Windstorm Insurance Association. And then there's the mother of all government-run insurance programs, the National Flood Insurance Program.
- **Self-insure.** This option means socking away money in banks or investments, to withdraw in case of emergency. You are, in effect, your own insurance company, paying "premiums" into your own bank account. This option requires both a healthy income and healthy discipline.
- **Go without insurance.** That's a dangerous road, but more people are taking it every year—involuntarily. As climate change produces more extreme weather, insurance as we know it may become "largely a luxury that might just be available for the rich," says Thistlethwaite.

Chapter 7

Protecting Your Children

YOU DON'T NEED TO WORRY THAT TALKING TO YOUR KIDS about climate change will upset them. Unless they're very young indeed, they're *already* upset.

"They know," says Caroline Hickman, a psychotherapist in the UK who researches young people's feelings about the climate crisis. "Children rarely initiate these conversations with adults, but they'd have to be living under a rock nowadays to not be anxious about climate change."

Meanwhile, the emotional toll that climate change takes on children is only the beginning. The great Venn diagram circles of climate change and children overlap in these four primary realms:

- **Gradual health effects.** Children are more vulnerable than you are. They spend more time outside, which exposes them to more ticks, mosquitoes, and heat. They've had less time to develop their immune systems, so they're more vulnerable to diseases and allergies. They have faster metabolisms, their body sizes are smaller, and their nervous systems are still developing, all of which magnifies the impact of heat, pollution, bug-borne diseases, and so on.

- **Extreme-weather events.** Every year, more children endure, with their families, devastating floods, hurricanes, and wildfires. There's danger when these disasters strike, but there's also trauma afterward, which leads to higher rates of depression, anxiety, post-traumatic stress disorder, and substance abuse.

- **Climate despair.** A child doesn't have to live through a weather disaster to be affected by climate change; the news that their future is uncertain is enough to be concerned about. Climate fears, wrote the authors of a 2018 study, "place children at risk of mental health consequences, including PTSD, depression, anxiety, phobias, sleep disorders, attachment disorders, and substance abuse. These in turn can lead to problems with emotion regulation, cognition, learning, behavior, language development, and academic performance."

That's probably not the future you imagined for your kids.

- **The baby decision.** As the population grows, the climate problem gets worse. The coming decades promise ever more devastating storms, floods, droughts, and heat waves; food disruption; and, eventually, mass migrations of people whose home regions are no longer habitable.

  No wonder that increasing populations of potential parents are posing a question that previous generations didn't ask anywhere near as often: Is it fair to bring a baby into this world we've messed up for children?

This chapter considers these topics one at a time.

The Health Effects

In every description of a climate-change consequence—wildfire, heat wave, flooding, air quality, tick- and mosquito-borne disease—the same sentence seems to crop up over and over: *The very old and the very young are particularly at risk.*

"Children are a uniquely vulnerable group that suffers disproportionately from these effects," writes the American Academy of Pediatrics. "Their immature physiology and metabolism; incomplete development; higher exposure to air, food, and water per unit body weight; unique behavior patterns; and dependence on caregivers place children at much higher risk of climate-related health burdens than adults."

Study after study illustrates that our children are already suffering from climate effects like these:

Heat

Extreme heat kills more people in the United States than hurricanes, lightning, tornadoes, and floods, and the risk is especially great for babies and children.

- **Pregnancies** go wrong more often in a hotter world. Heat increases the chances of premature births and stillbirths.
- **Babies** die more often as the heat rises. At the current rate of climate change, infant mortality rates will rise 5.5% for girls and 7.8% for boys by 2100.
- **Young children** wind up in the emergency room more often than adults during heat waves.
- **Student athletes** are especially susceptible to heat. A third of *all* heat-illness patients are teenage male athletes—usually football players.

  And according to the CDC, high-schooler emergency-room visits shot up over 130% between 1997 and 2006. And the number of young U.S. athletes dying from heatstroke has doubled since 2009.

Of course, none of this even touches on the indirect damage children experience when their *parents* suffer from heat illnesses or deaths.

Chapter 10 describes the steps for protecting your kids in a heat wave.

Air Quality

A headline-grabbing superstorm is one thing. But the declining quality of the air children breathe is longer lasting, more sinister, and mostly invisible:

- **Asthma**, the most common chronic childhood disease, affects 9.3% of all American children. A lot of it comes from breathing ground-level ozone, a compound that results when fossil fuel emissions get baked by heat and light—and that problem, as you'd guess, is getting worse. By 2050, ground-level ozone concentrations will rise 5% to 10%.

♦ **Hay-fever season.** As any climate scientist (or mosquito) can tell you, the warm-weather season is getting longer every year—which means the ragweed-pollen season is getting longer. And for the 9% of children who suffer from hay fever, that means more *days* of suffering. Already, your nose runs 27 more days a year than it would have if you were growing up in 1995.

It's not just the number of days, either; it's also the amount of pollen in the air. Believe it or not, ragweed produces more pollen in warmer weather, and when more carbon dioxide is in the air. No wonder the average pollen count is now twice what it was a century ago. The result: more kids having asthma attacks.

♦ **Wildfire smoke.** The 2020 West Coast wildfires added so much smoke to the airstream that the sky was hazy on the East Coast, 3,000 miles away.

Wildfire smoke is really bad for you—and guess who's particularly vulnerable? Right: the very young and the very old.

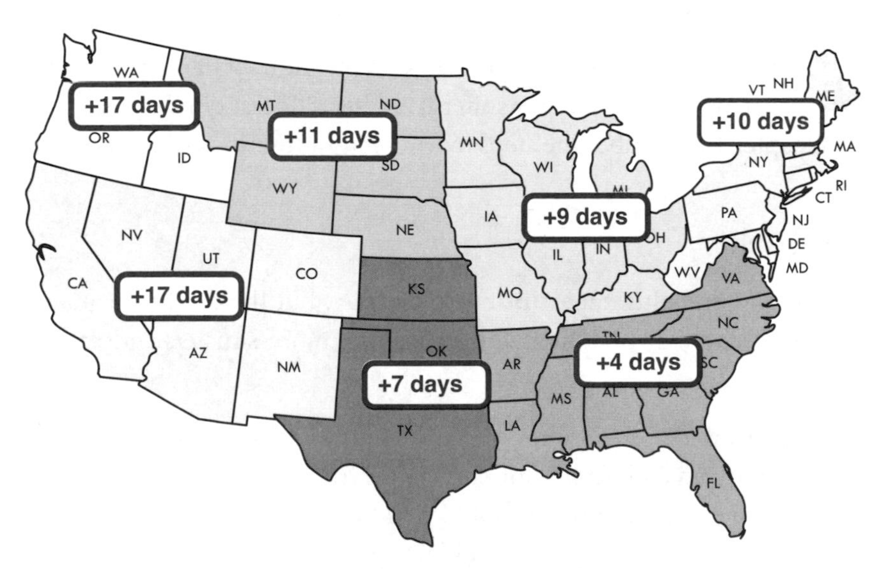

Figure 7-1. These numbers show how the frost-free season as of 2015—the warm-weather, hay-fever season—has changed in each part of the country, relative to the first half of the twentieth century.

To minimize smoke exposure, close up your house (doors, windows, and so on). Turn the air conditioner's fan to recirculate. In the car, same thing: Close the windows and vents, and set the vent control to recirculate.

Forget about those surgical masks; even the small ones don't work for children, because they're not fitted right. Humidifiers and breathing through a wet washcloth don't work for smoke, either. If there's smoke in your home, go to some other indoor location.

Infectious Disease

The changing climate delivers bacteria that make kids sick over two avenues: contaminated water and food, and bugs.

- **Diarrhea** kills 2,200 children *every day*, mainly by severe dehydration. That's more than the deaths from malaria, AIDS, and measles combined. It's the second-biggest killer of kids under five worldwide (after pneumonia).

  People get these diarrheal illnesses from bacteria (*E. coli*, salmonella, and so on), usually from unclean water and food—just the sort of things that follow the ferocious downpours and hurricanes in the new climate. Most people who get sick from contaminated water do so right after a freak rainstorm.

  Bacteria love warmer weather. Planet warms, bacteria thrive, more kids get sick. Put it all together, and you're looking at 48,000 more children dying each year from diarrhea by 2030.

- **Malaria, dengue fever, West Nile.** Chapter 14 should do an adequate job of scaring the hell out of you with facts about the new era of mosquito-borne diseases. Today, malaria kills 482,000 children under five every year; but climate change means that we're getting more mosquitoes in more places, and more children spending more time outside because the warm seasons are longer.

  Chapter 14 also makes it clear that if you bother to take precautions, there's no reason you or your children need to get sick.

Guiding Kids through a Disaster

It's traumatic for anyone to live through a wildfire, hurricane, flood, heat wave, or drought—but at least you, the mature and stable adult, have had enough life experience to develop a few coping skills. Kids are resilient, sure—but traumatic experiences can lead to lifelong depression and other setbacks.

Weather disasters are not rare, either. The number of "extreme-weather events" has tripled since the eighties. In the nineties, 66 million children a year lived through one of these devastating disasters; that number is expected to balloon to 175 million in the next ten years. As always, the kids in the poorest regions will be disproportionately affected.

In 2019 alone, natural disasters forced 25 million people out of their homes around the world; in the United States, the number of displaced people is about 1 million a year, brought to you by hurricanes and wildfires. That's a lot of families experiencing traumatic disruption.

The direct injuries from those events are bad enough: burns and smoke inhalation from wildfires, lacerations and puncture wounds from hurricanes, and so on. But there are also longer-term, more insidious, indirect effects.

They can include the trauma of being separated from family. After Hurricane Katrina, 5,000 families got separated, and it took six months for the last of them to be reunited.

A natural disaster may make *your* life difficult, but the trauma to children lasts far longer. Researchers have found high rates of post-traumatic stress disorder (PTSD) in children three years after hurricanes, floods, and other disasters. If your family must endure one of these weather nightmares, the playbook for recovery goes like this:

- **Reestablish stability.** "Provide as much comfort and security, and as many of the basic needs, as you can: food, warmth, shelter, love, and community support," says psychiatrist Sue Varma, assistant professor of psychiatry at NYU's Langone Medical Center.

  That might mean working on reuniting the family, offering the kids the chance to work or study, or getting in touch with any kind of familiar social systems that might provide meaning: religious, social, cultural, whatever you've got.

- **Care for yourself.** Some of the dangers to children after a disaster come from close at hand: their parents. The sense of instability and loss in the wake of a weather calamity leaves *you* psychologically fragile, too. "Substance abuse, domestic violence, and even child abuse may be correlated with increased stress post-tragedy," says Varma. "After Katrina, half of all adults experienced mental-health problems."

  The challenge is finding the bandwidth to take care of yourself—and your children—when the world seems to have fallen into chaos. Mental-health care is slowly coming into focus as a priority for disaster-relief organizations, but you may have to hunt to find it.

  Turn to any tools available to keep your own keel stable. That might mean therapy or medications for depression and anxiety. It also means limiting your use of alcohol and drugs, and seeking help from any kind of community.

  Group interactions can help tremendously. Local hospitals often offer group counseling, and getting together with other families can distract or engage kids while they're in the temporary housing.

  As you gradually reestablish your living situation, food, water, and shelter may be at the top of your priority list, but your emotional state should be next.

- **Consider the ages.** A kid's response to disaster disruption is age dependent. If you have a baby and you have to bolt from your home, try to grab a stuffed animal, blanket, or a toy that your kid loves. If you weren't able to, try to get a replacement.

  Older kids benefit from contact with any part of the social-support network that still exists, like family members or friends. For kids of any age, do whatever you can to maintain any pieces of their routine or activities that are still going on.

- **Let them help.** When people feel helpless in a bad situation, they're likely to become depressed and angry—including kids. Let them contribute. Give them tasks; listen to their ideas. For kids facing adversity, research shows that the chance to solve problems and take action is a huge help.

- **Check the app.** Download the free Help Kids Cope app for your phone. It's designed to help you talk to children before, during, and after a natural disaster.

Soothing Eco-Despair

Researchers have only recently begun studying the psychological effects of the climate crisis on adults; the psychological effects on *children* are an even more rarefied niche.

That's a shame, because record numbers of young patients are experiencing depression and anxiety about our climatic future. "I'm seeing massive increases," says UK therapist Caroline Hickman. "Not just directly from children, but also from adults who are working with children—teachers, parents, and museums."

In one survey, a *quarter* of children ages 10 to 14 reported that they believe that the world will end during their lifetimes. They feel about the threat of climate change much the way their parents or grandparents felt about the nuclear threat at the height of the Cold War: a nagging, overhanging sense of dread.

We're no longer talking about children who have endured a climate disaster. Just *knowing* about the climate crisis can bring them a world of worry.

Your kids may be even more freaked out than you are. That's partly because they have less experience discerning good information from bogus, and partly because they perceive the terrifying truth: *They'll* live their lives in the chaotic world we're leaving to them.

Your first step in helping your children adapt to climate change is observing them. You may see them having trouble sleeping or getting their homework done. They may get headaches, feel tired, or act clingy or immature. In teens, you might see changes in patterns of using alcohol, tobacco, or drugs. These are signs of a kid having trouble with anxiety.

Get Yourself Ready

Before you speak to your kids about climate change, make sure *you* understand the science. You can find good, plain-English explanations at sites like these:

- **National Geographic.** www.nationalgeographic.com/environment/global-warming/global-warming-effects/
- **NASA.** https://climate.nasa.gov
- **The US Global Change Research Program.** www.globalchange.gov

Get a grip on your own feelings about the changing world, too, following some of the techniques described in Chapter 1. Your children look for cues on how to react by studying how *you* react.

Bring Up the Subject

The most important step in helping your kids adapt to the psychological impact of climate change may feel screamingly counterintuitive: Bring it up.

Why counterintuitive? Because a good parent's instinct is usually to keep things upbeat, to create a happy world for their children. Why introduce something as upsetting as climate change?

First, you're *not* introducing it. In one study of fifth graders, 72% expressed pessimism about the planet's future. "What you're doing is waking them up to how they feel underneath," says Hickman.

Meanwhile, if children don't see you talking about the climate crisis, they're likely to feel even more troubled by that disconnect. "Why is the whole world talking about it, except at home?" they'll think. "It must be too scary to even talk about!"

A key benefit to having The Conversation is that, in the process, you may find out that your kid has heard all kinds of stuff that's not even true.

And, no, you can't leave climate-change teaching to the schools; fewer than half of U.S. schools teach anything at all about climate change. And when it is mentioned, it's often presented as a theory or as an undecided scientific debate. As with sex, race, and other delicate topics, *you* are the best presenter of climate change to your kids.

Laying the Groundwork

If your kids are very young—say, four and five years old—use your judgment about how much to say about the climate problem. Your primary task is to build a foundation by *exposing* them to nature. That can mean camping or hiking; if you live in a city, that can mean watching pigeons and squirrels in the park.

Many a child psychologist recommends using these outings as conversation starters; for young kids, discussions of plants, animals, air, sun, water, and earth are a lot more meaningful when they can *see* them.

That's especially true of the ocean. If your kid is lucky enough to see the immensity and power of the sea from the land, the idea of sea levels rising will be far more meaningful.

Or just watch some of those breathtaking David Attenborough documentaries, like *Planet Earth*, *Planet Earth II*, *Blue Planet II*, and *Seven Worlds, One Planet*. The photography and storytelling will blow *your* mind, too.

The point is that kids who grow up with an appreciation of nature grow up to care about the environment. They're the ones who become volunteers, environmentalists, and green citizens.

At these ages, you should focus on building some basic understanding of how nature and its cycles work. When you're outside, explain how plants "breathe in" the gases that we breathe out. When it rains, talk about where all that water goes—and where it came from in the first place. When you eat, make it real: that food came from somewhere and reached you somehow.

Eventually, you can talk about how energy, too, is a cycle. Turning on the lights isn't magic; it comes from somewhere. And you can mention some of the different ways we create that power.

You're teaching that "the environment" isn't some abstract, distant place. It's *their* world. You're also cleverly laying the groundwork for The Talk once they're ready.

Outlines for The Talk

Most problems that inspire your kids to approach you—bullying, unpopularity, romantic disappointment—will pass. But climate change is different, and you have to acknowledge that.

"'Everything will be okay' is the one thing you *cannot* say to children right now," says Hickman. "We know that things are not going to be great, we know that some countries will be better than others, we know that some places will be underwater—but we can't predict all of it at the moment. So we can't reassure them fully."

At the same time, "they also need to feel reassured, protected, and safe in the world," notes psychiatrist Varma.

The conversation, in other words, must strike a balance. Speak honestly to the kids, but without terrifying them. Your dialogue should be age appropriate and child appropriate. Don't use jargon without defining it.

When the day comes, your conversation might follow an outline like this:

- **Explain the cause of the problem.** For example: "For a long time, we've been burning oil and gas and coal to make energy—for heating our houses, making electricity, and flying planes. But burning all that stuff makes this bad gas come out—like invisible smoke. It wraps around the planet like a blanket and makes everything warmer."
- **Outline the problem.** "When the planet gets hotter, the storms get bigger, the world's ice starts to melt, and the oceans get fuller. And pretty soon, animals have a hard time finding places to live."
- **Propose action.** "It's a really big problem. There are a lot of smart people working on fixing it—and there's a lot that our family can do to help, too."

Hickman suggests emphasizing that you'll be there to see them through: "There will be some things that are okay, and some things that aren't, but what I can promise you is that we will not abandon you. We will be here with you, and we will talk with you, and we will share this problem together."

You'll notice that these scripts don't shy from suggesting that there's danger ahead, and that's on purpose. Kids will figure out that you've been sugarcoating things or being vague, and that'll backfire. It's scary enough to contemplate the world's deterioration without realizing that you can't trust your own parents.

For older kids, your chat should follow these principles:

- **Ask what they've heard.** Some of the things they worry about may be far-fetched wastes of anxiety. "They may tell you something so outlandish that it's easy to comfort them, to say, 'I hear what you're saying, but that's not happening tomorrow,'" says Sue Varma.

  Studies show that by fifth grade, much of what kids learn about the crisis comes from TV, movies, and YouTube videos, all of which may be delivering a mix of facts, semifacts, and falsehoods. They may be picking up distorted information even from teachers.

- **Talk facts, but avoid shock value.** Don't show upsetting photos or videos to elementary schoolers. Even for teenagers, you might want to prescreen news videos, so you'll know what they're going to see, and you can think about spots to pause and talk about what's on the screen.

- **If you don't know, research together.** If the kids ask a question that you can't answer, don't consider it a moment of embarrassment; look up the answer *together*. "They will learn from your humility and problem-solving abilities," says Varma. "How you acquire information is just as important as the information itself. It teaches them to be resourceful."

Keep It Going

Once your children realize that it's okay to talk about climate change, subsequent conversations can build on that trust. Talk about hopeful signs as they come along. You can watch videos together or explore websites, to keep the conversation going.

The web is teeming with great options along these lines. Many of these are offered as *teacher* resources, but, hey—since most schools aren't teaching climate change anyway, *you* are the teacher now:

- **WatchKnowLearn** (www.watchknowlearn.org). A directory of professionally produced videos—thousands—on all kinds of educational topics. Set the age of your audience, search for *climate*, and click one that looks interesting.
- **Alliance for Climate Education** (https://acespace.org). Lots of short videos, two to six minutes long. Excellent conversational snacking, all for a young audience.
- **BioInteractive** (www.biointeractive.org/classroom-resources). A huge array of videos, activities, and interactive displays for high school and college-age kids.
- **CLEAN** (www.cleanet.org) is the Climate Literacy and Energy Awareness Network, a vast collection of videos, articles, and other resources for middle school and high school kids.
- **Global Oneness Project** (www.globalonenessproject.org). Click "Climate Change" for about a dozen beautiful short videos.
- **Climate Kids** (https://climatekids.nasa.gov) is NASA's site for kids. Clear, updated, non-alarmist lessons on what's going on in the world.

There are hundreds of children's books about climate change; you can read them together with younger kids. No kid wants to read a bunch of depressing facts. The best of them, therefore, deliver the message sneakily, as a side dish that's secondary to the story.

Some examples:

- **Ages 3–7:** *The Problem of the Hot World* by Pam Bonsper. Five forest animals set out to find out what's going on with the planet.

  The Rhythm of the Rain by Grahame Baker-Smith follows a glass of water that a boy pours into a river—to the ocean, the clouds, the rain, to growing things.

  The Tantrum That Saved the World by Megan Herbert and climate scientist Michael Mann tells a rhyming story of motley climate refugees—a polar bear, Syrian farmers, and so on—who arrive at a young girl's home, looking for solutions to their plight.

Dr. Seuss's classic *The Lorax*, a parable about people exploiting nature beyond the tipping point, feels more current than ever.

- **Ages 7–10:** *The Magic School Bus and the Climate Challenge* by Joanna Cole. Everyone's favorite daffy teacher, Ms. Frizzle, explains climate change with humor and real science.

  Under the Weather by Tony Bradman is a collection of inspiring, climate-related short stories.

- **Ages 9–11:** *Basher Science: Climate Change* by Simon Basher. Using catchy illustrations, this book covers greenhouse gases, weather systems, renewable energy, and what we're doing about the problem.

  It's Your World by former First Daughter Chelsea Clinton is about current events and engagement—about climate change, yes, but also poverty, health, and other topics. She emphasizes actions that children and teenagers can take and highlights success stories of other kids' efforts.

- **Ages 12-18:** Novels like *Exodus* by Julie Bertagna and *Love in the Time of Global Warming* by Francesca Lia Block aren't explicitly about climate change—but by featuring teenage protagonists making their way in a ravaged world, they tell thought-provoking stories about the future.

  For nonfictiony reads, there's *A Teen Guide to Eco-Gardening, Food, and Cooking* by Jen Green (just what it sounds like), and *It's Getting Hot in Here* by Bridget Heos, a full-color, engaging, science-based primer on the causes and solutions of climate change.

Acknowledge the Anxiety

Once you've broached the No-Longer-Forbidden subject, you can begin to explore how your kids *feel* about it. If it turns out that they're afraid, worried, anxious, or worse, you can move on to the second phase: acknowledging, and even welcoming, their anxiety.

Any new situation produces anxiety. Going to school for the first time, going on a date for the first time, learning to drive—all of that is nerve-racking, and all of it is essential in growing and learning. The heightened anxiety probably served our ancestors well as they ventured into new, unknown situations and territories, because they'd be ready to fight or flee if danger arose suddenly.

The problem, says Caroline Hickman, isn't that your kids have these negative feelings. It's when they have negative feelings *about* those feelings. "You beat yourself up emotionally. 'What's wrong with me?' No, you *should* be anxious. It's a sign of being emotionally intelligent."

That doesn't mean allowing the anxiety to take over. Acknowledging it is only a first step; it still has to be addressed. Otherwise, the anxiety can grow wild, leading to panic attacks and phobias, and, eventually, the inability to function.

Conversation is half the battle. The other half is taking action.

Teaching by Example

Any action *you* take for the planet leaves an outsize impression on your kids: eating less red meat, riding Amtrak instead of flying, installing LED bulbs, adjusting the thermostat.

Those acts of engagement silently convey a sense that there's hope, illustrate a degree of control over the situation's outcome, and establish that individual actions matter.

Or not silently! Each step you take along these lines is an opportunity to talk about what people can do about the climate problem. (The free online appendix to this book, "Your Carbon Footprint," may give you some ideas and inspiration along these lines. It's available at www.simonandschuster.com/p/how-to-prepare-for-climate-change-bonus-files.)

And when you talk about future careers, don't just introduce fireman, pilot, doctor, lawyer. "You might discuss careers that help the environment (such as forest resource officer, hydrologist, geologist, agricultural technologist, biofuels engineer, solar sales representative, wind development associate, atmospheric scientist, meteorologist)," says Scholastic's guide to discussing climate with children.

A New Skill Set

Your kids probably won't need to forage for plants or build survival shelters anytime soon. But teaching them some survivalist skills now is worth considering for three reasons:

- ♦ Feeling mastery of these skills works wonders in relieving your kids' climate anxiety, even if they'll never need them. It's the same reason people feel more confident walking downtown at night after taking a self-defense class: The chances you'll use those skills are remote, but you *feel* better knowing that you're prepared.
- ♦ Taking survival-skills classes is fun for—yes, the whole family. We're not talking about falling in with roving anarchists; this is more along the lines of Boy Scout and Girl Scout merit-badge efforts. Taking a couple of camping classes, learning a few roughing-it skills, mastering a couple of knots, tasting your first wild berries—it all makes an incredible family-bonding experience.
- ♦ The cold truth is that your children *may* one day need these skills.

Dealing with Climate Deniers: Junior Version

What if you've done a great job raising an environmentally responsible leader-to-be, and then you get a report that they're running into young climate deniers at school?

If the other child insists that climate change doesn't exist or is part of a natural cycle, advise your kid to forget it—to walk away or change the subject. "They're either psychotic or they're just looking for a fight. Don't waste your time," says Hickman.

More often, other kids *disavow* climate change—they resist owning the problem. They may say, "Oh, it's not that bad" or "We won't see any effect for hundreds of years" or "Well, what about China? China has to fix its problems first."

In those situations, see the advice at the end of Chapter 1. Your child's job is not to beat other kids over the head with facts and numbers, but to appeal to the emotional centers, to listen to the other kid's thinking, and to give personal reasons—as personal as possible—about why the changing climate is a worry.

Take Action Together

Helping children adapt to their new, uncertain future involves many of the same thought processes as helping *yourself*. And the most useful tool you have, as you may remember from Chapter 1, is *doing* something about the problem. To somebody who's depressed or paralyzed by fear, any action, any size, feels good. You're still in a miserable situation, but at least you're not miserable *and* helpless.

Those steps can take the form of kid-appropriate undertakings like these:

- Change the light bulbs to LEDs.
- Bring the reusable shopping bags into the store.
- Build a "go bag" (Chapter 8).
- Kill the vampire power. All kinds of things keep using a trickle of electricity—"vampire power"—even in idle, standby, or "sleep" mode: microwaves, game consoles, TVs, cell phone chargers, computers, cable modems, cable boxes, garage-door openers, and so on. It's costing you, on average, $165 a year in wasted electricity, and the United States' total vampire-power consumption requires the output equivalent of 50 large power plants. Send your kids on a hunt to unplug stuff that doesn't need to be running 24-7.
- Get them used to leaving pee unflushed, to save water. Nursery rhyme for the new era: "If it's yellow, let it mellow. If it's brown, flush it down."
- Take them with you to the farmers market so you can talk about why locally grown food is much better than food shipped from far away.
- Blow their minds: Start a garden. Eat the results.
- Don't just tell them, "Go ride your bikes." Ride your bike *with* them, so they grow up with the bike as a natural option for getting somewhere.
- When you're shopping, let your kids say to the salesclerk who's reaching for the plastic, "We don't need a bag—thanks!"

- If your kids are on Instagram, have them follow eco-aware accounts like @youthgov, @nextgenamerica, @climateoptimist, @james_balog, and @RainforestAlliance.
- Work with your kids to write a personal note to your congressional reps. It's a great lesson in how the world works—and a note from a kid will get more attention.
- Add up your family's carbon footprint, using an online calculator like this one for younger kids (https://calc.zerofootprint.net) or this one, for middle schoolers and older (www.conservation.org/carbon-footprint-calculator#); both use a simple question/answer format. See if you can make changes that lower your carbon score.
- Plant a tree on your property, at the kids' school, or in a community garden.
- If, by the time they're tweens, your kids are environmentally aware and well-adjusted, congratulations! Consider setting them on the next path: becoming activists. The Alliance for Climate Education lists current environmental campaigns that your kids can join—and do something that's actually meaningful (https://acespace.org/ourwork/active-campaigns).

On the face of it, you and your children are doing all of this as part of your mitigation efforts. But you're simultaneously teaching your kids *adaptation* techniques. You're shielding your kids against depression, cynicism, and post-traumatic stress, because action erases feelings of helplessness at any age.

The Greta Effect

As an adult, it can already be hard to believe that you, an individual, can make a dent in the planetary climate problem. Now imagine how insignificant you'd feel if you were a *kid.*

If you encounter the "What power do I have?" moment, you may want to bring up Swedish teenager Greta Thunberg. She was 15 when she began her campaign of "school strikes for the climate" protests. The movement spread internationally, leading to mass student walkouts in cities around the world. In September 2019, the walkout involved over 4 million children.

She became famous, in part, through extraordinarily blunt and challenging messages to the older generations. "Adults keep saying, 'We owe it to the young people to give them hope.' But I don't want your hope," Thunberg has said. "I want you to panic. I want you to feel the fear I feel every day. And then I want you to act."

How much effect did this one kid have? She led a Montreal rally attended by hundreds of thousands, including Prime Minister Justin Trudeau. The secretary-general of the UN endorsed her message and her school-strike methods. The number of children's books about climate change doubled inside a year, which publishers called "the Greta Effect." Swedish Railway reported that ridership went up 8% in a year, in part because of the shame of flying that Thunberg had instilled in her message. And in Austria, the Greta Effect was credited with a tripling in support for the Green Party.

Figure 7-2. Greta Thunberg's famous sign says, "School strike for the climate."

You might also mention *Juliana v. United States.* It was a lawsuit filed by children, ages 9 to 20, against the U.S. government. Their argument: By supporting and even subsidizing the fossil fuel industry, and by refusing to take action on climate change, the government was denying them their constitutional rights to life, liberty, and property.

The case was filed in 2015. Incredibly, it has withstood dozens of challenges, petitions to dismiss, and other dilatory tactics by the U.S. government. By 2019, it had garnered amicus briefs (letters of support) from members of Congress, legal scholars, religious groups, women's groups, businesses, historians, doctors, environmentalists, and over 32,000 other kids.

In early 2020, a Ninth Circuit panel of three judges dismissed the case, writing that making the kinds of major societal changes the kids wanted should be entrusted "to the wisdom and discretion of the executive and legislative branches." The Juliana lawyers intend to appeal.

Behind the Scenes for Your Kids

You already know that schools may not be doing a great job teaching about the climate crisis. "Climate change is often not taught at all, skimmed over, taught in a hesitant way, or cast as a controversy," writes Mark McCaffrey, policy director of the National Center for Science Education. Indeed, about two-thirds of all U.S. students say that they were never taught about climate change in school.

The thing is, you're not helpless. As a parent and a taxpayer, your opinion matters. You can express your feelings about the lack of climate teaching to the school's leadership, or to the school *system's* leadership. You can also stand up for teachers who face political resistance from a hostile administration, parent cluster, or even other teachers.

If the issue continues to be blockaded by deniers or evaders, consider joining forces with groups like these:

- Climate Parents (www.climateparents.org)
- Moms Clean Air Force (www.momscleanairforce.org)
- Mothers Out Front (www.mothersoutfront.org)

- Ecodads (www.ecodads.org)
- 1 Million Women (www.1millionwomen.com.au). It's an Australian group, but its Facebook feed is bursting with good ideas for making changes.

These groups are dedicated to bringing climate-change conversation into schools, political decisions, and the public forum. They participate in rallies, raise awareness through social networks, and ask business and political leaders to act. You don't have to sign up with them or get involved officially; just reading their sites is plenty inspiring.

The Baby Decision

"I've always imagined I would be a mother," writes McGill University student Emma Lim. "There are all of these little things I can see myself doing: singing to my babies, listening to their stories, working through homework, baking, finger painting, going for nature walks, dancing in the kitchen. And it hurts almost like a physical pain when I realize I might never get to do these things."

Lim founded a movement called No Future No Children. Thousands of women joined her in pledging not to have children until governments ensure a safe future for them. At Birthstrike (www.birthstrikeforfuture.com), women make a similar pledge. Cofounder Blythe Pepino takes pains to distance her movement from the antinatalist movement (which argues that having offspring is morally wrong) or population-control movements (which have a sinister and racist history).

Instead, these more recent movements are allowing women to declare—in a way that they hope will draw attention to the tragedy of inaction—that they intend to be child-free until the world is rescued from the climate crisis.

These groups, in other words, are concerned with climate change's impact *on* children. But other potential parents cite another concern: the impact of *having* children.

It turns out that the most disastrous carbon dioxide gesture a human being can perform is simply existing. Sure, you can save maybe 1,400 pounds of CO_2 a year by giving up flying, and 500 pounds a year by giving up red meat. But those savings are minuscule next to the *total* emissions you produce each year.

For an average American, that's 20 metric *tons* of CO_2 every year. (One metric ton would fill a cube 27 feet on a side.) For a European, it's maybe 8 metric tons. At the lowest end of the scale, a citizen of Bangladesh is responsible for only about *one-quarter of one ton* of carbon dioxide per year.

Now, the birth rate in the United States has been sinking to another all-time low every year for a decade. But the *global* population is another story.

The global human population is growing by 80 million a year—the populations of New York City, Los Angeles, Tokyo, and Mexico City combined. "It's the highest annual number increase in human history," says climate educator and author Richard Heinberg. Not only do those new souls need feeding, sheltering, and caring, but they impose a continuing, massive toll on the climate.

Add all of these factors together, and you get one-third of prospective parents naming climate change as a reason they're holding back.

No wonder baby-decision therapy is a booming business.

"More millennials are saying, 'Yeah, this is a climate crisis. I'm not having kids,'" says Ann Davidman, whose potential clients now fill a waiting list. She's a "motherhood clarity mentor" and coauthor of *Motherhood: Is It for Me?*

The Two Decisions

The decision to start (or enlarge) a family is harrowing well before the climate question comes up. Parenthood psychotherapist Merle Bombardieri, author of *The Baby Decision*, likens it to being offered a job that requires a 20-year commitment—and you have to make your decision to accept it without meeting the boss, seeing the office, or having any idea whether you'll like the work. Oh—and you won't get a salary. In fact, you'll have to *pay* to do it.

And that's how hard the choice is before you bring climate change into it.

If you're struggling with the motherhood decision, it's important to identify the difference between what you *want* and what you *decide to do.*

"They're very separate processes," says Davidman. "If you try to solve them at the same time, you end up in gridlock and you feel tortured."

All of the external factors in the decision to have a baby are real, and they're important: the cost of having a child, your career, your age, your health, your relationship status, family pressure, your partner's opinions, and whether you already have a child.

But in figuring out what you really want, those factors are irrelevant. "They play a big role in what you're going to *decide to do*, but they play no role in figuring out what you *want*," says Davidman.

That's why her therapy begins with homework: Write down every one of those external factors you can think of—then seal them away in an envelope for a few weeks. The idea is to figure out what you want, internally, in the *absence* of all those external factors. Then, and only then, is it time to weigh those other considerations.

Maybe what you want and what you decide will align. Congratulations! You've won the bliss of certainty.

But maybe they won't.

"There may be grief," acknowledges Davidman. But at least you've made the decision with eyes wide open, having considered your deepest desires and expectations of your life. "You're choosing it and you know why you're choosing it."

The second decision is the answer to: If you want children, *why* do you want them? The reasons may be sound, or they may strike other people as irrational; it doesn't matter. The point is that you've considered the question and arrived at your own answer.

Baby-decision therapists like Bombardieri and Davidman both provide exercises to their clients: journaling, making maps of family relationships and dysfunction, visualizations, and so on. And both acknowledge that no matter which decision you make, there will be pangs of regret.

"There isn't one child-free person who isn't going to think, perhaps at a wonderful family celebration: 'I wonder what a child with our genes would have looked like. We would have been good parents,' and so on," Bombardieri says. "On the other hand, there isn't one parent who isn't going to regret their decision to *have* children on a day when their favorite rock stars are in town, and it's the one time in ten years, and the baby's sick or the babysitter's sick."

The word *decision* comes from *cis*, a Latin root meaning "to cut off"—for example, to cut off all other courses of action. As a result, Bombardieri says, "*Every* decision involves loss. So it's not 'Will I regret my decision?' It's 'Which decision will I regret least?' "

In her experience, you're far more likely to experience regret if your decision was made *for* you, passively: for example, if you had a child through a contraceptive error, or did *not* have one by drifting, indecisive, until you were no longer capable of becoming pregnant. Actively confronting and making the decision, one way or another, is the path of least regret.

Variations on a Family

In many cases, there may be a way out of baby indecision by thinking outside the "2.5 children per family" box. Baby-decision therapists often offer their clients less traditional ways to approach the problem. For example:

- **Become a conscious family.** Having a child doesn't have to mean a proportionate destruction of the environment.

  "You could have a family with *three* children who are doing more for reversing the climate crisis than a child-free person who's creating all kinds of problems," says Davidman. Your children, in other words, could be part of the solution. They could become activists, educators, or leaders who help the rest of the world on its decarbonization mission. They could live exemplary, low-carbon lives that have a ripple effect among the people they'll encounter in life.

- **Adopt.** Adopting a child lets you become a parent without adding a single new molecule of human-generated carbon to the atmosphere.

  "You have to think about what 'parenthood' means to you," says Davidman. For example, you may conclude that to you, being a parent means influencing the way a young person grows up, imparting your values, deriving joy from your kid's development and achievements, getting the satisfaction of being needed and loved—not necessarily passing on your own genes.

Bombardieri points out that the children don't all have to be *all* yours or *all* not-yours, either. You could have one child of your own, then adopt one or two. "That's one carbon footprint instead of two or three," she says.

- **Have one child.** Your choice of having children isn't a choice of two children or none. "People think, 'Well, if I have one child, I have to have two,'" says Bombardieri, "because an only child is selfish, miserable, immature, et cetera."

  In fact, though, copious research has shown that a typical only child scores higher on achievement tests, bonds better with their parents, and isn't any more prone to mental illness than children with siblings. (Only children do tend to be more overweight than kids with siblings, though, so do your best to foster active, healthy habits.)

The Final Analysis

Among the external factors that Ann Davidman recommends you set aside while you decide what you want, social pressure and family pressure to have children are sure to be the big ones.

"The prejudice is unbelievable," says Bombardieri. "The stereotype of someone who's going to be child-free is someone who doesn't want to do the work of parenting, wants to spend all their money on vacations, wants no responsibility, wants everything to be easy. It's changing, but it's still terrible."

Given the climate crisis, there's a certain irony in calling child-free adults "selfish." These are the people who are doing the *most* for society—both by keeping 9,400 tons of carbon dioxide out of the atmosphere (that's the tally of one child and descendants), and, in many cases, by freeing themselves to work on the planet's behalf. And our children's.

"Be your own judge," Bombardieri concludes. "You have a right to your own reasons to be child-free without being accused of selfishness, immaturity, or neurosis."

Therapists like these typically work with a client for a few months, then wish them well.

"When I say goodbye to people, I know what they want," says Ann Davidman, "but I don't always know what decision they'll make. They go away, and then sometimes I hear from them a year later. I get either a picture of their baby—or a picture of their dog."

Chapter 8

Ready for Anything

SURVIVALISM—THE HOBBY, THE PURSUIT, THE LIFESTYLE—has been around for decades. The headlines lately have focused on "doomsday preppers," the rugged individualists associated with underground bunkers, stockpiles of guns, and deep skepticism about the government's intentions. They prepare for scenarios they describe as TEOTWAWKI (the end of the world as we know it), WROL (without rule of law), and SHTF (shit hits the fan).

But in the new age of extreme weather, the *essence* of their philosophy seems increasingly reasonable: preparing now for emergencies to come.

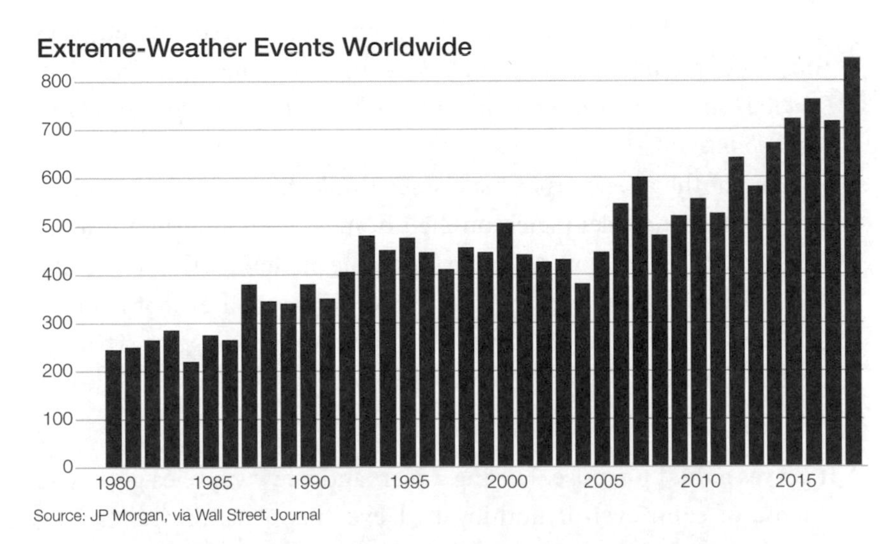

Figure 8-1. The number of hurricanes, wildfires, heat waves, droughts, heavy downpours, and floods has quadrupled since 1980.

That's because the number of massive, disruptive natural disasters has steadily been increasing. Every Katrina and Sandy, every arctic blast and polar vortex, every California fire season, displaces and disrupts hundreds of thousands of people. In 2019 alone, there were *14* disasters in the United States that cost over a billion dollars each.

And then, of course, the 2020 pandemic came; with grim smiles, the preppers pointed out that they would not be among the screaming Walmart customers fighting over toilet paper.

The problem with nature's systems freaking out is that they're usually followed by *human* systems breaking down: power outages, water contamination, and interruptions in the availability of transportation, groceries, water, internet, TV, cellular service, health care, and so on.

You may think, "Well, that's not going to happen to *me*." Trouble is, almost *everyone* has or had that thought, including the 25 million Americans who were struck by extreme-weather disasters in 2017 alone—8% of the population.

The arguments for doing some basic emergency prep are pretty ironclad:

- **It's cheap and easy.** As long as it doesn't take a lot of time, money, and energy to prepare, why not do it?

  "You don't have to become a MacGyver or a Green Beret," says Tony Nester, survival instructor and author of *When the Grid Goes Down: Disaster Preparations and Survival Gear for Making Your Home Self-Reliant.*

  Some of the attendees of his classes think they need 5,000-gallon water barrels and solar panels on their homes. They show up thinking they'll need to learn how to forage for edible plants and trap varmints.

  But for most people in the real world, he says, all of that is unrealistic overkill. Instead, he says, it's all about preparing for six basic needs: water, food, shelter, communications, medical care, and home security.

- **It serves a dual purpose.** Having a plan and a few supplies gives you a sense of calm even if nothing bad ever happens. Simply having thought everything through and socked away some basic supplies helps with your self-confidence, crisis or no crisis.

And if something bad *does* happen, you'll have a fighting chance at getting through the disruption relatively safely and comfortably.

- **It's good for the kids.** Your children also benefit from knowing that their family has thought through things that might happen. Once again, the benefit is both psychological and, when the worst comes to pass, physical.

The following chapters offer specific guidance on living through floods, heat waves, droughts, hurricanes, wildfires, and social breakdown. This chapter covers the broader concept of preparing for, and recovering from, *any* of those disasters. It's a prepping guide for any situation when the SHTF.

Apps to Install Now

You want easy and free emergency prep? Think you can handle installing an app on your phone?

In times of emergencies, the most useful tool in the entire world may be a charged phone. As long as the cell towers are still operating, you can call, you can text, you can keep updated, you can organize, you can look stuff up. Best of all, iPhone and Android apps can give you early warning for the wildfires, hurricanes, superstorms, tornadoes, and floods that are coming your way.

Early-Warning Alerts

In the 2018 wildfire that devastated Paradise, California, the greatest heartbreak may be hearing that people died simply because they didn't get the word about the emergency. They didn't realize a wildfire was nearby, never heard the evacuation orders, and simply died in their homes, trapped.

It's not that there wasn't a "Reverse 911" notification system; there was. But it was an opt-in service, and only 30% of the residents had signed up to receive the alerts. (Even *they* didn't all receive word of the fire. "Within the first hour of the fire, everything was out: electricity was off, cell towers were down, radio communications were down," says former Paradise mayor Woody Culleton. "Nothing worked.")

In theory, you shouldn't have to sign up to get emergency alerts. FEMA has put together a nationwide warning system called IPAWS. It may sound like an Apple puppy product, but it stands for Integrated Public Alert and Warning System. It lets the government blast out Amber Alert–style emergency warnings, called Wireless Emergency Alerts (WEAs), through every conceivable channel in the affected region: radio stations, TV channels (cable and broadcast), text messages, digital road signs, sirens, and so on.

But to send out a WEA, *local* disaster officials must *notify* the national system—and in the case of the Paradise fire, the local officials decided not to. WEAs didn't go out during the 2018 Sonoma County fires in California, either, and 24 people died.

Why not? Sometimes, local crisis managers worry that the WEA will

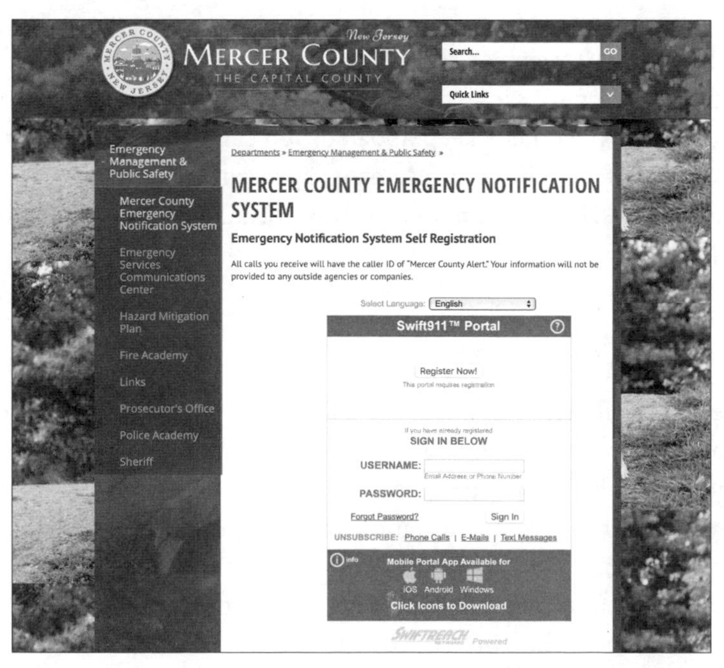

Figure 8-2. Here's your homework for tonight: Sign up for your city's emergency alerts.

notify residents in too wide an area, triggering unnecessary panic. Sometimes they don't know that the IPAWS/WEA system is available. And sometimes they just don't understand how it works.

You now know the moral of the story: Find out if there's a *local* public-warning system where you live, and sign up for it. Do a Google search for *[your city or county's name] emergency alerts* to find the sign-up instructions. Do it *today.*

School systems, religious organizations, and businesses may also have emergency-alert systems. These, too, are opt-in.

And this is critical: Make sure every family member's phone is signed up for these services, not just yours.

Early-Warning Apps

If you have a smartphone—that's about 80% of Americans and Europeans—you should take a moment, *now*, to install one of these emergency-alert apps. They're free, they take up little space on your phone, and they can sit there, unused and forgotten, for most of your life.

But when the day of disaster comes, they'll grab your attention with a notification—a sound, a vibration, a message on the screen (if your phone is charged and turned on). They'll buy you some time to get your act, and your stuff, together.

- **Emergency: Alerts.** This app, made by the American Red Cross, should be on every phone. It provides early warning about imminent disasters in your neighborhood: tornadoes, hurricanes, floods, tsunamis, thunderstorms, heat waves, wildfires, winter storms, and 27 other severe-weather and emergency events, natural and man-made.

  The app can turn your phone into an emergency beacon, with a flashing light and a siren. It shows you where Red Cross shelters are right now. Some of its information is available even if you have no internet signal, like instructions for creating a family emergency plan and what to do in an emergency.

  You can customize the warnings. You might enter the addresses of several places you care about—your home, your parents' home, your kid's school—and indicate which kinds of disaster you want to be alerted about for each one.

Get this app. Then you can forget all about it—until the day it saves you and your family.

- **Tornado, Flood, Hurricane.** The Red Cross also makes individual apps for these individual disasters. Each warns you when that type of extreme-weather event is headed you way, but each is also a mini-guidebook, filled with how-to information on these disasters and the aftermath.

  The Tornado app makes a loud siren sound when there's a tornado warning for your location; it plays even if you're not using the app at the time. The Flood and Hurricane apps also offer a customizable "I'm Safe" notification that lets you notify a predetermined list of friends and family, with one tap, that you're all right.

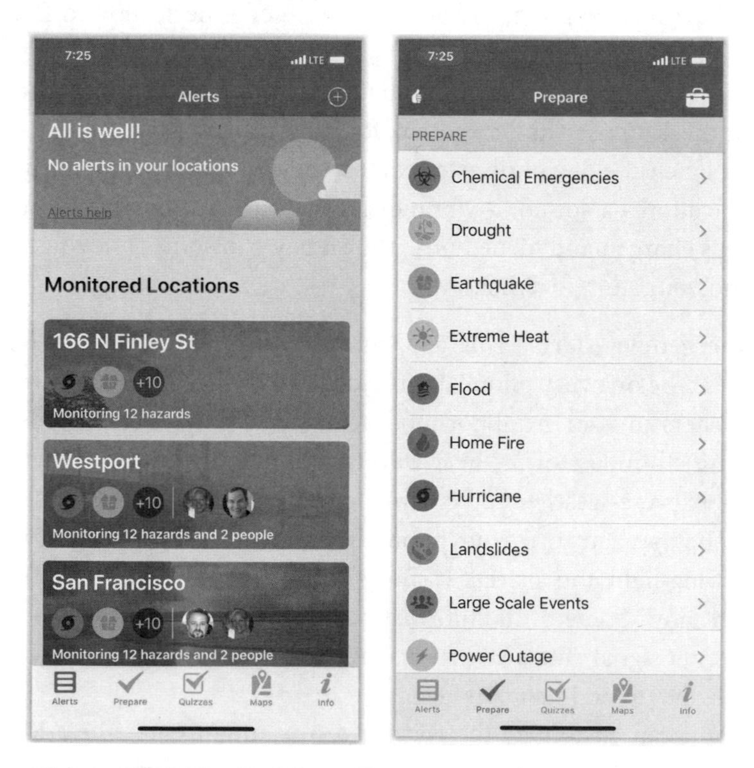

Figure 8-3. The Red Cross Emergency: Alerts app constantly monitors FEMA and weather status so it'll be ready to give you early warning of disaster.

- **Disaster Alert.** This app, by the Pacific Disaster Center, is the disaster peeper's dream. It shows a map of the entire country, with icons representing every disaster that's happening *right now*: floods, hurricanes, wildfires, earthquakes, droughts, volcanic eruptions, extreme cold, and so on. You can zoom in close enough to see exactly where the disaster is and tap the icon for details.
- **CDC.** The Centers for Disease Control and Prevention has its own app, too. It's teeming with useful, clearly written guidance about maintaining your health and safety during a disaster.
- **FEMA.** This app is brought to you by the Federal Emergency Management Agency, the U.S. government's increasingly busy disaster-

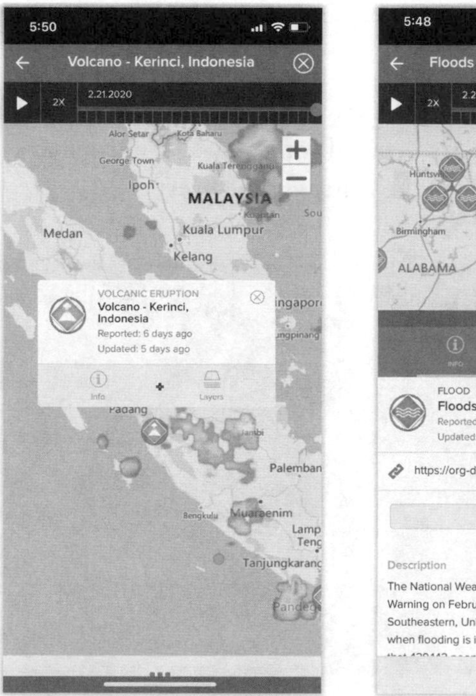

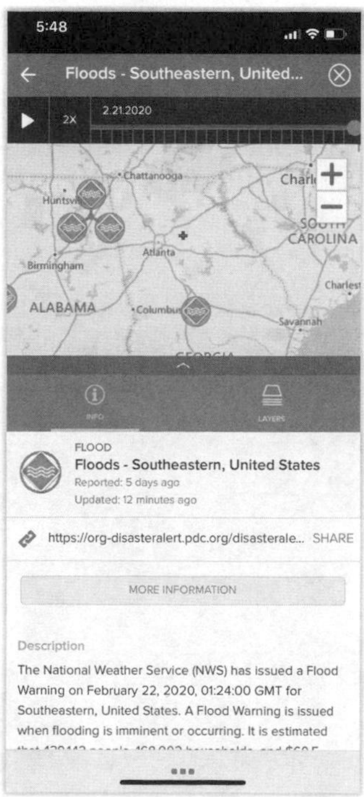

Figure 8-4. The Disaster Alert app shows you, with a startling jolt, that something terrible is always happening somewhere in the world. Usually many somethings.

recovery agency. It lets you know what kinds of help are available when bad things happen, including listing the closest shelters. The app even lets you share photos of the disaster, so first responders can get an idea of what the situation is near you.

Weather Tracking

In the olden days, you and your family might have huddled around a shortwave radio to track the progress of a blizzard, hurricane, or tornado. Apps do a much better job—much more targeted to your location, updated far more often, and far more visual.

- **MyRadar Weather's** star feature is its high-resolution radar imaging. With a glance, you can see whatever storm, hurricane, or blizzard is coming your way, hovering overhead, or about to pass. Weather forecasts, NOAA weather alerts, and nationwide temperature maps round out the package.
- **Weather Underground** offers excellent early-warning features for severe weather and offers detailed forecasts and radar maps, which make it easier to see when the worst will be over.

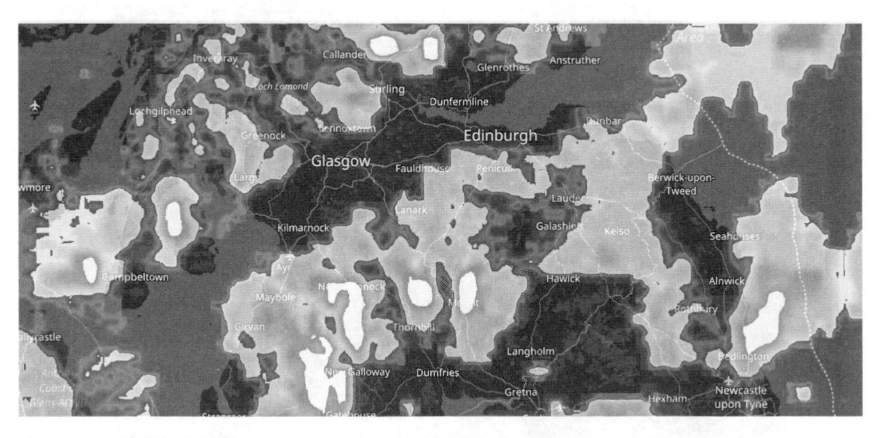

Figure 8-5. MyRadar gives you a satellite's-eye view of the storm.

Medical Info

It'd be great if everybody on earth went out and took a first-aid course, so everyone would be ready in times of disaster. But they won't, and you might not have time, either.

What you can do, though, is download these apps, so you'll have them to consult when you need them.

- **ICE (In Case of Emergency).** A simple, free app that stores your emergency-contact details, medical information, and a picture of your driver's license.
- **First Aid**, another American Red Cross app, is an excellent first-aid instruction guide and disaster-prep handbook. It comes with videos, quizzes, frequently asked questions, and lots of step-by-step guides to taking care of common medical problems.

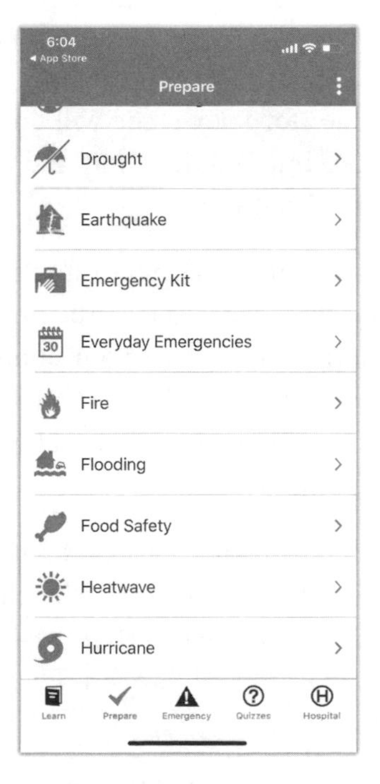

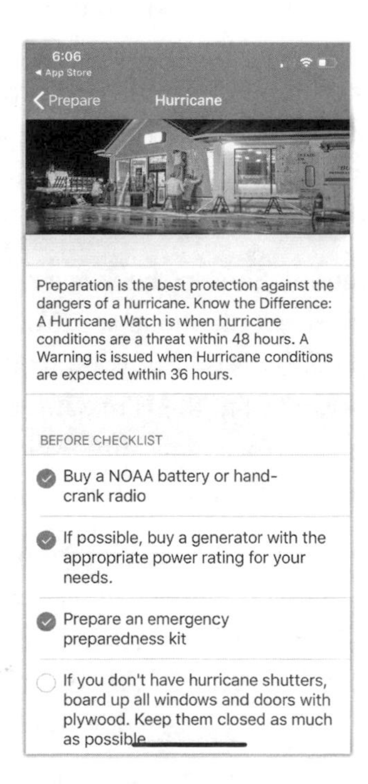

Figure 8-6. The Red Cross's medical experts can fit in your pocket, thanks to this first-aid app.

- **MedlinePlus** is a massive database of health articles, in English and Spanish. It's not *only* emergency-medicine articles, but they're in there, too. The WebMD app is similar.
- **Stop the Bleed.** Useful, illustrated reference for stopping bleeding—and saving lives.
- **Pet First Aid.** Veterinary advice for pet emergencies. Early-warning signs, first-aid guides with videos and photos, pet-hospital listings, interactive quizzes, and a Dog/Cat toggle.

Communication

You can bemoan people's smartphone addictions all you want. But in an emergency, if you can get online, apps like these can make all the difference:

- **SafeAndWell.** When you're riding out an actual disaster, do whatever you can to call up the https://safeandwell.communityos.org website, run by the American Red Cross. Here, you can list yourself as safe, along with messages like "Currently at shelter" or "Will make phone calls when able." Anyone who searches for your name will find your messages and learn of your status. It's a huge gift to family and friends to learn that they don't have to worry about you.
- **Facebook Safety Check** is a similar feature that's built into the standard Facebook phone app. You don't even have to take the initiative: If there's a disaster in your area, a notification wakes your phone and asks *you* to tap the "I'm Safe" button (if you've allowed notifications from the Facebook app). With a quick tap, you notify all of your friends and family on Facebook that you're okay.
- **Life360** is a real-time location-sharing app. You have to set it up on each person's phone in advance. Thereafter, you can see where everyone is on a map or get notified when they reach certain designated places (like home, work, or the hospital). If you have an iPhone, the built-in Find My app does the same thing.
- **Zello** is billed as a free push-to-talk walkie-talkie app; you can carry on real-time walkie-talkie chats with anyone else who has the app.

But there's a reason that Zello downloads surge after every major hurricane: It also works like a ham radio. During a disaster, you see a list of "channels," named for things like "Hurricane Whatever" or "Latest Wildfire"; if you tap it, you can push-to-talk, walkie-talkie-style, to everyone who's tuned in to that channel right now. It's a great way to get local information when you're feeling alone and scared.

- **Nextdoor** is like a private Facebook for individual *neighborhoods*—tiny areas a few blocks across. During most of the year, it's filled with lost-dog notices and "Can anyone babysit Saturday?" queries. But during a disaster, neighbors can post updates, cry for help, or, in some cities, view local police postings about evacuations, flood maps, open shelters, and so on.

Prepping Your Home

This may be a comforting bit of news: In wildfire and hurricane situations, you might be asked to evacuate. But in any other disruption—the power's out, you're snowed in, you've gotten 30 inches of rain, a distant wildfire has made the air outside dangerous to breathe, or a coronavirus has put entire swaths of the country in lockdown—you won't be evacuating. "You're going to be camping in your house or your apartment," says Tony Nester.

For real-world emergency preparedness, the idea is not to become a hunter-gatherer, roaming the apocalyptic landscape, fending off marauding gangs with a shotgun. Instead, you'll shelter in place. You'll be "bugging in" instead of "bugging out."

If a disaster has crippled the area, Nester says, you can dress in layers, sleep under a pile of blankets, cook on propane, keep updated from the radio or an app, read by candlelight or flashlight, wash by sponge bath, and collect rainwater. You can become your own first responder.

Your first prep step, therefore, should be equipping your *house* for surviving disruptions.

Planning for a home campout makes tremendous sense as a first step; in times of trouble, trying to drive somewhere might be a nightmare of traffic, downed wires, closed roads, and bad air. If you stay home, you already have a structure to protect you, you already know where everything is, and you don't have to pack or carry anything.

Best of all: Once things return to normal, you're already home.

Water

First thing you'll need: water.

"Doesn't matter if you're a Navy SEAL or triathlete, you can't go without water," says Nester. You hear all kinds of amazing stories of people going weeks without food, but you can't go without water.

If you're not outside, exerting yourself, you need around one gallon of water a day for drinking—per person. Make that two gallons a day if you plan to brush teeth, rinse dishes, take sponge baths, cook, and so on.

As noted in Chapter 3, the world is full of clever ways to install emergency water supplies in your home. Store-bought bottled water, unopened, keeps indefinitely.

You can also buy backup water supplies, like plastic jugs, big plastic barrels, or *inline* water tanks. Those are tanks that, ordinarily, water flows

Figure 8-7. A food-grade, designed-for-the-purpose water backup barrel should keep your water pure and ready for more than a year.

through, keeping it always fresh—but when the water supply gets cut off or contaminated, you flip a lever, cutting it off from the usual supply and making ready the most recent 40 or 80 gallons for you to drink.

The problem with filling containers yourself is that, as bacteria multiply, the water can go bad after a few months. The key is to prevent them from getting in there in the first place. If there *are* no bacteria, then the water in a sealed jug or barrel is safe for up to five years.

So before filling the container, sanitize it with bleach solution (a teaspoon of unscented bleach per quart of water), then rinse and let dry. Wash your hands before filling. Seal it tightly. Store somewhere out of the sun. Label the container "Drinking Water" and the date you filled it.

If disaster strikes and you haven't taken any of those steps, it's worth remembering the sources of clean water that you already have:

- **Your hot-water heater.** Right there, you've got 40 or 80 gallons of fresh, clean water, captured and ready to serve.

  You just have to get at it. First, shut off the valve that brings water in from the outside, so that contaminated water doesn't flow in. It's usually a lever on the incoming cold-water pipe at the top of the tank.

  Now, an important step: Turn off the heater's heat source, so that it doesn't try to heat an empty tank and ruin your expensive appliance. If you have a gas heater, turn the gas knob to the PILOT or VACATION position. For an electric one, turn off its circuit breaker, or just unplug the thing.

  If you try to drain the heater tank now, the water won't flow because it's blocked at the top; it'd be like your finger on the top of a full straw keeping the water from running out the bottom. To "lift your finger," locate the pressure-relief valve; it's usually a flipper switch high on the side of the tank, equipped with a down-pointing pipe that stops a few inches above the floor. Lift the pressure-relief-valve flipper switch so that its handle is pointing straight out. Only now can air flow in, to let the water flow out.

  Wait for the water inside to cool down.

  Finally, grab a bucket or some other container. Open the drain valve at the bottom of the tank—usually, that involves loosening the brass screw that's right by the spigot—and let the water flow!

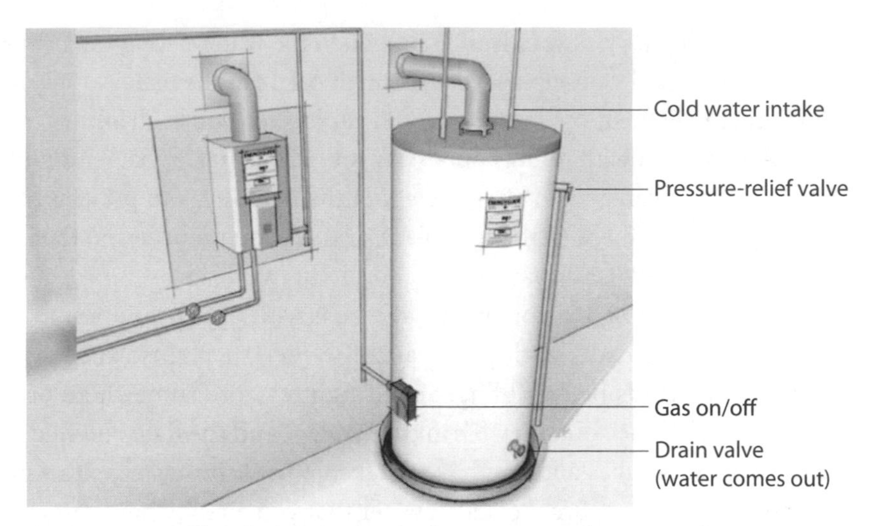

Figure 8-8. Your hot-water heater is an ever-ready source of drinkable water.

The first water that pours out may have rust and settled grit. Before drinking, clean it as described below.

(Once the crisis has passed, reverse all of these steps. Close the pressure-relief valve. Turn on the water intake on the cold-water line at the top. Make sure the heater is full *before* turning on the gas and power again.)

- **Toilet tanks.** Don't drink out of the toilet *bowl* unless you're a Labrador. But the *tank* behind the toilet holds several gallons of fresh water that hasn't yet flowed into the bowl. It's perfectly drinkable.

- **Your plumbing.** The pipes of your home are full of clean drinking water, too, poised to pour out of your faucets and showers on command.

  Open the highest faucet in the house, and capture whatever small amount of water comes out of it. Leave the faucet open. That way, the *lower* faucets in the house will flow on demand.

- **Ice in your freezer.** Presumably, those ice cubes formed while the water was still good, so they're yours for the thawing.

- **Canned food.** In a pinch, don't forget about the liquid in your canned fruits and vegetables. It's mostly water.

Your house may also contain a few sources of "gray water," not quite clean enough for drinking and cooking, but great for washing clothes, flushing toilets, watering plants, and taking baths:

- **The bathtub.** The age-old advice to fill your tub before a storm or blizzard descends is sound. You just have to remember to do it *before* the disaster hits; it won't already be waiting for you, like the water in the pool, the rain barrels, and the water heater. (Depending on your tub, your fastidiousness, and your gag reflex, you may not consider this gray water at all. You could drink it straight.)
- **Pool and hot tub.** These are big sources of water, all right—but you probably don't want to drink all the lovely sanitation chemicals you've poured in there.
- **Rain-catchment barrels.** These barrels catch rainwater coming from your gutters, as described in Chapter 3. This, too, is gray water.

You *can* use these gray-water sources for cooking and drinking *if* you first clean it, either by filtering or purifying it. Those are two different things.

Filtering removes sediment, bacteria, microorganisms, chlorine, lead, pesticides, and other contaminants from the water. A filter can take many forms:

- In an emergency, you can let the water sit in a bowl for an hour, so that the nastiest sediment floats to the bottom. Then pour the rest through a coffee filter.
- For more regular use, let the water flow through a made-for-the-purpose filter. There's a huge range of these things: $27 Brita water pitchers, $35 water bottles with built-in filtration in the straw, $50 faucet filters, $150 undersink filters, $700 whole-house filters, and so on. Most of them rely on cartridges that you have to replace periodically.

Purifying the water removes all of those impurities above *and* viruses, like norovirus and hepatitis A, which are far smaller than bacteria. You've got three options, listed here from most to least desirable:

- **Boil it.** The U.S. Centers for Disease Control and Prevention recommends bringing water to a rolling boil for one minute. Boiling kills everything: parasites, bacteria, and viruses. Let it cool before drinking.

  If you have a choice, boiling also beats any kind of cartridge-based filtering, because cartridges get used up.

  Boiling makes water taste kind of flat and gross. You can help that by pouring it into a new container and letting it sit for a few hours—or just add a pinch of salt for each quart you've boiled.

- **Bleach it.** Add six drops of bleach to a gallon of water, then let it sit for half an hour. You want *pure* bleach, unscented, where the sole ingredient is sodium hypochlorite. The additives in *scented* bleach—lemon, honey, whatever—are detergents that can make you sick.

  Bleach-sterilized water is fantastically useful. You can drink it, wipe down surfaces with it, sanitize your plates and utensils with it—all an excellent argument for equipping your home with a least a gallon of bleach *now*. (Another excellent argument: The empty cleaning-supplies aisles in stores during the coronavirus crisis.)

  And don't freak out that you're drinking tiny amounts of bleach. Bleach is sodium hypochlorite, exactly the same stuff your city has been using to purify your drinking water for decades.

- **Add tablets.** You can also disinfect water with a chemical purification kit like the ones backpackers use. A $10 package of tablets kills bacteria, viruses, and most parasites in 15 gallons of water, thereby making it drinkable.

  This is a fairly expensive option, but it's mobile. You can carry tablets with you if you have to leave the house.

  These tablets affect the taste of the water, by the way. Some tablets rely on iodine as their key ingredient, but it tastes vile; you'll have a hard time getting kids to scarf it down. The ones based on chlorine dioxide are easier to tolerate.

Figure 8-9. To use water-purification tablets, add one to the water, shake it up, and wait half an hour before drinking.

There's one big, fat, underlined caution here: None of these techniques get rid of oil, gas, lead, or toxic chemicals that may have seeped into the city's water if the system has fallen apart. If there's any chance that the water is contaminated that way, stick to bottled water or the water already in your home system.

There are some sources whose water you should *never* drink: radiators, boilers (that heat your home), and waterbeds. The first two contain anticorrosion chemicals, and waterbed water contains fungicide chemicals that you wouldn't enjoy.

Food

A disaster doesn't have to happen where *you* live to cut off your food supply. It could be droughts or deluges where the food is grown. It could be flooding in the packaging plants. It could be a diesel-fuel shortage that keeps the trucks from reaching you. It could be panic buying in grocery stores all over the region.

Fortunately, modern technology has brought us a terrific food-storage technology that lets food sit in your cabinets, as a backup, for months or years without spoiling. It's called cans.

You can buy just about everything in cans and jars: meat, fruit, vegetables, peanut butter. When you're stocking up at the store, some great options are cans of tuna, corn, black beans, green beans, diced tomatoes, pears, pineapple. You can get chicken and ham in cans, too, as well as hearty soups and stews.

Dried foods in boxes are also good to have: rice, oatmeal, beans, pasta, dried soup mix, coffee, and tea. And don't forget pet food. (Not for you—for the pet.)

This is all regular grocery-store food. Plenty of companies sell "prepper food," too: turkey tetrazzini, chewy beef stew, and other dishes that

Figure 8-10. They may look like ordinary groceries, but you're looking at a long-term "camping at home" supply station.

are high in calories and comforting in an emergency. But prepper food costs a lot and doesn't taste great, and for camping-at-home use, you don't need it. Some of it may even spoil sooner than the canned and dried staples listed here.

Build up at least two weeks' worth of emergency rations, and you've checked off this box.

Paper plates, too, can be handy when the water supply is cut off. And backup supplies of sugar and salt are handy for seasoning any of your backup foods that are a little bland.

During an extended power or gas-line outage, you'll also need some way to cook. Maybe you already have an outdoor grill; it might do nicely in a pinch for all kinds of heating up. Alex Wilson, founder of the Resilient Design Institute, also likes the BioLite outdoor cookstove (BioLiteEnergy.com), a $200 portable firepit that uses an electric fan to circulate air over coals or wood. The result: a cooking fire with half the

Figure 8-11. The BioLite FirePit lets you cook over charcoal or wood without smoke. You control the heat of the fire using, believe it or not, a phone app.

fuel needed for a campfire—and no smoke. The fan unit's battery can even charge your phone.

But making things hot isn't your only worry when the power's out. You also won't have power for refrigeration.

Food in the fridge stays cold for four hours without power, and food in a full freezer stays frozen for 48 hours. You can prolong those times like this:

- **Keep the fridge doors closed.** Every time you open the fridge or freezer, you let warm air in.
- **Shift to the freezer.** Move leftovers, meat, and milk from the refrigerator to the freezer. It'll stay cold longer.
- **Pack food together in the freezer.** That is, eliminate gaps between the packages. Once again, it'll stay frozen longer.
- **Stuff the fridge with ice.** It can be a block of ice, some *dry* ice, or even frozen gel packs. All of those will keep the air colder in the fridge.

If the power is off for four hours or less, and you haven't opened the doors, then everything inside is definitely safe to eat. Any food package that still has ice crystals on it is okay to eat, too.

But if the temperature inside the fridge has gone above 40°F for more than two hours (the FDA suggests using a freezer thermometer to check)—or over 90° for *one* hour—toss the perishable stuff, like fish, eggs, and meat. It could make you sick.

Heating and Cooling

When it comes to survival, the priorities—water, food, temperature, medical care—are identical whether you're foraging in the wilderness or enduring a power failure in a big city. "If you have a fractured tibia or you're hypothermic, your body doesn't care whether you're in downtown LA or Yosemite," Tony Nester says.

If you're "camping at home" when the power's out during very cold weather, for example, he suggests designating one room—the snuggest in the house—to be the warm one. It might be the corner room in your basement, or whatever room gets the most sunlight during the day—usually on the south side of the house.

Then seal off its openings with clear plastic sheeting and duct tape. Round up the whole family in that room, set out sleeping bags, start up a small heater, and camp out. If it's a propane heater instead of electric, crack a window to avoid carbon monoxide buildup.

Nester suggests borrowing a philosophy from Japanese culture: Heat the individual, not the entire home.

Layer up like you're outside: hat, scarf, sweater, down jacket. Get the stove going. Eat warming, calorie-dense foods, so that the metabolic heat of digestion will warm you: soup, rice and beans, Snickers bars, peanut butter and jelly.

For sleeping, put a bottle of warm water at the foot of your sleeping bag or in your coat.

If hot weather is the problem, you can use the same philosophy. You'll be coolest in the basement or on the north side of the house. Stay hydrated, don't move around a lot, and use sprayers and misters to keep your skin cool through evaporation.

If it's too brutally hot to sleep, you can use what Nester says is an old pioneer trick: Wrap yourself in a wet cotton sheet.

The Special Cases

As you prepare your home for an emergency situation, don't forget anyone who's:

- **A baby.** A baby's got special supplies that need stocking, too: formula, diapers, baby food.
- **A pet.** Pet food, medicines, leash, carrier or crate.
- **Medicated.** Do you have two weeks' worth of emergency blood-pressure medicine, diabetes medicine, heart pills, asthma inhalers? How about contact lenses and solution?
- **Addicted.** If you're a smoker, or if you can't function well without caffeine, the last thing you need is the stress of withdrawal when you're

living through trauma. Set up a stash that will tide you through a couple of weeks at home.

Prepping Your Family

Part of preparing for climate change is preparing for extreme weather, and part of *that* is thinking through the worst before it happens.

How will you find each other when the world is in chaos? How will you communicate? Where will you go if you can't go home?

And don't say, "Oh, I'll text everybody." In times of disaster, the first thing to go down is often the cellular network.

Your phone's cell signal comes from cell *towers*, evenly spaced across the nation's populated areas. Those towers rely on electricity.

Some cell towers have backup generators that kick in when the grid power goes out, but they can run for only a few hours without recharging or refilling with fuel.

The cellular companies also have portable cell towers that they can roll into place. But how do you refuel a generator or wheel a mobile one into place if something stops you from getting to the downed tower—like, oh, a *wildfire*?

That's why, during California's 2018 Camp Fire, 17 cell towers went down. In October 2019, after Sonoma County's wildfires, a quarter of the towers were nonfunctioning.

In short, the cell companies have told the FCC that their backup systems "are not effective for mitigating the disruption to wireless communications when our facilities are damaged by fire."

Fires aren't the only problem, either. The cell networks also went down in the aftermath of the 9/11 terrorist attack, when cut underground cables shut off service for days.

Hurricane Katrina took out about 1,000 cell towers, rendering voice calls spotty and cellular internet dead for weeks. In 2012, Hurricane Sandy killed a quarter of all the towers in ten states. And in 2017, Hurricane Harvey knocked out 70% of all cell towers near Houston.

Even if the networks are working, certain family members might not have their phones with them, or they may not be charged.

The point is, communicate a plan now, for when you can't communicate later.

The Plan

When weather hits hard, the world goes to chaos, and you're feeling overwhelmed, are you likely going to manage a reunion of everyone in your family? Fat chance, says Tony Nester. "It's hard enough getting everybody together for a Sunday barbecue."

The far easier solution is to make a plan in advance. Chances are good you haven't done that; fewer than half of Americans have even discussed household emergency plans.

The more details you've worked out in advance, the less time you'll waste fumbling when something bad is happening outside. Acting quickly means you'll be able to save more people and more stuff—and you'll recover faster afterward.

First step: How will you get out of the neighborhood? Obviously, you know how to drive into town from your house *now*. But in an emergency, all kinds of things might be blocking your path: floodwaters, fire, emergency vehicles, or just a never-ending line of stopped traffic. You need *alternative* paths.

Consult a map and trace out your options. What if it's dark and smoky? What if a foot of rain has fallen, or four feet of snow?

Keep in mind, too, that if the power's out, the pumps probably won't be working at gas stations.

Second step: Figure out four meeting spots in case the family members aren't together and can't communicate when the disaster strikes:

- **An indoors spot.** Where's the safest place in the house? Depends on the disaster. If it's a hurricane, tornado, or windstorm, it's a space without windows, like an interior bathroom or even a closet. The basement is a great option, if you have one. If it's flood, you'll want to go upstairs.
- **A neighborhood spot.** In a wildfire, well, your main priority should be *leaving* the house. But if everybody gets split up during the confusion, then what? The entire family should know to gather at a desig-

nated nearby spot. "The mailbox on the corner" or "that weird knotty tree in front of the Caseys'" works fine.

- **A town spot.** What happens if disaster strikes during the workday or school day? Who will pick up the kids from school? Which parent will pick up which kid? And then where will you go if you can't reach home?

  Designate a couple of meeting points on different sides of town, to spare each family cluster a cross-city drive in traffic mayhem. It could be the library, the Walmart, a church or a temple, or a relative's house.

  Designate a *backup* meeting point, too. "If we can't get to Dad's house for any reason," you might decide, "we'll all meet up at the Home Depot; if that's blocked off, meet at the Safeway," or whatever.

- **An out-of-town spot.** Finally, choose an out-of-town spot. That will be your meeting point if an evacuation order comes down when you're not together. Maybe it's a hotel, a landmark, or a relative's house.

  FEMA also recommends choosing an out-of-town relative or friend to be your emergency coordinating contact. Sometimes, it's easier to make long-distance calls than local ones when disaster strikes and desperate calls are jamming the lines.

Once you've figured out the logistics, here's the key: *Write them down.* Write them or print them out on a card. Put a copy on the kitchen bulletin board, in your go bag (page 285), in the car's glove compartment, and in your desk at work. Store a copy in the Notes app on your phone, then share that Note with everyone else in your family.

You can find an emergency cheat-sheet card to fill out at the back of this book, or you can use FEMA's version at http://j.mp/2D3AHqQ.

But do it, and make sure every family member has it. Because when you're under disaster duress a year later, you might not remember who's supposed to pick up whom, and where to go, and where to go after that. You *will* remember, though, that you've worked through this and written it all down.

Finally, go over all of this stuff with your kids. They should know the plan. They should also know who'll pick them up (and who's *allowed* to pick them up) if they're not at home. They should know the warning signs of an impending hurricane, tornado, or wildfire. Give everyone jobs to do when the chips are down. Even little kids should know how to dial 911.

You should also hold family practices of all of this stuff. For young kids, you can make it feel like a game. But the point is to have walked through certain tasks—grabbing the go bag, heading to the meeting places—before disaster descends.

FEMA recommends reviewing all of this information with the whole family once a year. The family plan won't do anybody any good if people forget that it exists.

Prepping Your Pets

At https://secure.aspca.org/take-action/order-your-pet-safety-pack, you can order a free ANIMALS INSIDE! sticker. It lets emergency teams know that there are pets inside, and how many.

Then, if you take your pets with you when you leave—as you should—you're supposed to write EVACUATED on the sticker.

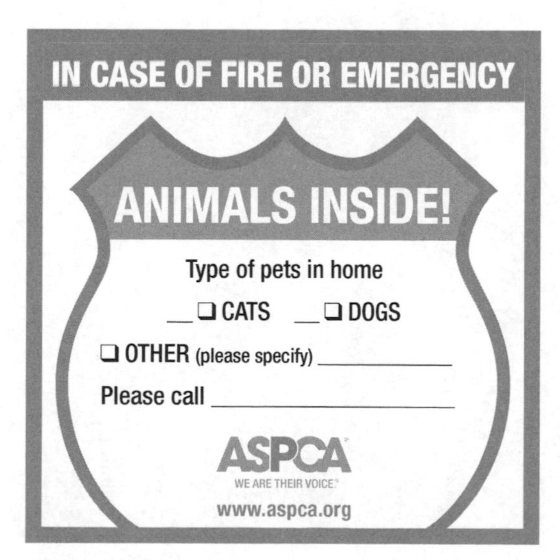

Figure 8-12. This sticker on a front house window lets rescuers know that animals are inside.

Think through what you'll do if you evacuate with your pets. Who'll be responsible for which pet, and where will you bring them?

Friends and relatives beyond your neighborhood are the best bet. But your vet can also suggest kennels. Some local hotels might accept pets, especially in a crisis. Many animal shelters will, too. In every case, though, you need to do your research ahead of time. Sites like petswelcome.com and pet-friendly-hotels.net are directories of pet-friendly hotels.

The ASPCA also recommends designating a "foster parent" for your pet: a neighbor you trust enough to equip with keys to your place, somebody who's home while you're at work. The idea is that if a disaster crops up when you're out of the house, this person can rescue your animals—and, if anything happens to you, will even be willing to adopt them.

Somebody with pets of their own is a good candidate; you can make a reciprocal offer to *them*.

Finally: How is anyone supposed to return your pet to you if you get separated in a disaster? At this moment, does your pet have an ID tag, bearing its name, your phone number, and urgent medical issues?

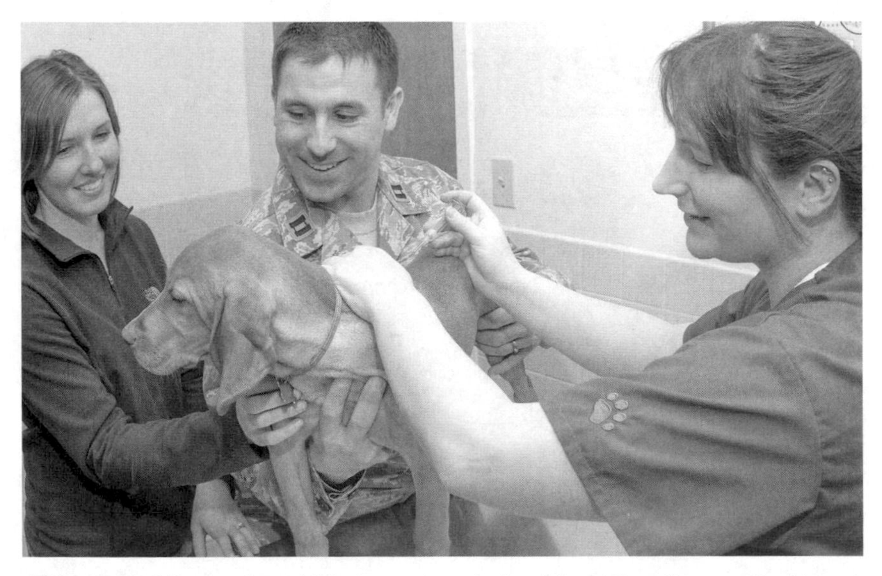

Figure 8-13. A microchip, shown here being injected under a pet's skin, gives you some hope of getting your pet back after a disaster separates you.

Your dog, probably. Your cat, probably not.

In either case, the ASPCA recommends an implantable ID microchip. For about $45, your vet can painlessly inject this tiny thing, the size of a grain of rice, under the animal's skin.

This tiny chip has no battery and requires no power. It does, however, reflect an ID number to a special reader wand that's passed over it.

If someone finds your pet and turns it over to a vet, your town's animal-control department, an animal shelter, or the ASPCA, they'll use that reader wand to look up your contact information on a central pet database. (There may be a monthly charge for services such as continuous monitoring, but you don't have to pay it; ID lookups are free.)

You'll have a fighting chance of getting the pet back.

Prepping Your Phone

It's important that your phone stores all the important phone numbers:

- Family members (including your out-of-town coordinator person)
- Emergency services, the Red Cross
- Hospital, doctor, pediatrician, veterinarian
- Insurance companies
- Utility companies
- Employers/supervisors
- Schools
- Church or temple
- Social-service providers
- Your homeowners association, if you have one
- Your plumber, roofer, carpenter, electrician

But having the numbers in your phone is not enough. In times of catastrophe, the phone could be dead, waterlogged, or unreachable. This may sound preposterously old-school, but you also need those numbers *on paper.*

Specifically, you should stash a printed cheat sheet of these numbers on the fridge, in the car glove compartment, at your workplace, and in the bedroom bureau—because you don't have them all memorized now, and you won't then.

It's not a bad idea to create a contact group on your phone, either, consisting of all family members' numbers. That way, a single text message can reach everybody else, and nobody wastes any time. Give it a test to make sure it works.

Finally, your smartphone offers an Emergency screen that first responders or helpful strangers can access even when the phone is locked. But it won't do any good until you've filled it out with your medical details and an emergency contact.

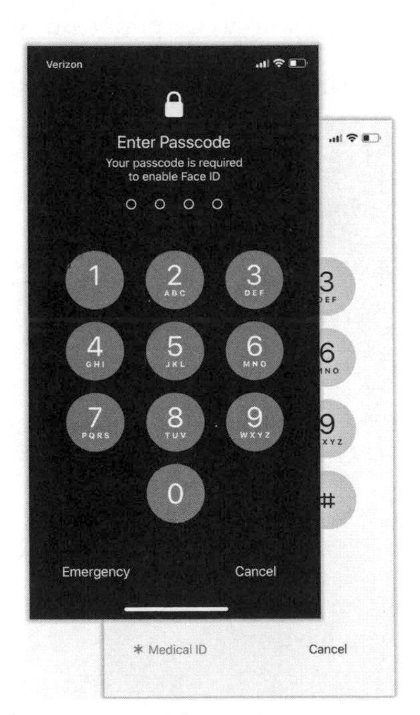

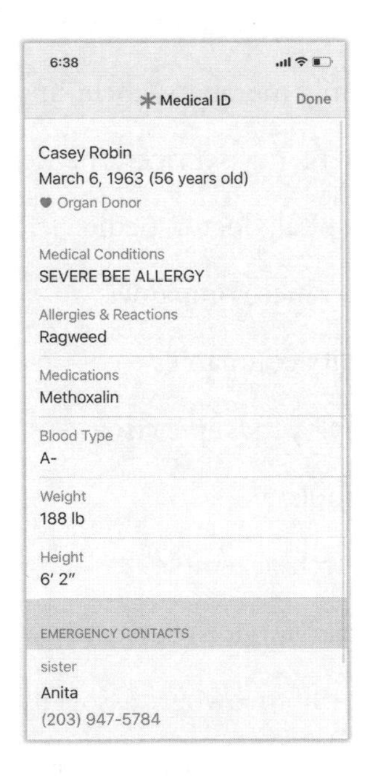

Figure 8-14. Left: An emergency responder can wake your phone and, even if it's locked, tap Emergency (lower left) and then Medical ID. Right: Your important medical information appears.

- **iPhone.** Open the Health app. Tap your icon at upper-right, then tap Medical ID. Here, you can record your medical conditions, allergies and reactions, medications, blood type, organ-donor status, and an emergency contact.
- **Android.** The steps vary according to your phone model and Android version. In general, you set up the Emergency Information screen in Settings, either in About Phone, Users, or User & Accounts. If you have the Google Safety app, use that.

No matter which kind of phone you have, all anybody has to do to find that screen is to wake your locked phone (by swiping up, for example) and tap Emergency > Medical ID, or Emergency > Emergency Information.

Knowing the House

This weekend, confirm your knowledge about the essential workings of your house.

Do you know how to turn off the gas? Can you shut off the power to the whole house? Do you know where the circuit breakers are? Do you know how to shut off incoming water to the house, so that if it's contaminated, it won't poison the water you've already got?

When a tree has fallen on some electrical wires outside your home, or you're seeing sparks, or you're smelling gas, you'll want to know.

Find out the answers to these questions and write *them* down, too. It wouldn't hurt to make a cheat sheet, laminated at a Fedex Office shop and hung inside your garage, basement, or utility room and reproduced on a page in your phone's Notes app. You and your family members won't have to remember what's on that cheat sheet—only that there *is* one.

The Home Inventory

If you're smart, you'll *inventory* your house. And you'll do it now, while the weather's not unleashing its rage. Photograph everything you own, especially jewelry, art, electronics, furniture, appliances, expensive clothes, and collectibles. Record how much you paid for each thing, and what.

In the event something bad happens to your house, you'll be incredibly grateful you bothered, for one towering reason: insurance.

"You're standing there in your ashes with your insurance adjuster, and there's nothing left of your home, and you naively think they're going to write you a check for your policy maximum—because, hey, they can see that the house is gone, right?" says Linda Masterson, author of *Surviving Wildfire*. "Unfortunately, it just doesn't work that way."

Believe it or not, the insurance company will be unconvinced by your desperate *recollections* of the stuff you once had. They'll pay you to replace only the stuff you can *prove* you had. You have to show them exactly what you had and what condition each thing was in *before* the disaster hit.

As a bonus, a home inventory also comes in handy when completing your tax returns or writing your will.

You can organize your master list by room or by category. Pick your favorite approach:

- Use a phone app like Encircle (free) or Nest Egg ($4), which are designed for quickly creating home inventories.

  You can snap several photos of each belonging, along with receipts and paperwork. The app makes it simple to search and sort your stuff, assign them to rooms of your home, and calculate a total value. Nest Egg also tracks your warranties, lets you scan bar codes for quick inventories of your books, DVDs, and wine, and auto-recognizes many objects to save you data entry.

- Walk through the house capturing video on, for example, your phone. As you spot an item of value, walk up to it, hold the phone still, and narrate what you know about it.

- Take digital photos. Import them onto your computer's Photos app. Annotate each one in its name or description box.

- Print photos. Write the details on back of each printout.

Don't forget to cruise your garage, attic, basement, patio, and yard for other stuff to inventory.

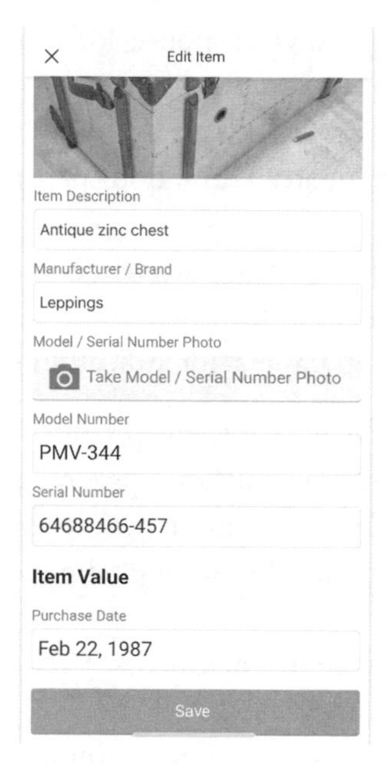

Figure 8-15. Encircle is a free app that makes the home inventory simple.

Then, when it's over, ensure that your inventory is backed up somewhere outside the house, as described next.

Finally, keep the inventory updated! Every time you acquire some new physical belonging, make sure the inventory knows about it. And if it's worth more than $5,000, you're required (and supersmart) to let the insurance company know about it. Your premiums may go up, but your *payout* also goes up if there's a disaster.

Data and Documents

All kinds of files, both physical and digital, need rescuing when all hell is breaking loose. There's no reason at all to suffer their loss; you could take care of creating backups before you go to bed tonight.

- **Proof of identity.** If your house is totaled, or if you're not allowed to go back to it (remember Katrina?), how will you prove who you are?

How will you prove your relationships to your spouse, kids, and even your pets?

You need copies of your essential people records, like birth certificates, marriage or divorce certificates, adoption papers, and child-custody papers.

You'll need personal ID, too: passport, driver's license, Social Security card, green card, military service identification.

And, for your dog or cat: ownership papers, identification tags.

- **Proof of homeownership.** If you expect to get help from your insurance company or government disaster-assistance programs, you'll need evidence that your house is yours. That could include your lease or rental agreement, mortgage documents, home equity line of credit, and property deeds.

  If you have a car, you'll need your loan documents, VIN (vehicle identification number), registration, and title.

  You should have on hand evidence of your financial life, too, like utility bills, credit-card statements, student-loan paperwork, alimony and child-support records, and bank records (checking, savings, debit cards, retirement, investment accounts).

- **Insurance paperwork.** When it comes time to make a claim, you'll definitely need copies of your insurance policies—homeowner's, renter's, auto, life, and flood—and any documents you have that goes with them, like appraisals and your inventory (photos and lists) of valuable belongings.

- **Proof of income and investments.** It can be useful to have copies of your pay stubs, government benefits, alimony and child-support records, and investment accounts.

- **Tax statements.** Do you have access to your federal and state income-tax returns? How about your property and vehicle tax statements?

- **Estate planning.** You should know where to put your hands on copies of your own will, trust, and power-of-attorney documents, too.

- **Medical documents.** Round up backups of your health insurance, dental insurance, and Medicare/Medicaid/VA health benefits. You'll also need lists of your prescriptions, immunizations, allergies, medical equipment and devices, and pharmacy information. You may also have a living will, medical power of attorney, contract with a caregiving agency, or documentation of a disability.

- **Passwords.** Finally, what about the passwords to all of your accounts? Are they stored somewhere, in some format, that can't get burned or flooded?

All right: You've rounded up all these records. Wasn't that an enjoyable weekend? Now you have to figure out how to disaster-proof all of it. You have two options:

- **Protect paper copies.** If you've made paper copies, you can store them in a fireproof, waterproof storage box or safe. You can buy a good one for about $35 on Amazon. (Note that these boxes aren't fireproof forever; a typical low-cost one might resist flames for 30 minutes.)

  You can also store them in a safe-deposit box at a local bank. That costs money, but it's off-site and heavily protected from fire or flood. Just make sure you've specified who's allowed to have access to it if something happens to you.

  Stash copies of the most important stuff—ID, ownership, insurance stuff—in a plastic folder to keep in your go bag, described later in this chapter.

- **Back up digital copies.** If you have digital copies of all of those records, this part is not rocket science: *Store the backups off-site.* They won't do you any good when your laptop burns up in a fire or drowns in water.

  Off-site could mean "on a flash drive at the office," "on a flash drive in the glove compartment of the car," "on a flash drive at Mom's house," or "all of those."

  It could also mean storing them on your Google Drive, iCloud Drive, Microsoft OneDrive, or Dropbox. All of those services place a "drive" icon on your computer's desktop. Any files you copy onto it appear like magic on any computer you use, anywhere in the world—but in reality, they're stored in a heavily protected data center hundreds or thousands of miles away.

Best of all, these services are free for your purposes—backing up a couple gigabytes of important paperwork. If you have more than 5 gigabytes (iCloud, Dropbox, OneDrive) or 15 gigabytes (Google Drive), you'll have to pay a monthly fee. But for storing a bunch of documents, the free levels should cover you.

Photos

Ask a hundred people what possessions they'd rescue first from their burning or flooded homes, and 99 of them will probably say, "My photos and videos."

Most damaged goods can be replaced. Your photos, however, usually can't. Permanently losing your photos is *really* painful.

It's also unnecessary. Backing up your entire digital photo/video collection to an online site is free and easy.

- **Amazon Photos (photos.amazon.com).** Do you subscribe to Amazon's Prime service? One in 1 in 3 American households do. In that case, you're welcome to back up your entire photo collection—no charge. These are the full-resolution originals.

Figure 8-16. Once your photos are backed up online, you can view them, arrange them, edit them, and share them on the web. You can even search them according to what they're pictures of.

- **Google Photos (photos.google.com).** This service is similar but absolutely free to everyone. Note that Google slightly compresses the photos to save space on its hard drives; these are not precisely identical to the original files, although it's unlikely that you'd see the difference. (If that bothers you, you can pay for additional storage through a program called Google One. For example, $20 a year buys you 100 gigabytes of storage—and no photo compression.)

And above all: If anything bad ever happens to the photos you've stored on your phones and computers, you can download them all from the site with a couple of clicks.

If you have photos and videos you care about, and you're *not* using these free services to back them up for you—are you crazy?

All right. But what if your photos aren't digital? What if they're prints, stuck in frames, albums, or old drugstore envelopes?

Then you should get them digitized. Soon. This has nothing to do with climate change—or not *only* to do with climate change. Having only one of *anything* precious, with no backup, is an invitation to heartbreak.

Getting photos scanned and returned to you in digital form (on DVDs, for example, or as downloads) is fast and relatively cheap. ScanMyPhotos.com, for example, charges about 16 cents per photo; if you have 500 photos, that's $80.

You can also buy a photo scanner for as little as $70 and scan the photos yourself. That's a lot more work than letting someone else do it, but you get the nostalgic pleasure of *looking* at each photo as you scan it.

You can get your old videotapes converted to digital, too, through companies YesVideo.com and Southtree.com, or at stores like Walmart, CVS, and Costco. (YesVideo does the actual work for those stores.) It's not cheap—maybe $8 a tape—so cherry-pick the important ones to send them. You probably don't need to get your old VCR recordings of *Cheers* transferred to digital formats.

Either way, do it. As the old saying should go, there are only two kinds of people in the world: Those who've learned to back up their photos and videos—and those who *will*.

Town Plans

When Tony Nester was visiting his children's school one day, he learned, to his amazement, that the local school system had a detailed disaster plan in place.

"If we have to evacuate the school, we're going to get all the kids over to the Home Depot a half mile away," the school principal told him. "I've already worked things out with the manager there, and he'll take care of us until all the parents or grandparents can come and get the kids. If that area's compromised, then we have a river-rafting warehouse a half mile in the other direction, and I know the owner there. We'll camp out there, overnight if necessary, until the last kid is sent home."

In other words, you're not the only person who's been thinking about disasters. Law enforcement, hospitals, and city managers routinely train for natural disasters, mass-casualty events, and pandemics.

As part of your prepping, it's worth visiting your city's website or even calling or visiting, to find out what kinds of plans are already in place. Knowing will not only make you more relaxed about the future, but might tie in with your evacuation plan for getting your kids.

In fact, depending on the size of your community, you don't have to just *read* your community's hazard-mitigation plan; you can help to shape it. Get involved in your town's disaster-recovery planning, resilience planning, and climate-adaptation planning. You can, for example, push the town to adopt the latest building codes in your area as a minimum standard for new or renovated buildings.

The Neighbors

What's the difference between a truly resilient home and a glorified bunker? Relationships with your neighbors.

Nobody knows that better than climate educator Richard Heinberg, who lives in Santa Rosa, California. In 2017, he was awakened at 3:00 a.m. by neighbors pounding on his front door. He recounts what happened next:

> I opened the door. The entire northern horizon was lit up in orange flame, and the wind was headed in our direction.

So we packed up, turned on the radio, and drove to the closest shelter. We found a place to park for a few hours, listened to our car radio, and tried to think of contingency plans. We had no idea if our home was still standing.

In late afternoon, I cautiously drove home to assess the situation. Smoke and ash were everywhere, and the power and gas were off in our neighborhood, but our home was OK.

Over the next ten days, everyone was relying on their friends. We got to know our neighbors as never before. The stuff in our refrigerators was spoiling, so we'd get together with our neighbors to cook stuff. We had electricity sometimes when other people didn't, and vice versa. It was ten days of smoke, ash, and sharing of food and conversation over candles.

About a week later, the power came back. The smoke lingered days longer, after which Janet and I drove a mile north to the Coffey Park neighborhood. It looked like an atom bomb had exploded there—complete devastation in all directions, as far as the eye could see. We felt lucky, but the experience was deeply sobering.

The point is that *other people* may be your greatest resource in times of weather crisis. "It's really important to get to know them ahead of time and establish that basic level of trust," Heinberg says.

An atheist, he even suggests joining a church or a temple, if only for cultivating a network of trusting friends and neighbors. "The evidence is really clear that that's a good thing to do," he says. "It's how people get through disasters. People in their church will help them out when other people won't."

You don't have to become best friends with every neighbor. But learn their names. Wave as you drive by them. Get to know them, even if you live in farmland and the next house is some distance away.

There are all kinds of ways to start building relationships. Organize a potluck-dinner series. Set up an annual block party in the summer. Come up with ways to work together on projects, like seasonal cleanup walks or creating a community garden.

If you live in a wildfire-, hurricane-, or tornado-prone area, hold an annual neighborhood meeting to make basic emergency plans. Figure out who will check in on older or ailing neighbors who may not be able to fend for themselves. Find out, together, what your town or city's government has drawn up as its emergency plan. Locate the closest shelters, cooling centers, and hospitals.

Use flyers, or Nextdoor.com or the Nextdoor app, to spread the word of these gatherings. Nextdoor is something like a Facebook for neighborhoods. It's ideal for publishing news of an emergency-planning effort.

Those contacts can make the difference between getting by and struggling, or even life and death.

Take a Class

You might expect that treating your family to a survival-skills class would heighten their anxieties about disaster, that it would somehow take an amorphous concern and make it more real.

In fact, though, picking up some outdoorsy skills has the opposite effect: It provides a sense of *control* over our fates. These classes are also something new, so they're inherently interesting; they're outdoors and active; and they're a lot of fun. It's cool to discover that you can eat some of the plants that are growing wild nearby, or that you can turn a poncho into a tent, or how to patch up a gash when you're far from a hospital.

Find out in advance about the instructor's approach, though. You don't really need to learn about tanning hides and making obsidian spearheads. The object is to cope with a few days of chaos, not to prepare for a *Mad Max* apocalypse.

Above all, says Tony Nester, "Survival's meant to be empowering. Stay away from the fear-based schools, the ones that emphasize doomsday scenarios."

Leaving Home

In the climate-crisis era, all kinds of things might force you to leave home: wildfire, hurricane, flood, tornado, landslide. Man-made disasters like chemical spills, nuclear accidents, and even terrorist attacks might require you to leave home in a hurry, too.

In those situations, it's time to forget that camping-at-home business and get out—or, as survivalists say, *bug* out.

To bug out means to scram. To get out, fast. The term arose as military slang during the Korean War (possibly from the British slang *bugger off*, which is a vulgar way of saying "go away").

In emergency terms, it means to get away from the disaster to save your skin. Bugging out is a whole different skill from the whole camping-at-home thing, which you might call bugging *in*.

The Go Bag

A *go bag*, also called a bug-out bag, is a backpack or small duffel bag packed with basic supplies to see you through three days away from home—long enough to get somewhere safe when your home isn't.

Each member of the family should have one of these bags. That's because, first, no single bag can hold food, water, and supplies for a whole family for three days. Second, you might get separated. Third, each person will need different things—medicine, spare glasses, whatever.

Keep your go bags in a closet by the door to your home. Everybody should know that they're there. When bad things are happening, and you have to get out, *now*, you can grab the bags and run, safe from the agonizing worry that you're leaving home without something you'll need. Because you've thought ahead, you can toss your bag into the car and get going inside of five minutes. Everybody else will either have to leave home empty-handed and panicked, or stay behind, losing time as they throw together the things they'll need.

What you pack in your go bag depends on you and the kind of threat you might experience. If you're in California, you might think "wildfire"; if you're on the Texas coast, you might think "hurricane." There's no one

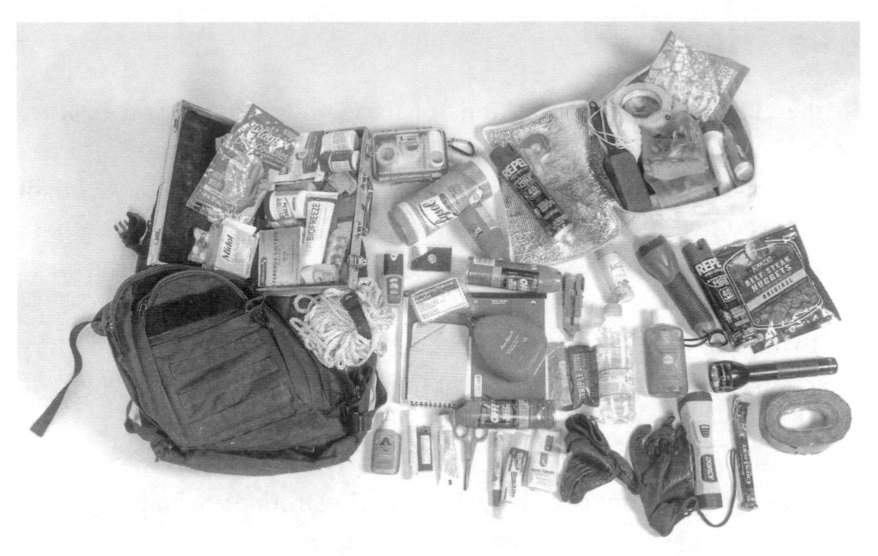

Figure 8-17. A go bag, ready to go. Not shown on top: a note to remind you of last-minute things to grab before you ditch.

universally useful go bag. (This is one reason that professionals generally don't care for the complete, ready-to-use bug-out bag kits sold online.)

Building a go bag not only makes a good bonding activity for the family—for kids, it can be a scavenger hunt—but also leaves all of you feeling more comfortable and confident in life. Being ready is a great anxiety reliever.

To build a basic go bag, start with a backpack or rolling bag. You'd be wise to avoid camouflage print or paramilitary designs that scream *THIS IS A BUG-OUT BAG!* because when things fall apart, that kind of bag might be a target for desperate thieves.

Here are the essentials:

- **Water.** You won't be able to carry around three gallons of water, which is what a person needs for three days. It's too bulky and heavy to carry in your bag.

  Instead, plan to purify the water you find on the run, using the hiker's purification tablets described earlier in this chapter. They're compact, they work, and they remain potent for at least five years.

  Beware the "water straws," which purport to purify any water as it's traveling up to your mouth. "They're pretty useless," Tony Nester says. First, they offer no way to transfer the water you're purifying into a *container*; they require you to sit there, hunched over your water source, sucking through that straw. Second, these straws clog easily. And finally, water proceeds through them painfully slowly. "I'm not kidding," says Nester. "Sometimes it takes two minutes of sucking on that thing to get the equivalent of an eight-ounce glass of water in your system."

- **Food.** You want stuff that you can eat without cooking, and that can sit in there for a year without going bad. There aren't many truly delicious options that fit those two categories, but when times are desperate, you might be willing to compromise.

  You can, of course, buy survival rations or MRE (meals ready-to-eat)—special synthetic bars and envelope packs, originally developed for the military, that last 25 years. That stuff is expensive, though, and, according to survival instructor Tony Nester, tastes like chalk. "The color will drain out of my kids' faces if they are ever confronted with an MRE again," he says.

  Instead, throw in a jar of peanut butter and some packets of salmon. Or look into the meal-replacement bars often preferred by bodybuilders, like MET-Rx, Quest Bar, or Pure Protein. The MET-Rx bars have, for example, 28 grams of protein, 400 calories, and a lot of fat.

  These aren't intended to tide you over for weeks. "After about two days, you're going to be having dreams of apples and steak talking to you," Nester says. But in a bug-out situation, when you have no access to a stove or groceries and you're on the move to someplace safer, a block of calorie-dense nutrition is exactly what you need.

- **Clothing.** You'll need extra socks and underwear, especially if you get wet, have to walk a long way, or both. If you pack them into ziplock bags, they'll stay dry *and* give you more spare bags.

  The go bag should also contain some sneakers or broken-in walking shoes. You don't want to be condemned to making your escape with whatever may happen to be on your feet, like flip-flops, dress shoes, or high heels.

Finally, include a change of clothes. If you're in wildfire country, it should be cotton clothes that won't melt onto you if they catch fire. No matter where you are, it should be long pants and long sleeves, for warmth and protection.

If you're in storm country, take exactly the opposite advice: *avoid* cotton clothes (like T-shirts, jeans, and sweatshirts). When they're wet, they suck heat away from your flesh. It's perfectly possible to die of hypothermia when the temperature is as warm as 70°F.

In general, therefore, wool, polyester, and fleece are the way to go.

- ♦ **Toiletries.** You'll feel 1,000% better if you're able to brush your teeth and wash your hands. Bring toothbrush, toothpaste, baby wipes, and a bar of soap.

  Technically, you can survive without toilet paper—but if you bring it, take some off the roll and flatten it to make it less bulky.

  If you might need tampons or pads, bring those. (No snotty remarks, gentlemen: Tampons and pads absorb a lot of blood. They're great to have if someone's wounded.)

- ♦ **Portable radio with NOAA reception.** NOAA is the National Oceanic and Atmospheric Administration, the government agency that tracks and warns about hurricanes, tornadoes, and other storms. If you have to leave your house, you'll be happy to have some way to monitor the storm's progress.

  For $15, you can buy a portable, rainproof, hand-cranked radio that gets NOAA weather bulletins—and has a built-in flashlight. For $25, you can get a model that also has a built-in solar panel, a loud sonic alarm, and a built-in battery-power bank, which can recharge your phone when the power's out (or you're out).

- ♦ **Flashlight.** A cheap, bright LED flashlight is fine. A headlamp—a flashlight on an elastic headband—is even handier.

- ♦ **First-aid kit.** You can buy prefab kits online that cover the basics: gauze, antiseptic wipes, and bandages to handle cuts and scrapes; scissors or EMT shears to cut clothing or bandages away; painkillers like Tylenol or Advil; and maybe a squirter syringe to clean out deep wounds. The more you pay, the more stuff you get (and the more you'll have to carry).

The review site Wirecutter recommends the Adventure Medical Kits Mountain Backpacker Kit ($30), which is compact, well organized, and even accompanied by a handbook of instructions.

Survival author Tony Nester recommends supplementing its contents with Imodium (for diarrhea), Benadryl (for severe bug-bite reactions), tweezers (for splinters), and something for dry, cracked skin, like Bag Balm.

- **Hand sanitizer.** When you're in transit, you may not enjoy the luxury of running water and flushing toilets, and you definitely don't want to get violently ill.

- **Baby wipes.** They're incredibly useful for multipurpose cleanup jobs. And when you can't get to a shower for a while, they make you feel so much better.

- **Whistle.** It's tiny, cheap, lightweight, and loud enough to get people's attention. It's a way of saying "Over here!" without having to yell your throat out for an hour. That could come in handy if you're pinned, trapped, or wounded and need to get a passerby's or rescuer's attention, or if you've gotten separated from your family in the dark. Keep the whistle clipped to the outside of your pack.

- **Plastic bags.** Ziplock bags come in handy in a thousand situations, like keeping water out (your phones when it's raining) and keeping water in (carrying water).

- **Charging cable.** Put a spare USB charging cord for your phone into the bag. When the chips are down, you can recharge your phone from a car somewhere—but not without that cable.

- **Three-day supply of medicine.** Bring a copy of the prescription, too, in case you need a refill while things are still messy. You can ask your doctor for a refill prescription for this purpose.

  Remember that medicines expire; this is one item you'll need to replace periodically in your go bag. Or tape a note to yourself onto the bag that reminds you to grab whatever *current* medicines you need before heading out.

- **Glasses or contact lenses.** In a classic *Twilight Zone* episode, a painfully introverted and farsighted bookworm survives an atomic blast that devastates the earth. His one consolation: He now has all the time in the world for his favorite activity, reading, since the city library has withstood the blast. But no sooner does he realize that his wish has come true when—*twist!*—his glasses fall off and shatter. He's at last surrounded by books, but he can never read them.

  You can avoid that cruel fate by packing extra glasses, maybe an older prescription. If you wear contacts, you'll need some of those, along with the solution.

  If you wear hearing aids, think about batteries for them, too, or even a spare pair.

  Oh, and pack a pair of sunglasses if you live somewhere they'd be useful.

- **Cash.** If the cellular network is down, credit-card systems will also be down. Stuff some small-denomination bills into your bag. Some change might be handy if you find a vending machine.

- **Documents.** Once you're away from your house, you may find yourself among throngs of other people trying to get into shelters, get help, and establish that they belong. You'll want to be able to prove who you are, that you've had your shots, that those family members are yours, and that your home is yours.

  Ownership or rental documents come in especially handy when you return to your home after a disaster and discover fellow disaster refugees occupying it. You'll be able to show the authorities that *you* are the rightful resident.

  Keep duplicates of your essential papers—birth, marriage, and divorce certificates; custody or adoption papers; passport; insurance documents; medical records; military service and discharge papers; mortgage, rental, or house deeds—in a plastic folder in your go bag.

- **Photos.** If you're any kind of a decent parent, you've already got photos of your spouse and kids on your phone. They come in most handy as in "Excuse me—have you seen . . . ?" situations. But having printouts in your go bag is useful, too, so that you can post "Missing" notices on walls and phone poles.

- **Games, toys, books.** If you've got children, pack something small that can engage them when, for example, you're spending days sitting in a shelter. Heck, something small, like a deck of cards or a book, might help *you* stay sane, too.

If you have a baby, that little go bag will need diapers and formula. You should make a bug-out bag for your pets, too. It should contain at least three days' worth of food for them, and three days' worth of whatever medicines they need. It should also contain an envelope or ziplock bag that holds photos of the animals, for "Have you seen them?" situations, along with copies of any ownership and inoculation paperwork. You may need to show them to the officials at a shelter if you intend to bring the pets with you.

FEMA has also prepared a web page that lists prep steps and resources for people with disabilities. It's at www.ready.gov/disability.

Depending on the size of your bag, the size of your budget, and the size of your inner Daniel Boone, you can escalate to a rougher-hewn level by adding these more hard-core bug-out ingredients:

- **Duct tape.** You have to admit: Duct tape is useful in a thousand emergency situations. Fixing your car, crafting a sling, holding a door open (or closed), fastening your flashlight into place, whatever. A roll is small and doesn't weigh much; take it.

- **A multi-tool.** A good, old-fashioned Leatherman or Swiss Army multi-tool is a compact, inexpensive addition that can save your bug-out bacon in all kinds of ways. At the least, it should have a knife blade, a serrated blade, pliers, a can opener, and a couple of screwdrivers.

- **Fire starter.** A plain old Bic disposable lighter lets you start a fire without any training or equipment. Depending on how paramilitary you're feeling, you can also buy waterproof matches, windproof lighter kits, or ferro (ferrocerium) rods, which produce sparks when you strike them with almost anything, like a knife.

Cotton balls make great kindling. Once they're in a plastic bag, you can squish all the air out of them, resulting in a package that weighs almost nothing and occupies almost no space in your bag.

- **A tarp or poncho.** A waterproof sheet like an old military poncho or a camping tarp can make an excellent shelter. But you can also use it to keep yourself warm and dry, to collect rainwater, to drag something heavy, to keep the bugs off when you're asleep, and so on. Maybe toss some parachute cord (plastic rope) in there, too, for even easier tarp-draping.

  At a minimum, stuff some trash bags in there. They, too, can keep you dry and serve a hundred other functions.

- **Gloves, safety glasses, dust mask.** This kind of protection gear might be useful if you think you might be picking through wreckage, either to get away or to help out after the disaster. Keeping your hands protected, in particular, is a great idea if you'd prefer not to get cut and infected.

- **A pocket mirror.** You can use a mirror for signaling to rescuers over long distances, for checking an injury on your back, or just for checking for spinach in your teeth. The plastic ones are more rugged than glass, but not as bright.

- **Crayons.** You can use crayons to write on anything—to leave notes for your family members or rescuers who come along later, for example—but they also work well as candles or as kindling for a fire. Each one burns for about half an hour.

As a garnish, tape a sheet of paper to the handles of your bag, listing the few essentials that you can't prepack. It might say, for example, "Meds. Wallet. Charger. Documents bag."

If your home is in danger of being lost, as it might be in a wildfire, add truly irreplaceable keepsakes to this list that you'll want to grab as you run. They might include, for example, your love letters, your stamp collection, and your Oscar. Let your kids designate a thing or two for themselves, too.

The Get-Home Bag

The Navy SEALs have a saying: "Two is one, and one is none."

It means that if you have only one backup, you're still at risk. When it comes to critical supplies, redundancy is essential. "One of anything critical is a poor system," notes Tony Nester.

Suppose, for example, that you've spent a delightful weekend carefully preparing a go bag. It's equipped, it's ready, it's sitting in the front closet. Then one day disaster strikes—while you're at work.

The go bag back at home won't do you much good.

Now you understand the argument for a smaller, secondary pack known as the get-home bag, which you keep in your car or your office.

This bag is simpler; it's intended to last you only long enough to get home, which might have to be on foot. Pack some protein bars, a couple of bottles of water, a first-aid kit, sneakers or walking/hiking shoes, a tarp or poncho, and a change of socks and T-shirt.

Finding Water

The old prepper mantra has it that you can survive three weeks without food, three days without water, three hours without shelter from extreme heat and cold, and three minutes without air.

In a city that's falling apart, air usually isn't a problem, you can improvise shelter, you can barter for food—but you can't create water where none exists.

If you can find a hose spigot on the side of a house or building, that's your first choice. It's clean and ready to drink.

But since you've shrewdly packed your go bag with water-purification tablets (right?), you don't have to worry about finding pristine, crystal-clear water. Lakes, ponds, and streams are fair game. So are swimming pools and, on golf courses, even water hazards.

As part of your family emergency-planning conversation—the same chat where you talk about your meeting places when the chaos is breaking—spend a moment to identify some water sources beyond your own home.

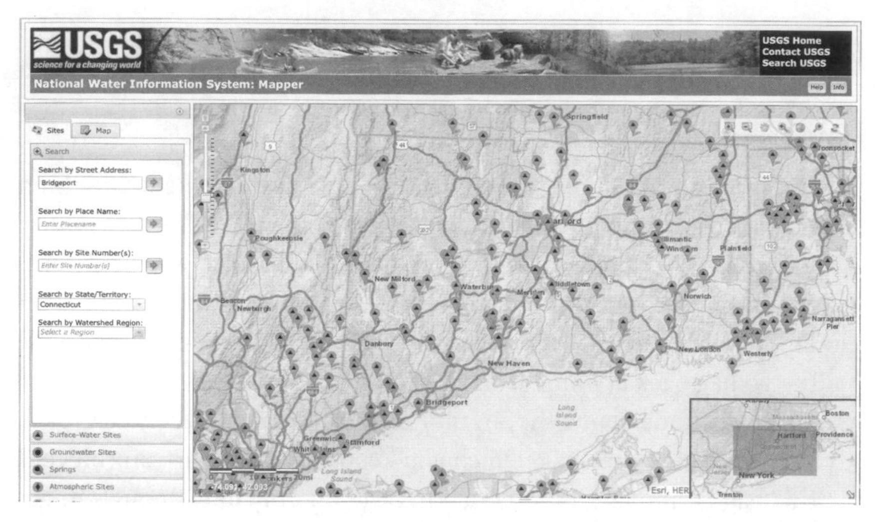

Figure 8-18. The government's water map shows you all the bodies of water near you.

For inspiration, open the Google Maps app on your phone, or maps.google.com on your computer, and scan your neighborhood for blue patches in public areas. Or visit the government's Water Resources of the U.S. page (https://maps.waterdata.usgs.gov/mapper/), a searchable map that displays and identifies every pond, stream, and river that the government knows about.

Remember that water always flows downhill. If you're on your own outdoors, look for the spots where hills meet. That's where water will wind up.

And one more tip: Standing water is always riskier to your health than moving water.

Your Phone in a Disaster

You may remember that in the aftermath of the 9/11 terrorist attack in New York City, people couldn't make phone calls—but they could send texts.

That's typical in the immediate disaster aftermath. If the cell towers are still working, they're usually swamped as everybody tries to call everybody else simultaneously.

Text messages, on the other hand, consume almost zero bandwidth on a congested cell network, so they may go through when voice calls won't. Better yet, a smartphone automatically keeps trying to send them until they go through.

Facebook and Twitter messages are good bets, too. They, too, require little data—they send quickly even when the cell networks are slammed—and as a bonus, they can reach lots of people, not just one person at a time.

If you must make a voice call, keep the conversation short and to the point, for the sake of minimizing network congestion. If you don't get through the first time, FEMA recommends waiting 10 seconds before redialing; otherwise, the connection between your phone and the nearest cell tower hasn't had time to clear, and you're contributing to a clogged network.

However, one brief call is worth making: If you spot somebody who needs immediate help or some danger that requires immediate attention, call 911. Don't assume that somebody else has phoned it in.

It may be some time before you can charge your phone again, too, so do everything in your power to draw out its battery life. Turn the screen brightness down. Turn on Low Power Mode (iPhone) or Battery Saver (Android). (To find the switch, open Settings > Battery.)

Keep in mind that most cars made since 2010 have USB jacks in the dashboard or seat console, which you can use to recharge your phone. Thank goodness you included your USB charging cable in your bug-out bag!

Getting Help on the Road

Once you've had to leave home, what are you supposed to do next?

If you have an urgent medical problem, you can use the old standby: dialing 911.

If you're looking for a shelter, you can use any of these methods:

- Send a text message to 43362 (which is *4FEMA*—get it?). In the body of the message, type *SHELTER 10024* or whatever your zip code is. They'll text you back with the address and contact information of nearby shelters.

- Visit www.redcross.org/find-help/shelter, the American Red Cross's shelter-location website.
- Use the Red Cross's Emergency app, or FEMA's smartphone app. Each contains a shelter-lookup feature.

A shelter is usually a school, library, community center, or fairground. It's open 24-7 to anyone who needs a place to wait, to get protection from the weather disaster, and to get food, water, and medical attention.

Staying Positive

The longer you're away from the stability of home, the harder it is to stay positive. Feeling anxious and fearful is a normal and healthy reaction to a distressing situation, but you can take some steps to fight the lowest lows:

- **Be with people.** A solo expedition is infinitely more frightening and upsetting than one with other people around—beyond the sharing of food and supplies. Conversation, empathy, and feeling a part of something are critical mental-health aids.
- **Occupy your brain.** If you are alone, give your brain *something* to do. Sing, whistle, go get something.
- **Create something every day.** Carve something. Draw something. Construct something. Write a letter, a poem, a journal entry. Reorganize your pack. Creativity is a known antidote for depression.
- **Sleep.** Everything feels 50 times more overwhelming and depressing when you're exhausted. When you're in a bug-out situation, you probably won't get much rest. But when the chance presents itself, take it. You hereby have permission to sleep as often, and as much, as you can.
- **Stay clean.** A daily sponge bath, baby-wipe wipe, or hand-sanitizer session can give your mind a subtle signal that you are still civilized, that you still have some control over your situation, that you're taking care of yourself.

- **Stick to routine.** Prisoners of war often cite the importance of a daily routine to help keep them from going crazy. It might go like this: breakfast, cleanup, make progress on foot, lunch, nap, gather supplies. Or whatever.

The Aftermath

When the violent weather finally passes, you can begin rebuilding your life. You've got a thrilling insurance-company adventure ahead of you, as described in Chapter 6.

You may wind up in temporary housing for a while, for which you'll need temporary supplies for your temporary lifestyle. In that case, you've got some calls to make:

- **The post office.** They can hold your mail for pickup for up to 30 days; after that, they'll forward your mail to a different address for up to six months.

- **Your cable company**, so they'll stop billing you for TV and internet you're not using. You should notify your water, gas, and electric companies for the same reason.

- **The credit-card companies.** Linda Masterson, wildfire survivor, discovered her credit cards frozen when she needed them most—when she was setting up shop in her temporary housing. She was buying things that the credit-card company thought she already owned, a pattern that the banks' computers interpreted as fraud.

  "We appreciated their vigilance and supersmart tracking systems," she says. "But since we couldn't charge anything until we got it straightened out, it added stress at a time when we really didn't need any more." Call the card company at the first opportunity to explain what's happening.

- **The banks.** If your house has been destroyed, you want your mortgage company to know. Your regular bank should know that you've moved, too. Switch to electronic statements and avoid any hassle.

- **The tax assessor.** Presumably, your damaged or destroyed home is worth less now than it was before; therefore, you should pay less in property taxes for it. But your rate won't drop unless the assessor knows about it and reassesses it.

Here's one other idea to contemplate: As you come to understand what aspects of your life insurance will and will not pay for, you may realize that your finances are in sadder shape than you'd thought. Natural disasters often disrupt your ability to work, and therefore your ability to generate income.

If your region has been declared a federal disaster area, you can apply for disaster-relief grants to get back on your feet; these usually take the form of low-interest loans. You can apply at www.disasterassistance.gov or by calling (800) 621-3362.

Then there's always GoFundMe.com, the crowdfunding website. If you genuinely need money, there's nothing at all to stop you from creating a "campaign" for yourself.

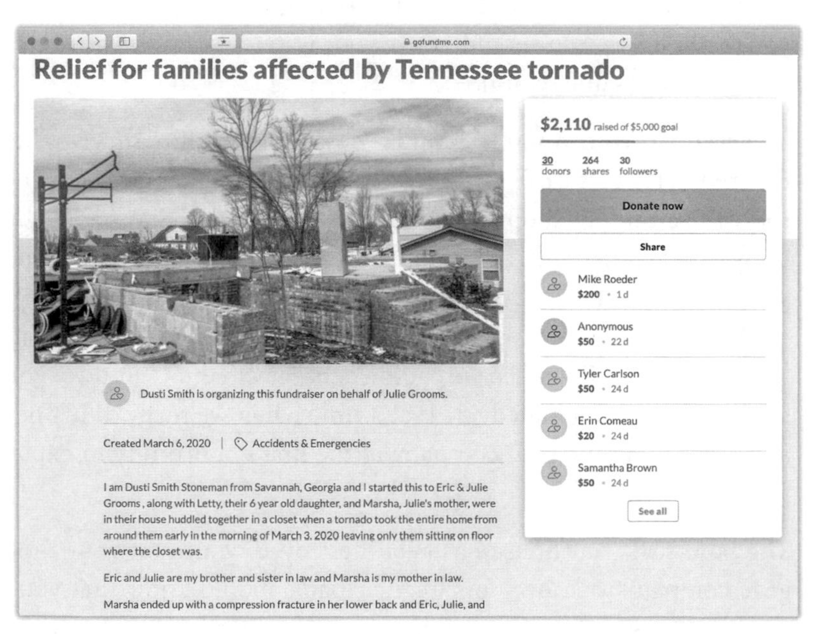

Figure 8-19. A GoFundMe page lets anyone appeal to the public for donations.

Pets and Disasters

When you return home after a disaster, you can always explain to your kids what's happening and what your plan is to restore normalcy. It's much harder to explain all that to your cat.

When you return home with your pets, they may be clearly upset, disoriented, and agitated. The landmarks and smells they once knew are all scrambled now. For that reason, the ASPCA recommends that you don't let them roam loose outside; they can get lost. Even inside the house, keep dogs on leashes and cats in carriers while you look around to scope the place out. Block off nooks and crannies in unstable parts of the house to make sure your cat doesn't hide there.

In the weeks after your return, your animals may exhibit signs of stress: behavioral changes, peeing and pooping, aggression. Be patient with them; restore as much of their routine as possible. If there's no improvement, consult a veterinarian.

It's free, and it's easy to do, but these campaigns don't raise much money unless the public hears about them. That generally happens in one of two ways: either your story gets some media coverage, or you have a big following on Facebook, Twitter, or in real life.

The important thing is to tell a great story—who you are, what happened to you, why you need the money, what it will be used for—and back it up with photos and frequent updates.

GoFundMe is worth considering if you're panicking about money after a disaster. The worst thing that could happen is that people don't contribute much, but at least you do get to keep anything they do donate.

Mental Health

Eventually, stability will return. It may take a year, but sooner or later your home and your daily routine will return to normal, even after a big disaster. But the psychological damage can last far longer.

After her house burned down, it took weeks before Linda Masterson and her husband could sleep through the night. They'd close their eyes and see their home on fire.

Depression and post-traumatic stress syndrome following disruptive disasters is staggeringly common. Twenty-one percent of victims were still suffering from PTSD a year after Hurricane Katrina, and people with suicidal thoughts more than doubled. For anyone who was already experiencing mental-health problems, the anxiety and stress of a hurricane, flood, or wildfire makes it worse.

Children, in particular, can suffer from the effects of a natural disaster. The world, the life, and the routine they were just getting to know has been dismantled; the home they equated with safety isn't the same anymore. Chapter 7 offers the latest psychiatric guidelines on helping them through.

As for you: The recipe offered in Chapter 1 is twice as important now, as you're rebuilding. Resurrect as much of your old routine as you can. Spend time with other people; social involvement is a huge antidote to sadness. Do everything in your power to get sleep. Drink a lot of water. Get some exercise every single day; it's critical for mood elevation. Accept that these feelings of depression, helplessness, and grief are entirely normal and, in fact, healthy.

Finally, if you can't shake the anxiety, depression, flashbacks, anger, or grief, get some counseling. Talk to someone who's interested in helping you out. Call the government's Disaster Distress Helpline at (800) 985-5990; trained crisis counselors answer the phone, 24 hours a day, 365 days a year. It's free, confidential, nonjudgmental, and completely anonymous.

If you'd rather text than talk, send the message "TalkWithUs" to the number 66746. You'll get a reply promptly.

In both cases, you'll be offered counseling for emotional distress, help recognizing that distress and how it might be affecting you and your family, tips for coping, and referrals to local crisis centers, who can take it from there.

Chapter 9

Preparing for Flood

WORLDWIDE, FLOODING KILLS MORE PEOPLE THAN ANY other weather disaster—around 8,000 people a year. It's also the most common natural disaster in America; seven out of ten times that the president declares a disaster, it's because of flooding.

Each year, there are more floods and *worse* floods in more places—and it's going to get worse.

Where Floods Come From

You might think that flooding is a result of storms—and, yes, that's one source. But in fact, floods arise in several different ways. All of them are becoming more frequent and more dangerous, all because of the greenhouse effect. The earth warms, which melts the planet's ice. It also warms the ocean waters, which increases evaporation, which makes storms more powerful and produces more rain. You get the idea.

Oceanographer John Englander, a sea level expert and author of *Moving to Higher Ground: Rising Sea Level and the Path Forward*, puts all flooding into five primary source categories. Here's a tour.

Sea Level Rise

As the earth warms, the sea levels rise. This cycle has repeated itself many times over the millennia—about every 100,000 years, in fact, corresponding to the ice ages.

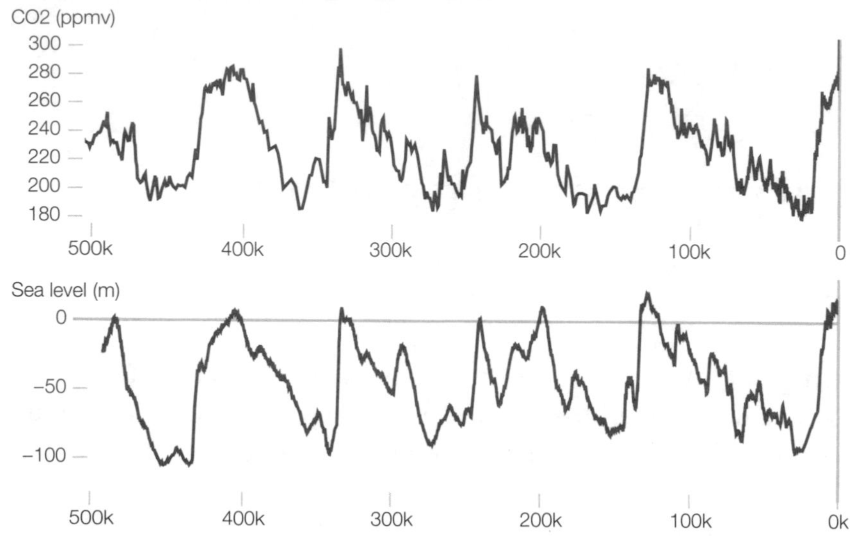

Figure 9-1. Top: Carbon dioxide levels over the last 500,000 years. Bottom: Sea levels over the same period. Notice anything similar about those graphs?

What's disturbing, though, is the far right end of the graphs in Figure 9-1. The sea level rise has never happened before during human existence—and it's never shot up this fast.

You may have heard the explanation boiled down to "Well, the planet's ice is melting, so there's more water," and that's essentially true. But in fact, *three* effects are contributing to sea level rise simultaneously:

- **The great land ice sheets are melting.** Incredibly, 99% of *all freshwater on earth* is locked up in the massive ice sheets that cover Greenland and Antarctica. That'd be kinda bad if it melted, huh?

  That, of course, is what's happening. Trillions of gallons of water are melting into the oceans, and the water level's going up.

  Melting *sea* ice—icebergs—is another story. Melting sea ice doesn't actually make the sea level rise, for the same reason a melting ice cube doesn't make your drink overflow. Any floating ice has already had its effect on the water level.

  Melting sea ice does accelerate global warming, though. It represents a switch from huge, reflective white areas (ice) that bounce the sun's heat back into space—to huge, dark areas (ocean water) that *absorb* the heat. (The tendency for white surfaces to reflect the sun's radiation is called *albedo*, but that won't be on the test.)

- **The glaciers are melting and calving.** The world's mountain glaciers have always melted a little bit in the summertime and then grown back with the winter snows. But now, they melt a *lot* in the summer, and the winters are too short to snow them back up.

  They're also doing a lot more calving, which is when huge walls of ice break off their edges, plunge into the sea, and melt. The result either way: rising sea levels.

Figure 9-2. Land ice is melting all over the world—in this case, Grinnell Glacier Overlook, photographed in 1940 and 2013 in Glacier National Park, Montana.

- **Warm water expands.** As you may have learned in fifth grade, water expands when it's warmer. Here's the second reason the sea levels are rising: Ocean water simply takes up more space as it heats up.

  So far, this effect has accounted for a *third* of all sea level rise. But as those ice sheets start melting faster, the thermal expansion percentage will shrink proportionally.

Since 1900, sea levels have risen 8 inches, and NOAA calculates that they could rise as much as eight more feet by 2100. In fact, the sea level will continue rising for decades, because there's so much heat already stored in the oceans.

"We've passed the tipping point," says John Englander. "Even if we cut carbon dioxide emissions to zero, we're still going to have extreme sea level rise. And for about 30 years, you can't even *slow* it, because the heat's already in the ocean." In short, he says, "We need to begin adapting."

Now, a foot of sea level rise doesn't sound like a lot. But the vertical rise isn't the freak-out statistic—it's the resulting *horizontal* spread. For every foot the sea level rises, it spills inland hundreds of feet—into our homes, businesses, transportation networks, sewage systems, and farmland.

As recently as 1971, cities like Annapolis, MD; Sandy Hook, NJ; Washington, DC; Atlantic City, NJ; Charleston, SC; and Wilmington, NC, never flooded more than five days a year. Now they flood at least 20 days a year.

For the first time in human history, our shorelines are moving inward, inland. Wallops Island, Virginia, home of a billion-dollar NASA launch base, has been losing 12 feet of shoreline a year. The famous Colonial-era homes of the Point, a historic neighborhood in Newport, RI, may be lost entirely. Hawaii's famous Waikiki beach will vanish by about 2035. Two-thirds of Southern California's beaches will be gone by 2100.

"The shoreline is the most important line in the world," says Englander. "It divides what's underwater and what's valuable coastal real estate. It hasn't changed much in all of human civilization—5,000 years. But it's going to change greatly."

By 2050, over 300,000 American coastal homes will flood about 26 times a year. Fifty years after that, 2.4 million homes, worth over a trillion dollars, will be partly or entirely lost to the sea. Not from *storms*, mind you—simply from the rising levels of the calm oceans.

The floods in some cities come about not only because the sea is rising but also because the *land* is *sinking*. That's partly because exploding populations have been sucking out water from the ground for their own use, in effect deflating the land.

In the United States, the fastest-sinking places include New Orleans (2 inches a year), Houston (2 inches), California's Central Valley (2 inches a *month* in some places), and the San Francisco airport (1 inch a year).

In some megacities, the sinking problem is especially devastating, because the populations are so huge and the sinking is so fast. Jakarta, Indonesia, is sinking 6.7 inches a year; Shanghai, China has sunk more than *six feet* since 1921.

Storm Surges

But everyday flooding isn't the only deadly result of higher sea levels. In bad weather, there's an even more devastating effect: the storm surge.

When high winds approach the coastline, they push a huge mound of water onto the shore. The ocean floor slopes upward to meet the land, forcing that hill of water even higher. And the confining contours of a harbor, bay, or waterway can squeeze it taller yet.

The resulting mountain of ocean—the storm surge—can be many feet higher than high tide. In Hurricane Katrina, the storm surge was *28 feet* above sea level.

A storm surge can travel along rivers, pushing as far as 100 miles inland and causing devastating damage along the way.

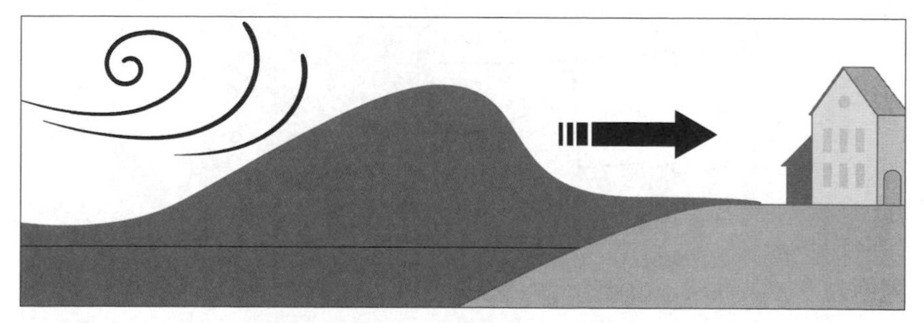

Figure 9-3. A storm surge is a huge mound of water, pushed toward land by a storm and pushed upward by the shallower seabed.

If today's sea level is higher even when it's *calm*, you can imagine what kind of elevated starting line it gives to the storm surges.

"Those storm surges are going to be riding in on a higher background sea level," says Tom Knutson, whose department creates computer climate models for NOAA. "That creates greater flooding problems, even if the storms themselves don't change at all."

Water, as you know if you've ever had to carry buckets of it, weighs a lot. That's why storm surges do such stunning amounts of damage. They destroy roads and bridges, take down buildings, erode beaches, and kill people. The effects can reach *miles* inland from the coasts.

Heavy Rainfall

At this point, you may be thinking, "Good thing I don't live near the coast!"

But flooding is increasing inland, too—*especially* inland, in fact. In the decade ending 2017, eight of the top ten most-flooded states were inland. Arkansas—Arkansas!—took the lead, with 17 federal flood disasters.

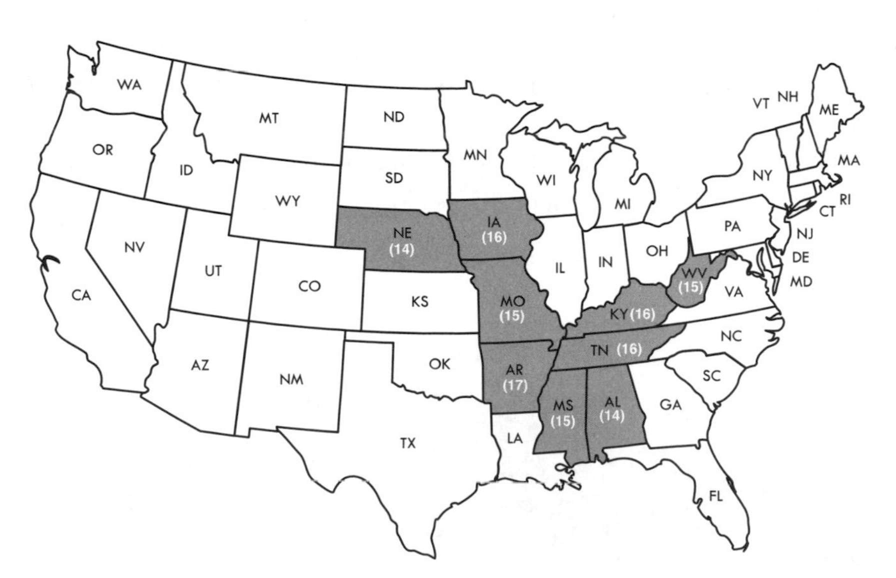

Figure 9-4. Of the ten most-flooded states, eight aren't on the coasts of the country.

How could that be? Because the world is getting rainier. In the United States, there was more rain in the 12-month period ending June 2019 than in any year ever recorded.

As the greenhouse effect warms up the world's oceans, lakes, ponds, and streams, more water evaporates. More water vapor in the atmosphere means more water that has to fall back down on us. The result: bigger storms dumping more rain, no matter where you live.

In low-lying regions, a common result is rivers overflowing their banks. All that water spills outward into whatever fields, roads, and buildings happen to be nearby.

Runaway Runoff

Heavy rains wouldn't matter so much if all the world were a sponge. Unfortunately, once the ground gets saturated and can't absorb any more liquid, the next few billion gallons of rainwater have nowhere to go but sideways: across the ground, downhill, or into overflowing streams and rivers, looking for a first floor or basement to pour into.

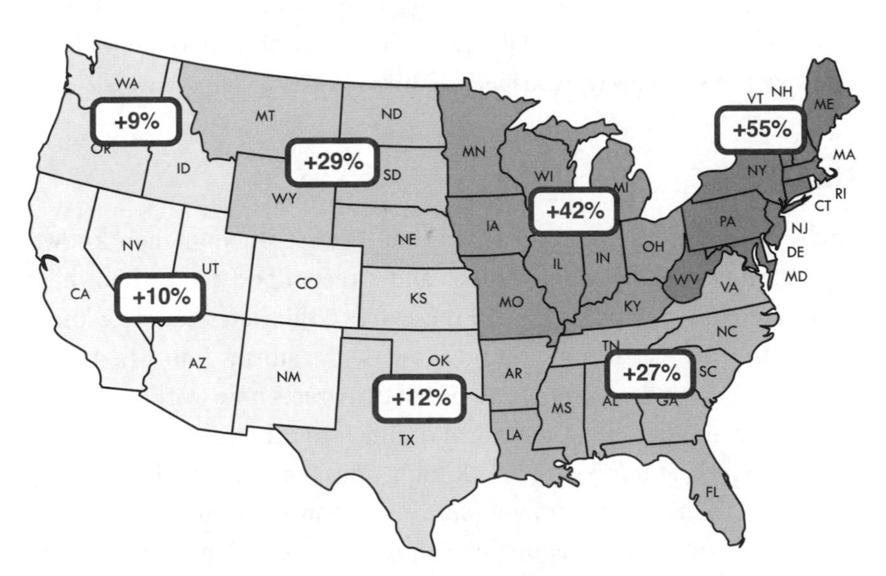

Figure 9-5. Here, you can see how much rainier each part of the country grew from 1958 to 2016.

Runoff leads to exciting flood mutations like these:

- **Flash floods** are defined as floods that develop in less than six hours. At lunchtime, you walk down Main Street; by dinnertime, people are rowing down it.

  Slopes are prime flash flood suspects, because gravity does its thing to the runoff water, and valleys collect it all. The National Weather Service has seen six-inch-deep mountain creeks turn into raging rivers ten feet deep—in an *hour.*

  In cities, flash floods are especially common. Thanks to all the pavement, there's nowhere for the water to go, so it just shrugs and improvises a new river or lake on the spot.

  There may not even be any logical connection between where the *rain* was and where the flash flooding occurs. The water from a deluged region can pour furiously into a part of town that didn't even get rain.

- **Dry Wash.** In the climate-change era, we're getting longer, hotter periods of drought followed by longer, wetter periods of rain. During the dry spells, the ground gets hard and dry; during the rain that follows, the water can't soak into the ground. As in a paved city, the result is spontaneous artificial rivers that, in Western states, have been known to wash away unsuspecting hikers.

King Tides

Lately, coastal cities are experiencing the bizarre phenomenon known as *sunny-day flooding*, where homes and streets flood even when there hasn't been any rain or storms. In these areas, the water can flow knee-high in the streets. In some Florida towns, the storm drains that were designed to move excess rainwater out of the streets now work in reverse, spilling ocean water *out* of the storm drains into the streets.

These are *king tides*: super-peak high tides, occurring a few times a year when the earth, moon, and sun line up just so, multiplying their weak gravitational pulls enough to drag the ocean high up onto the shore. They're the biggest of three kinds of tides:

The Real Name of the King Tide

"King tide" is the popular term for six-times-a-year, super-mega-high tides along the coasts. It is not, however, the proper term.

Scientifically, these über-tides are called *perigean spring tides*. Once you break it down, that term makes perfect sense.

About once a month, there's a full moon, when we see the moon as a full, bright circle. At that moment, the celestial bodies are lined up in space like this: sun—earth—moon.

There's also a *new* moon once a month, where the moon looks completely dark to us. In that case, the lineup is: sun—moon—earth.

Both of those alignments produce especially strong gravitational pulls on the oceans, and especially high tides result. Those are officially known as *spring* tides—named not for the season spring, but because the tide *springs forth*. We get spring tides twice a month.

There's one more factor to go, though. When you were a kid, somebody probably taught you that the moon circles the earth. They were simplifying things a bit, though; after all, you were a kid. In fact, the moon *ovals* the earth. Its path is an ellipse, not a perfect circle, and it takes about 27 days to complete.

Twice a month, therefore, the moon comes closer to the earth than usual. That moment of nearness is called the *perigee*. And a closer moon exerts a stronger gravitational pull.

You can probably see where all of this is going. What would you call it if you *combined* the extra tidal pull at the moon's perigee *with* the extra pull of the spring tide? Why, you'd call it the perigean spring tide, of course. And that's exactly what scientists call it. We get perigean spring tides between six and eight times a year.

Nobody's going to correct you if you call them king tides. But now you know what scientists call them—and why.

- **Daily tides.** As you probably know, the sun and the moon exert a gravitational pull on the oceans as the earth spins. Every 12.5 hours, they tug the water level higher up onto the shore, a period we call high tide.
- **Spring tides.** About twice a month, at the full moon and the new moon, the high tides are especially high. Those are called *spring* tides (nothing to do with the season; see the box on 309).
- **King tides.** The moon exerts its maximum gravitational effect when it's closest to the earth. About six times a year, its nearness coincides with spring tide *and* the daily tidal pull. The result: high tides that are *especially* especially high. These are the king tides, the ones that cause the deepest sunny-day flooding.

As the climate warms, king tides are growing higher, washing farther inland, and lasting longer. Since the base sea level is higher, then of course the tides have a higher starting point. Or, as the EPA puts it: "Sea level rise will make today's king tides become the future's everyday tides."

Flooding's Future

If you have any doubt that the cycle of warmer oceans/rising sea levels/more intense downpours/more flooding is underway, here are a few stats to get you started:

- In the United States, we now get intense rainstorms 30% more often than in 1900.
- In 2017, deaths from flash floods more than doubled the 10-year average.
- In Texas, by the end of the century, the annual chance of a Hurricane Harvey–level rainstorm (20 inches of rain from a single storm) will have increased *1,800%*.

- The global cost of flood damage, now $96 billion a year, will quintuple by 2030.
- Florida's problem isn't just seawater coming from offshore. It's also coming from *under the ground*, thanks to the porous limestone that forms much of its land mass. No number of seawalls can block that kind of incursion.

The areas of the United States susceptible to flooding is increasing, thanks to climate change and new construction near rivers and coastlines. FEMA expects river floodplains to grow by 45%, and coastal floodplains to expand by 55%, by 2100.The problem isn't just that the rains are getting more intense and the sea levels are rising. It's also that we keep creating dangerous areas and *moving into them*. We cut down trees and clear land for roads and buildings, which increases our exposure to flooding. In fact, we're moving into floodplains twice as fast as we are into other areas. Half of the U.S. population now lives in coastal counties.

In low-lying countries, all of this is much, much worse. Bangladesh, for example, is expected to lose 10% of its land to the sea in the coming decades, forcing 18 million people to look for new homes.

What's Wrong with Flooding

What we're worried about isn't a little water in the basement. Climate-crisis flooding is a longer-lasting, more sinister threat.

First, flooding kills people. Around 90% of flood victims die by drowning, but electrocution, shock, physical trauma, heart attacks, fire, and overexertion are also on the list. Incredibly, more than half of flood victims die in their *cars*, trying to drive across flowing water. The water lifts the vehicle off the ground, pushes it sideways, and flips it.

Then you've got the sewage problem. In 700 American cities, the sewer system, fed by toilets, is *also* the drainage system, fed by rain overflow. That unified liquid is supposed to flow beneath the city into water-treatment plants for cleaning. But in heavy rains, the plants are overwhelmed by the sheer quantity of water. The plant managers have no choice but to dump untreated, raw sewage into the public waterways.

Figure 9-6. "Floodwaters" is often not quite the right term. It's water plus sewage, fertilizer, chemicals, and fuel.

In Chicago, a storm that dumps 2.5 inches of rain or more in a single day is enough to send geysers from manholes in the streets, backing up into basements, and flowing into Lake Michigan. Nationwide, our treatment plants dump more than 850 billion gallons of raw sewage into the public waterways every year—enough to fill over a million Olympic-size swimming pools.

A lot of people get sick as a result. After Milwaukee's 1993 rainfall—the heaviest in 50 years—100 people died and over 400,000 got horrific diarrhea that lasted for days. In 2000, a five-day deluge in Walkerton, Ontario, left the drinking water with such high levels of *E. coli* that 2,300 people got sick and seven died.

Microbes aren't your only contamination concern during floods, either. As water from overflowing riverbanks or shorelines sluices across the land, it picks up agricultural waste, fertilizer, heavy metals, oil and gasoline, and other nastiness. It's what scientists call "black water": hazardous to humans. In New Orleans after Hurricane Katrina, people called it toxic gumbo.

This nasty stuff can find its way into your house; into fishing spots and swimming holes; and, worst of all, into the drinking-water supply.

The awfulness isn't over once the floodwaters recede, by the way. At that point, you begin a very slow, expensive cleanup process—but now you also have to contend with mold and toxic chemicals, which present their own health risks and expenses.

If there's one tiny bright spot in this dismal picture, it's that you can prepare. There's a lot you can do, often inexpensively and without much inconvenience, to protect yourself from the worst of the floods in the new era.

Prepping in Calm Weather

First question: How likely are you to get flooded?

It really doesn't matter where you live; flooding these days is a risk anywhere. Even inland, "mile-high" cities like Boulder and Denver, Colorado, routinely get flooded.

You can look over the government's official flood-risk maps at https://msc.fema.gov, although you may find them confusing without the secret decoder box on page 201. For a clearer and more realistic risk reading, type in your address at https://floodfactor.com. Unlike FEMA's site, FloodFactor.com's calculations take into account data like sea-level rise, rainfall patterns, and flooding along smaller creeks. As a result, it lists twice as many properties at risk than the government does.

As you're now aware, the definition of the "floodplain" is changing fast these days. The floodable areas are rapidly growing. If you like, you can sign up to get a free email notification if your neighborhood's flood risk (as determined by the government) changes. Register for it at https://msc.fema.gov/portal/subscription.

Buying a Home

When you're house shopping, you might think that it'd be a simple matter to make sure you're not about to buy into a nightmare floodplain. Surprise: Privacy laws specifically *prevent* you from looking up a property's flood-insurance claims in FEMA's database.

Well, surely there's some federal law that requires the seller to disclose a home's flood risk!

Nope—29 states have some flood-disclosure requirements, but in the other 21, there's nothing at all. Sellers have no obligation to tell you if flood insurance is required, if the house is in a floodplain, or if it's had flood damage. The property may flood every single summer, and you'd have no way of knowing. The states without disclosure requirements even include coastal states like Virginia; Georgia; Alabama; and, incredibly, Florida, where 30% of homes are in the floodplain.

You may discover the truth about a home only at the last minute of the buying process, when you fill out your mortgage paperwork. The *bank* cares a lot about flood risk and will dig up the flooding history before granting your loan. But that can be an extraordinarily inopportune time to discover that the home you've worked for weeks to buy is a flood magnet.

Remember, too, that if the property is in what FEMA designates as a floodplain, you won't be allowed to get a mortgage unless you have flood insurance.

Building (or Rebuilding) a Home

In the flood era, builders and renovators have all kinds of sneaky ways to make a home more floodproof. They can raise it. They can choose building materials that won't turn sodden and hopeless when they get wet. They can install the water heater, furnace, and air handlers above ground level.

Chapter 3 offers a complete rundown of the ways you can build or retrofit a home to make it floodproof—with plentiful reminders that you'll save money on flood insurance if you take these steps, and the serendipitous news that the government has money available to help you pay for them.

House Fortifications

Most of the preparation you have to do for flood aligns nicely with the advice in the preceding chapters. When floods come to call, you'll be in excellent shape if you're ready with early-warning apps, a bug-out bag, a family evacuation plan, and copies of important documents (Chapter 8). You'll be grateful that you've made an inventory of your belongings and looked into flood insurance (Chapter 6). And if you're stuck at home because the roads are flooded, you'll appreciate having thought ahead about emergency power, water supplies, and check valves to prevent sewage from backing up into your house (Chapter 3).

A few emergency-readiness steps, though, are unique to floods:

- Keep your gutters and downspouts clear and unclogged. That means hauling out a ladder a couple of times a year and putting on gloves. When the rain comes down hard, you want your home's drainage system to *work*.

- If there are cracks in your basement walls, seal them up with waterproofing compound to prevent water from seeping in.

- Buy a waterproof storage box for your birth and marriage certificates, passport, custody and military papers, and so on. Of course, you already have copies of these in your bug-out bag (right?), but this way the originals will be safe. If you have time to evacuate in your car, rather than on foot, you could grab this box and toss it in the back, so you'll have all the original papers.

- Buy a boat. If you live in superstorm areas, this tip might not be as far-fetched as it sounds; in Katrina/Sandy/Harvey situations, where roads are under four feet of water and your house isn't usable, a floating es-

Figure 9-7. A waterproof, fireproof document case like this (left) costs about $40 on Amazon; the inflatable raft (right) goes for $120, and includes a hand pump and aluminum oars. (Note: In case of superstorm, the models' expressions may be less happy.)

cape like this could mean the world to you. On Amazon, you can buy a cheapo two-person inflatable life raft, with oars, for $50, or a nicer one for $120. An actual rowboat costs about $500.

- An indoor water alarm is a $15 battery-powered gizmo that goes off loudly if water starts pooling wherever you've put its sensors. It's a cheap way to give you an early warning before real damage occurs.
- Waterproofing your basement is not quite so cheap, but it can be a blessing if your neighborhood floods a lot.

  The process usually involves hiring a company. Most companies dig a French drain around the perimeter of the house—either just outside the foundation or inside, along the margins of the basement floor.

  The trench contains a perforated pipe. When water comes in, these drains intercept it; a sump pump shoots the water out and away from the house. For an added cost—and a worthy one—you can equip your sump pumps with battery backups and dying-battery alarms.

  Neither trench is visible after installation; it's covered over. But the indoor approach means that you don't have to dig up your flower beds, steps, and walkways on the outside of the house. The inside piping is also less likely to clog over time.

Your Evacuation Plan

Chapter 8 covers the general topic of family emergency planning in detail. But in the context of flood evacuation, your extreme-weather dossier should also include answers to questions like these:

- How do you shut off the power and gas to your house? Those are important steps before you head for the hills.
- Which roads are likely to flood? If you have to get home in a bad storm—or get *away* from home—knowing those flood patterns are important.
- Where's higher ground? What's the fastest way to get there?

As the Flood Begins

In general, you get to hear about a hurricane, storm, or intense rainstorm in advance. The news will be filled with ominous warnings, everybody will be talking about it, and there'll be a run on bottled water.

But there's a difference between a flood *watch* (a flood is possible) and a flood *warning* (somebody has actually spotted a flood or flash flood). Either one is a reason to pay attention.

If there's a flood *warning* nearby your home, take these steps, in this order:

- Start monitoring the situation on the radio or a free phone app like American Red Cross Emergency. Listen for the locations of shelters that your town is setting up.
- If you have pets, keep them upstairs and confined where you can keep tabs on them. Call the shelter and ask if pets are welcome.
- If you've got livestock, like horses, get busy moving them to higher ground.
- Gas up, or charge up, the car. Evacuation traffic is always frighteningly slow, and gas stations may not be operating.
- If you have cars or boats you won't need for evacuation, or a trailer home, move them to higher ground. Your town may announce good locations for them.
- Seal the cracks around your exterior doors against encroaching waters by laying a plastic tarp against each one, and then weighing it down with sandbags. Many towns give away sandbags free as the emergency approaches, but you can also buy them yourself. Stores like Home Depot and Lowe's sell self-sealing heavy-duty plastic bags ($19 for 50); you have to provide your own sand.

  If the water's coming from a certain direction, you can also use sandbags to build a temporary dam. According to the Red Cross, two people can create a 20-foot–long sandbag wall, a foot tall, in about an hour.

- Unplug your appliances. Don't touch the metal parts of anything that's electronic and wet, either. The shock you'll get from them can be much more severe than the usual shock from a household outlet, because the electronics may contain a capacitor (a little component that stores extra power).
- Check over your go bag. Make sure it's up-to-date; that medicines haven't expired; and that it includes whatever important, grabbable items you wrote on your last-minute list.
- Move anything valuable to high, safe places, like the attic or the top shelf of an upstairs closet: jewelry, electronics, your flat-screen TV, fancy clothes, and scrapbooks and albums. Even try to move your nicer furniture as high as you can get it. Once muddy, contaminated water enters your ground floor, you can kiss most of your valuables goodbye.
- Fill some bottles or jugs with water, so you'll have something clean to drink. Fill the bathtub and sinks with water, so you'll have some water for flushing, cleaning, or (after purifying) cooking or drinking.
- Turn off your propane tanks; they're a fire risk.
- Scope your yard for things that could get ruined or float away if your property ponds up: patio furniture, sports gear, propane tanks. Strap them down or bring them inside.

If You Have to Leave

The unanimous advice from flooding experts goes like this: If you get word to evacuate, do it—now. And even if you haven't heard any official warning, if you *think* you should evacuate, do it—now.

Your checklist for evacuation should go like this:

- Start listening to your NOAA radio or one of the emergency apps described in Chapter 8. You'll want to know about road closures, evacuation notices, boiled-water alerts, and where to go for higher-ground shelters.

- Run around unplugging appliances. If you're wet or in water, don't touch the appliances themselves. You could be in for quite a shock.
- Shut off the power to your house. You generally do that by shutting off the main circuit breaker on your main panel, usually above the smaller branch circuit breakers.

  Why? Well, suppose you forget. The power goes out in your neighborhood, and your house gets flooded. Later, when the power comes back on suddenly, your wet electrics can short out, spark, or start a fire.
- Prep your photos. Of course, because you've read Chapter 8, your digital pictures are all backed up online. But what about framed photos, pictures in albums, or boxes or envelopes full of photos?

  Do a quick run through the first floor, your basement, attic, and garage, looking for photos. Check bookshelves and coffee tables for albums; check the walls and mantels for framed photos. Get your kids to help you.

Figure 9-8. The main circuit breaker is the off switch for all the power in your home. It's usually the centered top switch on your breaker panel.

If you can't take the photos with you, put them into two layers of plastic—ziplock bags that then go into trash bags, for example. Pad sharp corners of frames so they don't puncture the plastic. Label everything. Carry the bags to the highest floor and store them on the highest shelf of a closet.

According to the internet, the dishwasher is a great place to protect your photos from floodwaters. Unfortunately, dishwashers are usually on the ground floor—the first to get flooded—and they're fed by pipes that back up in a flood, spewing water into the dishwasher and drowning your photos. Just this once, something you read on the internet was wrong.

- Grab your go bag, your waterproof document case, and whatever you've listed on your last-minute-grab list and get out of there.

Don't Cross Floodwaters

Once you're on the road, here's a quick and easy way to avoid dying: Don't cross floodwaters.

Every single year, people die trying to cross floodwaters. Some are on foot, some in cars; it doesn't matter. *Don't cross floodwaters*, not even a flooded intersection or a highway dip.

Why not? Because:

- **Rushing floodwaters sweep people away, even in their cars.** Six inches of rushing water can knock you off your feet. One foot of rushing water can float your car (the tires act like rafts). Eighteen inches can lift an SUV or truck. Once you're afloat, the current carries the vehicle, turns it at crazy angles, and often flips it over. You don't want to be inside when that happens.

  As noted earlier, that's how over half the fatalities happen in floods: People drown in their cars.

- **You can't see what's under there.** You can't tell how deep it is. You can't tell how fast the water is moving. You can't tell if there's even a *road* under there; it could have washed out.

  Meanwhile, the dark, flowing water may conceal holes, gaps, rocks,

Figure 9-9. It doesn't seem like something as heavy as a car could float—but drop it into floodwaters, like these in Iowa, and you'll get your proof.

tree branches, and all kinds of sharp and heavy things that it's picked up on its journey. There could be live downed wires in there.

Your car can stall in the middle of it, or get hooked on something you can't see, and you'll be stuck.

- **The water itself is likely to be contaminated** by raw sewage, chemical pollutants like motor oil and antifreeze, fertilizer, oil, gas, and other ingredients. Researchers have found over 100 kinds of bacteria, viruses, and parasites in floodwaters that can make you sick.

  Don't let your kids near floodwaters; and if you get wet in it, wash your hands at the first opportunity.

As you read this, you may be saying to yourself, "Okay, okay! I won't drive through floodwaters. How stupid do you think I am?"

You're forgetting about social influence. Once you see somebody *else* successfully drive through floodwaters, you're probably inclined to believe that it's safe. In one 2009 study, more than 50% of people who drove across floodwaters said they did it because they'd seen someone else make it across.

If you understand the logic behind "Don't cross floodwaters," then you'll probably get this part, too: Don't try to go around barricades that emergency workers have set up to *keep* you out of floodwaters. In every major flood, people think they can get away with that—and they die in the process.

Don't drive across a bridge if there are rushing floodwaters beneath, either. Floodwaters often chip away at the footings of bridges, making them unstable.

But mainly, don't cross floodwaters. Or, if it's easier to remember, use NOAA's handy almost-rhyme: "Turn around. Don't drown."

If you learn nothing else from this book—well, then it was a pretty worthless book. But at least you'll remember: Don't cross floodwaters.

In Your Car in a Flood

You already know Rule 1. But what if it's too late, and you find yourself in your car, deep in water?

Your general goal should be to get out of the car as fast as you can. That, however, may be easier said than done:

- **If you can't open the door** because the car is partly submerged, climb out through the window. As long as the car isn't completely submerged, power windows still work.

  And if you can't get out of the window for some reason, open the door and get out. If the door won't open because of the water pressure on the outside, wait for the car to fill up *even more*, up to your neck, so that the pressure equalizes. At that point, you'll be able to open the door.

  While you wait, get rid of anything that might hold you back, including your seat belt and jacket. Unlock the doors. Turn on the headlights and flashers to get rescuers' attention. Take deep breaths to stay calm.

  Once the door's open, leave your stuff behind. Hold your breath and swim.

  Once you're out, work your way to higher ground. Use a stick, if possible, to gauge what's under the water in front of you before each step.

- **If your car is stuck in moving water**, FEMA suggests that you open the window and climb onto the roof to await rescue—provided the car isn't actually floating.
- **If you get swept into rushing water**, lie on your back with your feet pointed in the direction of the current. That way, your feet strike any floating objects instead of your head. If you can find something to use for flotation, like a branch, grab it. If you encounter big obstacles, clamber over them. Never try to swim under.

  Your goal is to find either the shore or something you can hold on to, like a tree branch or a rooftop.

When You're Stranded

If you manage to get someplace that's up out of the water, like a tree or a roof, stay there. Wait until rescuers come to get you; don't try to swim for it.

If you're trapped in a building, go to the highest floor possible. The roof is good, too. But be careful of the attic; if there's only one way in and out, you could get trapped there as the water rises.

Try to get people's attention from a window. Go onto the roof only if you absolutely must—for example, if that's the last available high point—because roofs are often sloped and spiked with sharp protrusions.

Deaths from Flood Tourism

Now that you understand how much danger floodwaters can conceal, you might be amazed to learn that a huge percentage of people who sustain injuries and deaths entered those floodwaters *for fun*.

"Flood tourism" is when people gather on dangerously unstable riverbanks and bridges to gawk and take pictures, or boat on or drive through floodwaters for the purpose of posting YouTube videos. More people are making these recreational floodwater excursions every year, and more of them are dying. In Australian flood fatalities, the percentage of people who died in flood tourism escapades quintupled between 1950 and 2008.

"There is significant overrepresentation of male fatalities," notes one researcher diplomatically, citing "a male to female death ratio of 4:1."

Figure 9-10. Flood fun on YouTube. Bad idea.

After the Flood

Sooner or later, the floodwaters recede. You can stop thinking about survival and start thinking about putting your life back together.

The aftermath, unfortunately, is awful. Chunks of roads and bridges can be missing. Your furniture and belongings might be waterlogged and contaminated. And the mold—oh, man.

And then there's the invisible stuff: the sewage and chemicals in the local water supply. Gas leaks. Live power lines fallen from the poles. Short circuits in appliances that got wet. Where hillsides got saturated, mudslides can sluice into buildings and across roads, picking up debris along the way and burying stuff.

The bottom line: Flooding is the gift that keeps on taking.

Keep listening to the news, and keep checking your emergency app. You need to know which roads are still out, whether the water is okay to drink, and when the power will be back on. You still should avoid walking into floodwaters, and you should still keep out of the way of emergency workers.

Keep in mind, too, that any spot of ground that was flooded is now waterlogged and saturated. The instant more water arrives—some more rain or runoff, for example—it will flood again instantaneously.

Contaminated Water

In times of flooding, the water-treatment plants that are supposed to purify your drinking water often get overwhelmed, and the water supply gets contaminated.

During Hurricane Harvey in 2017, for example, 800 sewage-treatment plants wound up spilling 2 million pounds of human waste into the environment. Researchers afterward found unsafe *E. coli* bacteria levels in every water sample they took, and officials were still finding human feces washing up on Texas beaches seven months later.

In a big city, it's not unusual for hundreds of thousands of people to get stomach bugs, respiratory infections, rashes, eye/ear/nose/throat irritation, and awful diarrhea after a flood.

In shelters after a major storm, where a lot of people are living in close quarters, more serious diseases break out. We're talking dysentery, cholera, and hepatitis A, all of which are horrible. If you find yourself in that situation, get obsessive about washing your hands or using hand sanitizer like Purell.

If your local authorities consider the water too contaminated to drink, they'll issue a boiled-water alert to radio stations, TV stations, and newspapers, as well as on their own websites and social media accounts. In some cities, you can sign up to get these notifications by text message, email, or phone call. You'll get another one when the water is safe once again.

Until you've heard for certain that the local drinking water is safe for drinking, therefore, purify your tap water before drinking or cooking, following the procedures described in Chapter 8.

Bugs and Critters

After a flood, there's a lot of standing water, and nobody loves standing water like mosquitoes. They breed best when there's a lot of water around. If the flood has left ponds near your home, the CDC recommends that you use DEET or picaridin when you go outside (Chapter 14).

People aren't the only ones flushed out of their homes in a big flood, either. So are pets, reptiles, rodents, ants, and other critters.

You may see a lot of wildlife where you usually don't. After some of the recent hurricanes, for example, people went back to their homes as the waters were receding and found not just mud and branches in their hallways and living rooms, but *snakes*.

Keep clear of displaced animals and help them keep clear of *you* by making a lot of noise—bang the floor with a stick, for example—as you venture back into your home.

Returning Home

If there was an evacuation order, don't go back until you get official word that it's safe to go back. Best not to hover around your home in the meantime, either, while emergency workers are trying to do their jobs.

Actually, be careful going into *any* building. Often, water leaves buildings with weakened floors, water-weighted ceilings, and sodden walls, to say nothing of waterlogged electrical systems. Get the all clear from

Figure 9-11. After a flood, homes like this one after Hurricane Harvey are clearly unsafe to enter.

somebody in authority before you enter any building that shows signs of damage.

While you're waiting, let your friends and family know you're safe, using one of the apps or websites described in Chapter 8. Keep listening to the local news, or the NOAA Weather Radio, for the latest. Keep your pets and kids away from standing floodwater.

The Red Cross recommends leaving your kids with a friend or relative when you visit your home for the first time after the disaster—first, because the house might not be safe, and second, because seeing the home damaged can be upsetting for them.

Once you've determined that it's safe to return, approach the building cautiously. Look for signs of damage before you go in, like loose power lines, foundation cracks, and missing support beams.

Broken glass, overturned cars and grills, ruptured gas lines, and wandering animals are likely to make your return less exuberant than usual. If emergency personnel have put colored tape across your doors or windows, don't be a hero (or an idiot): Don't go inside.

Try to arrive during daylight, so you don't have to turn on lights (and thereby tempt the short-circuit gods). If you need illumination, use flashlights, not candles, lighters, matches, or anything else with a flame. There could be leaked gas everywhere.

If the door is jammed, don't force it, because it might be supporting a bunch of building that could fall on you.

Once you're inside, you may see that you've got quite a job ahead of you. Mud damage is a nightmare. Water damage is horrifying, and contamination from floodwaters is the *worst*.

Here's your returning-home checklist:

♦ **Mind your sagging floor and ceilings.** They're probably waterlogged. If they're disturbed, they could collapse. You can build a bridge across a sagging floor with thick plywood or boards. A sagging ceiling will probably have to be replaced.

  If the floodwaters got up pretty high, be wary of leaning or pushing against *any* surface, actually. Stairs and walls can be unstable.

♦ **Don't turn on the gas, the power, or any electrical appliances that got wet** until they've been checked for safety by an electrician. If you'd have to step into water to turn the power on or off, don't do it—call an electrician.

The Centers for Disease Control feels so strongly about this point that it uses both boldface *and* caps: "**NEVER** turn power on or off yourself or use an electric tool or appliance while standing in water."

- **Beware the gas leak.** If you've been out of the house for a while, step in only long enough to throw open the doors and windows. You want to let it air out for half an hour before spending any time in there, just in case there was a gas leak.

  If you *smell* gas or hear hissing, turn off the main gas valve, if you know where that is. Then open all the windows and get out. Call the gas company and the fire department. Don't turn on anything electrical; it could spark and make your whole house go boom. Go back in only when you've been told that the leak has cleared.

- **Take a lot of pictures.** You'll need them to show the insurance company. Take them in good light from different angles.

Power

Water + live power = danger, pain, or death.

That formula applies in all kinds of ways.

- **Power lines.** Don't touch downed power lines (you, too, kids!). Treat every downed power line as though it's energized and dangerous, and report it to the power company. Don't drive into a flooded street where downed power lines are visible, either. Yes, people do that.

- **Turning on the power at home.** If the basement is flooded high enough to cover power outlets, power cords, or your utility panel, don't even go in there.

  If it *was* flooded but the water has receded, don't turn the power back on until an electrician has approved your whole electrical system.

  If it's safe to get to the panel box, look over the circuit breakers. If one of them has tripped, don't turn it back on; it may indicate damaged wiring in the walls, which you'll need to have checked out.

Figure 9-12. Downed power lines are dangerous because they're down—within reach of people who can get electrocuted.

- **Other utilities.** Don't use your faucets, toilets, or sinks if you spot any signs of a ruptured sewer line. Call a plumber.
- **Appliances.** Don't use any appliance or electrical machine that got wet in the flood; people get electrocuted this way after every flood. No appliance is safe to use until it's *completely* dry, inside and out, and an electrician has checked it out.

  Your dishwasher, washer and dryer, stove, microwave, and water heater *may* be salvageable if they got wet; your water heater, air conditioner and heater, fridge, and freezer will probably need replacing. In general, the longer your appliances have spent wet, the less likely it is that you'll be able to salvage them.

  And here it is, one more time: *Never turn on an appliance, or turn your power on or off, while you're standing in water.*

  If you think somebody's been electrocuted, by the way, don't touch them until you're sure they're not still touching the source of the electricity. If they are, *you* might get the current, too. Call 911, turn off the power or separate the person and the gadget, and use CPR if the heartbeat or breathing is weak.

- **Generators.** If the power's still out, but you have a generator, don't celebrate just yet. You should not turn the thing on if any part of it is wet. If there's any chance that it got wet, an electrician should check it out before you power it up.

  Meanwhile, one of the leading causes of death after a storm or flood is carbon monoxide poisoning from people using generators, camp stoves, or charcoal camp grills indoors.

  If you have one of those machines, don't run it indoors or in the garage with the doors closed. It should be on a dry spot that's at least 20 feet outside of any door, window, or vent on your home. The CDC even recommends buying a carbon monoxide detector (about $14) that can run on battery power or battery backup, just to make sure.

The Cleanup

If your home was flooded, you've got your work cut out for you. You'll have to spend the first couple of hours (or days or months) cleaning up. This isn't just running the vacuum and straightening a few pictures. There's a whole art to tending a flooded home:

- **Wear gloves and rubber boots** to minimize contact with toxic water, pulled nails, and snakes.

- **Be careful lifting furniture.** Wet furniture weighs a *lot* more than dry furniture, and the last thing you need right now is a pulled back.

- **If it's sodden, then it's rotten.** That's not an actual FEMA saying, but it's catchy. Anything that has sopped up water and can't be dried completely in 48 hours has to be thrown out. Soaked carpets, mattresses, couch cushions, stuffed animals—it all has to go.

  If you have flood insurance, hang on to everything until the insurance claims adjuster comes to see it—even stuff you intend to get rid of. Put the damaged stuff into its own pile.

  If you have to rip up and throw away soaked carpet before the adjuster's visit, at the very least save a two-food-square piece of it and the padding underneath. The adjuster will want to see it.

This is a painful exercise, especially if you loved some of that stuff; the sheer waste on display after a flood is heartbreaking. At least, having read Chapter 6, you know that the insurance company will buy all new stuff for you.

- **If it's plastic, that's fantastic.** That means that it didn't soak up water, so you'll probably be able to reclaim it. Same with anything else made of a hard, nonabsorptive material like glass, metal, ceramic, porcelain, or sealed wood. Items in this category: counters, some appliances, some flooring.

  Wash with soap and water, and then disinfect with bleach solution: eight drops of unscented bleach per gallon of water. Then dry thoroughly.

  Don't let your kids play with their toys until you've cleaned them with a bleach solution (the toys, not the kids).

- **Can it dry? It's worth a try.** Anything with a prayer of drying out in a couple of days might be salvageable.

  Wash curtains, clothing, sheets, blankets, and comforters in hot, soapy water. Bleach them if possible. Then dry them, iron them, and see how they fared. Or just get them dry-cleaned, if you can afford it.

  For rugs and furniture: Flush them with clean water, shampoo them, then air-dry. If a bigger carpet got soaked, you can steam clean it.

  For mattresses, pillows, and chairs: If they're not totally soaked, let them dry in the sun. Then spray them with a disinfectant like Lysol.

- **Food.** If the power was off for more than four hours, then the food in your fridge may not be safe. Toss any food that looks, smells, or feels weird. The stuff in your freezer is okay without power for up to 48 hours.

  You're usually told to throw out any food that came in contact with floodwaters, too. But in fact, you can salvage any food that got wet if it's in a can, jar, or pouch. Take off the label, if you can, wash the container in hot soapy water, rinse in drinkable water, wipe off with our friend the bleach solution, then relabel the container.

But if the container has a screw cap, twist-off cap, or snap top, toss it, because there's no way to thoroughly disinfect it. Anything in a cardboard container has to go, too—even those waxed ones like juice boxes or milk boxes.

- **Baby food.** Clearly, you don't want to feed your baby formula made from contaminated water. In order of preference, the CDC recommends: breastfeeding, ready-to-drink formula, formula made with bottled water, formula made with boiled water, formula made with treated water.

  By the way: There's no way to sanitize bottle nipples or pacifiers that got wet in floodwaters. Throw them away.

- **Wells.** If you get your drinking water from a well, don't use it until you've had it tested and disinfected. If you're not sure, contact your local or state health department, which you can find with a quick Google search.

- **Make temporary repairs.** Don't rebuild any damage until your flood-insurance adjuster arrives to make an inspection! You want them to see the full result.

  It's fine to make temporary repairs to make your place livable, though, like putting a tarp over a hole in the roof, plywood over a broken window, bracing a wall, and hauling away trash and debris. But since flood insurance will pay for all of this, save your receipts for all the materials and keep taking pictures.

- **Dry out with fans.** Fans can help dry your place out—if, of course, you've got power and it's safe to use appliances. Aim them *out* the window, to avoid blowing dust and mold around.

- **Pump out the basement slowly.** If you empty a flooded basement all at once, the sodden, heavy ground outside might apply so much pressure that the basement walls collapse inward. Slow it down. Pump out no more than a third of the water each day.

- **Insurance coverage.** If you, like almost everyone else in the world with flood insurance, have a policy from the National Flood Insurance Program, here's what they'll pay for: the full replacement cost of the home and everything fastened to it (cabinets, plumbing, central air, etc.), up to $250,000. They'll pay for hauling away debris from inside the home, too. If it's a vacation home, you'll get only its current cash value.

They'll also pay you the cash value (the "used" price, not the "brand-new" price) of all your possessions in the house, up to $100,000—except for whatever was in the basement.

NFIP policies don't reimburse you for mold damage that you could have avoided; food that goes bad because the power's out; yard cleanup; physical money and stock certificates; structures outside the house like decks, patios, pools, hot tubs, landscaping, fences, and wells; cars; and living expenses while you're rebuilding.

Photos, Books, and Documents

Chapter 8 recommends that you get your photos digitized and then backed up and stored off-site. If it's too late for that, and you've now got wet prints and slides—well, you've got some work, and some disappointment, cut out for you. As you work, obey the two golden rules:

- **Don't touch the wet image.** You'll smear the dyes and ruin them forever. (If they're already smeared, there is, alas, nothing you can do to save them.)

- **Dry them soon.** The longer you wait, the greater the permanent damage, like mold. After 24 hours wet, the photos may not be salvageable.

  And once they've been wet and then *dried*, they're usually permanently stuck to each other, or to whatever frame or album they're in, and ruined for good.

The most urgent step is to stabilize them—by drying them. To get them ready for drying, remember these four special cases:

- **If the photos are muddy**, first rinse them in a tub, bowl, or baking dish into which you're running clean, cold water. Don't let the faucet water strike the photos; the goal of the running faucet is to keep the water constantly refreshed with clean water. Don't rub! Handle the photos by their edges, and keep dipping until the water dripping off the photo is clear.

- **If photos are stuck together**, they may not be salvageable, especially if they've dried face-to-face.

  But you do have a chance—by freezing them. That process halts the deterioration and prevents mold from advancing. Later, as they're thawing, you may be able to peel them apart. At that point, lay the separated photos out to dry, as described below.

- **If they're in albums**, first dip the entire album, still closed, in cold water, to rinse away the mud and debris.

  Lay the album on a towel and open it to the first page. If the photos are held in place by corner holders, carefully lift the photos out with a knife, and lay them aside to dry.

  If the pictures are in clear plastic sleeves, cut away the sleeve around the photo, giving you a better shot at peeling it away. Lift the photo off the backing page with a knife, if necessary.

  And if the photos are in a clear sleeve back-to-back, use a knife to carefully peel them apart; if they're too wet to separate, freeze them first.

- **If they're in frames**, the first step is to scan them, or take digital photos of them, while they're still recognizable as images. If the photo gets ruined when you try to separate it from the glass, at least you'll have a record of what it looked like.

  If you can afford to send the frame to a photo-conservation company, do it. You'll have a much better success rate than trying to separate the glass and the photo yourself. (That's another reason to take a digital pic of the photo: The restorers will have something to work from.)

  If you do want to go it alone, your goal is to get the pictures out while they're still sopping wet. As soon as they start to dry in the frame, the photo sticks to the glass, and you'll never be able to fully rescue it.

  Here's the technique: Remove the glass and photo together, as a unit, and hold them under running water from the faucet. The stream of water helps you peel the photo off the glass.

- **Rescue prints first, negatives later.** Negatives are slower to deteriorate when wet.

Figure 9-13 Spread out wet photos and let them air-dry.

Drying them comes next. If there's no opportunity at the moment, separate the pictures with pieces of waxed paper, and then put them in your freezer to halt the deterioration in its tracks. Later, you do the drying business.

Here's how that goes: Spread the photos out to dry, faceup, on something absorbent like cloth or unprinted paper towel. Every couple of hours, replace the cloth or paper towel with fresh, dry ones.

Fans can help them dry faster. The photos may curl (especially in the sun), but you can flatten them later by moistening the backs with a sponge and then letting them sit, separated by pieces of acid-free paper or photo blotters, under books for a couple of days.

You can rescue slides by rinsing them and then dipping them into a commercial slide cleaner.

If important paperwork got wet, you can use similar techniques. Carefully rinse off any mud; lay the pages out to dry (or hang them from clothesline); use a fan to speed up the process. If there's no time for that, freeze them. Later, you can thaw them and then dry them with a blow-dryer.

As for books: Put a sheet of paper into the book, one per 20 pages or so, and then let the books dry. Change the soaked paper sheets every few hours.

As you work, remember that books and papers can get moldy even if they weren't actually *in* the floodwaters. Sitting around in stagnant high humidity is enough. That should be your incentive to move them someplace drier at the first opportunity.

Mold

Any building that's been flooded for more than a couple of days probably has mold. In general, you'll see it and smell it; it has a musty scent.

Mold is the worst. It's a fungus that likes to grow on damp surfaces—and after a flood, there are a *lot* of damp surfaces, and they're very hard to get dry. After hurricanes Katrina and Rita of 2005, about half of all homes in greater New Orleans got mold contamination. In 100,000 homes, the mold was so bad that the owners had to discard everything porous in their entire homes, including the sheetrock of their walls.

Mold is especially partial to cellulose-based materials like wallpaper, ceiling tiles, cardboard, insulation, and wood; but you'll often find it on walls, carpets, clothes, boxes, furniture, books, toys—just about anything

Figure 9-14. Mold after a flood. It's nasty.

that got wet. It might be growing in spots you don't see right away, like under your carpet, inside the walls, or above your ceilings.

There are thousands of different mold species, but they all spread the same way, using a creepy science-fictiony mechanism: microscopic, single-cell "reproductive units" called spores. They float invisibly through the air, seeking new damp spots to colonize.

You really don't want mold in your house. Breathing the spores, cells, and tiny mold fragments can irritate your lungs, throat, and nose. Some of those microfloaters are toxic, especially to people who already suffer from allergies, asthma, respiratory problems, or weakened immune systems.

The side effects of living around mold include runny or itchy nose, watery eyes, itchy throat, or sneezing; in some people, the effects can include breathing problems, coughing, headache, bronchitis, insomnia, fatigue, fever, and other fun stuff.

Mold grows incredibly fast. The most urgent business after your house has been flooded, therefore, is to dry it out. That means removing everything that got wet. It also means setting up dehumidifiers or, at the very least, fans pointed out the windows.

Hiring a Company

If you discover big patches of mold, a professional remediation company is definitely the way to go. (Google *mold remediation dallas* or whatever your town is.) The *average* cost is between $1,000 to $4,000, depending on the size of the building and the amount of mold—but really, the sky's the limit. If there's mold all through a house that was flooded, the remediation could cost $30,000 or more.

Fortunately, the flood insurance you were so smart to buy in advance will generally pay for the cleanup.

Incidentally, don't take anybody up on their offer to "sample your mold." That's a scam. There's no point in knowing what species of mold you have. You clean them all the same way.

Doing the Job Yourself

If you must tackle the mold removal yourself, settle in for an unpleasant job. Children, people with asthma, and people with weakened immune systems should stay clear of the house, no matter how eager they seem to participate.

Put on some good music, tell the boss you won't be in for work, and dig in:

- **Dress for the job.** Wear gloves—don't touch moldy things with your bare hands. Wear long sleeves, long pants, and waterproof boots. You'll need fully enclosed safety glasses—not the kind with air holes—and a disposable face mask whose package says N-95 protection or better. (As you may have learned during the COVID pandemic, that notation means that it blocks out 95% of particles as small as 0.3 microns, which is 225 times smaller than a human hair.) Those floppy surgical masks aren't good enough.

- **Air the place out.** Open all the doors and windows while you're working. Even when you're done for each night, leave as many of them open as you feel safe doing.

  Open all the *interior* doors, too, for better airflow, and whatever doors, stairs, or openings go to the attic. If you've got power, turn on the fans in the bathrooms and on the kitchen hood.

- **Wash everything.** If the place is covered with mud, that's got to come off first. Only then does it make sense to start the de-molding process.

  For the first round, use hot soapy water and a bristle brush. Then comes the mold killer: either diluted bleach (one cup per gallon of water) or a spray-bottle mold killer sold just for the purpose. Immediately wipe dry, blow-dry, or fan-dry the surfaces to make sure the mold doesn't grow right back.

  Don't mix cleaning products, especially bleach and ammonia, which creates toxic vapors and even more problems for your life.

- **Be brutal with discards.** If some object got wet, and you didn't clean and dry it within 48 hours—or you *can't* clean and dry it—you have to throw it away. It *will* get moldy.

  It's really hard to get the mold off ceiling tiles and sheetrock. You'll probably have to toss them. If you lift the carpet and see big patches of mold on the backing, that'll have to go, too.

- **Rent a wet-dry vacuum cleaner.** Use this special industrial liquid-sucking vacuum (a "wet vac") to slurp up all the puddles. (Home Depot charges $24 a day for the rental.) This part is urgent; you want to dry everything in the home as quickly as you can. Mold can't grow without wetness.

- **Open cabinets in the kitchen and bathroom.** Completely remove all of the drawers. Wipe them clean, and leave them out until they're dry.

- **Set up fans and dehumidifiers—if it's safe to use power.** You can rent huge industrial-looking dehumidifiers that are designed specifically for drying out a home after floods.

  It's important, if you use fans, to point them out the windows. If they're aimed anywhere else indoors, they'll just blow mold spores into new nooks and crannies of your home.

  Avoid using air-conditioning and forced-air heat for the same reason.

- **Don't repaint** until you've made absolutely sure that your water problem is solved, the house is totally dry, and the mold is completely gone. If there's any hidden mold left, it'll grow right back despite the new coat of paint.

- **Don't bring it with you.** At the end of each day, take a shower and machine wash your clothes in hot water, and spray disinfectant on your brushes. The last thing you want to do is contaminate wherever you're living *now* with mold.

Figure 9-15. Let these rented dehumidifiers and blowers run for a few days. They'll dry out the air, and the dry air will dry out the walls and floors.

Chapter 10

Preparing for Heat Waves

IF YOU WANT A SNEAK PEEK AT THE FUTURE OF THE HOTTER United States, stop by Phoenix, Arizona, which is one of the fastest-warming cities in the country. Almost a third of its 2019 days baked residents at over 100°F. It's often 94°F even before the sun rises.

During the summer, life in Phoenix shifts. People do their workouts and errands before sunup. Construction crews begin work in the middle of the night. The local zoo opens at 6 a.m. and closes at 2 p.m., for the sake of both the visitors and the animals.

And during heat waves, planes can't even take off, because the air is too thin to provide aerodynamic lift.

Figure 10-1. In Phoenix, heat waves make the air so thin that planes can't take off.

In the United States, most years, heat is the biggest killer of them all. In 2018, heat killed 50% more Americans than floods did, and three times as many people as hurricanes.

But heat waves are killing people all over the planet. France, 2003: over 14,000 people. Moscow, 2010: over 11,000. Europe, 2019: Thousands of schools were closed, social workers rushed around to rescue older people before they died of heatstroke (about 900 elderly died in the United Kingdom, 1,400 in France), and entire villages were evacuated.

The bottom line: In the 25 years ending 2003, heat exposure killed more people than hurricanes, lightning, tornadoes, floods, and earthquakes *combined*.

Temperatures have always varied and always will, season to season and day to day. There have always been heat waves. And not every place is heating at the same rate. The farther north you go, the faster temperatures are rising.

But in the big picture, the trend is clear: On average, each decade since 1980 has been hotter than the previous one, and the decade 2009–2019 was the hottest ever recorded on this planet.

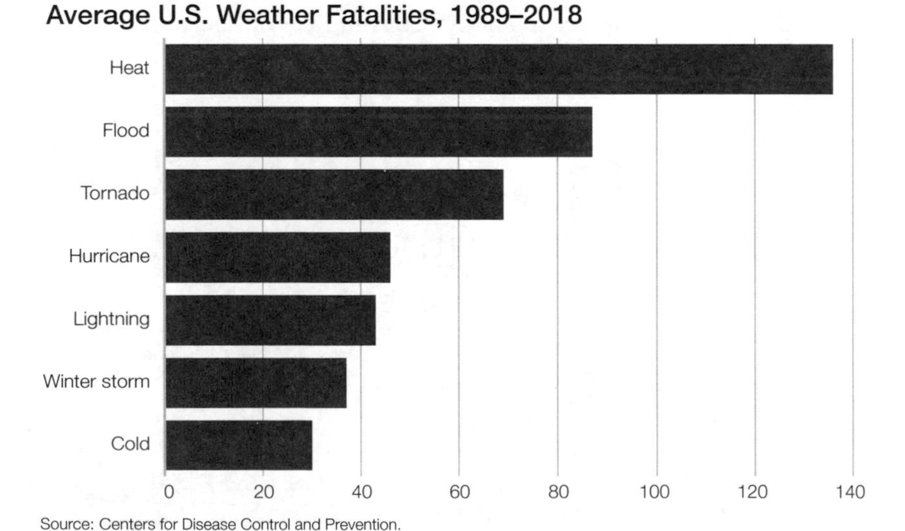

Figure 10-2. Of all the extreme-weather monsters, heat is the deadliest.

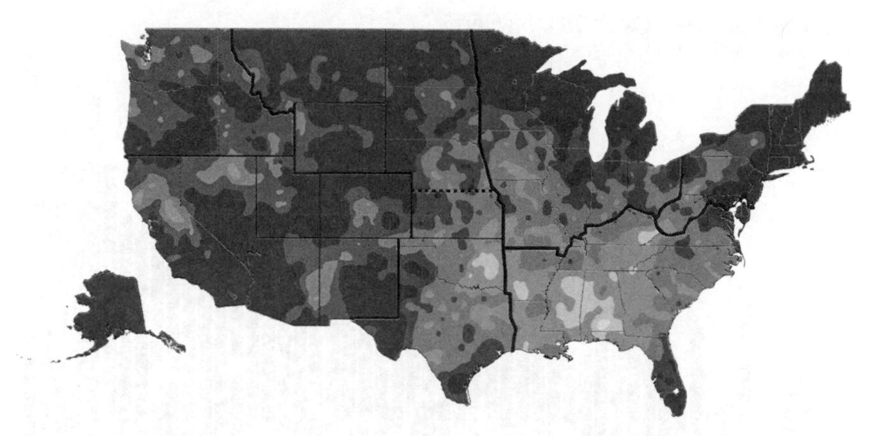

Figure 10-3. In the United States, the Northern and Western states have warmed the fastest since 1991—look at Alaska! The Southeastern states have heated up but not nearly as much. (Deeper blue spots have heated up more.)

You could fill an internet with examples of heat records being broken recently (all temperatures in this chapter are in Fahrenheit):

- The last five years have been the hottest ever measured on the planet.
- July 2019 was the hottest single month ever recorded.
- Since 2001, we've experienced 19 of the 20 hottest years ever recorded.
- The hottest single temperature ever reliably recorded on this Earth was in 2020. It was 130° in Death Valley, California.
- In the summer of 2020, Siberia suffered a blistering heat wave. ("Siberia" and "heat wave" are not terms that occur often in the same sentence.) On June 20, one Siberian town hit an all-time Arctic record of 100.4°. In *Siberia.*
- Germany, Belgium, Luxembourg, the Netherlands, and the United Kingdom all set new national heat records in 2019. France got the highest temperatures ever recorded there: 114.6°. Even Helsinki, Finland, way up north, recorded its highest-ever temperature: 92°.

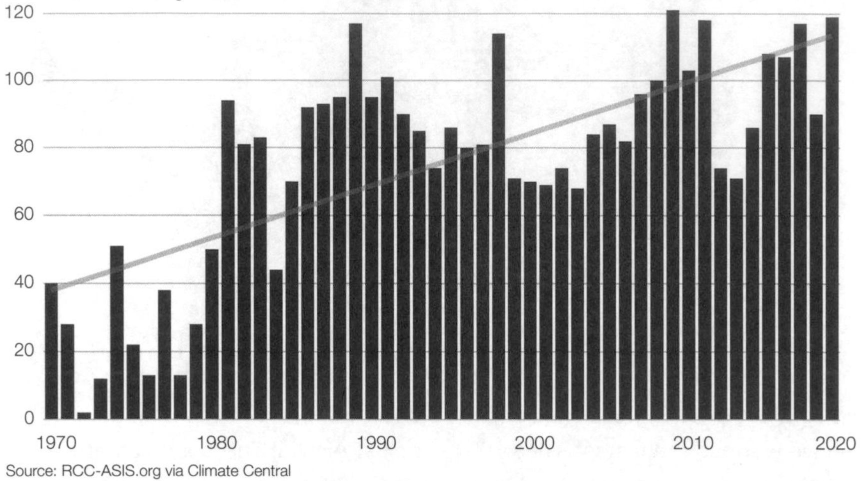

Figure 10-4. This graph shows how many days Miami gets each year over 90°. Hint: There are a lot more days like that lately.

- The 2019 heat wave in India gave Delhi *its* hottest day ever recorded: 118°. Since 2004, the country has experienced 11 of its hottest 15 years.
- In U.S. cities, 2,655 new daily high-temperature records were broken in September 2019 alone. In some cities, new heat records were broken for at least 10 days straight.
- in 2020, Phoenix had 53 days of at least 110°. (The previous record was 33 days.)
- Every year, Miami now gets 75 more days above 90° than it did in 1970.

But these records are made—and guaranteed—to be broken, over and over again. Computer models predict that:

- From 2050 to 2100, heat waves will triple in frequency.
- By 2070, Tampa will swelter with 40 more "above 95°" days than it does now.
- Tucson, Arizona, will experience heat waves that last 28 days longer than the longest ones now.

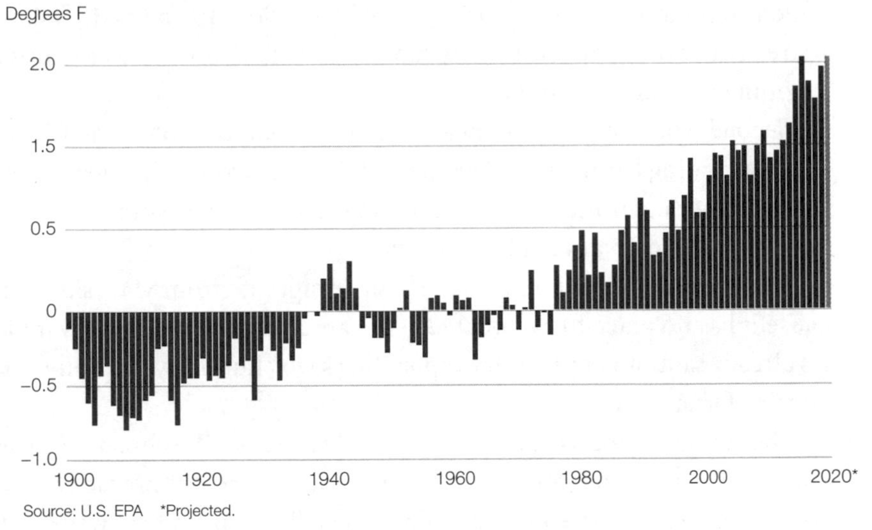

Figure 10-5. The 0 line represents the twentieth-century average worldwide temperature, as measured by both land-based and sea-surface sensors. The downward bars indicate below-average temperatures; the upward ones indicate above average.

"But wait," you might be saying, furrowing your brow. "If the whole world is warming, won't the winters be warmer, too? Won't that save lives from cold?"

Yes. Fewer people are dying from cold in the warmer era. Hurray! Unfortunately, that's not enough of an effect to offset the increased fatalities from heat.

How Heat Affects Us

Heat dries us out, weakens us, impairs our thinking, makes us more argumentative. It also stresses our hearts, which might not be an immediately obvious side effect. To understand why, it helps to grasp how your body ordinarily gets rid of heat.

First, it *radiates* heat, handing off heat to the cooler air around it. But if the air *isn't* cooler, then you don't get cooler, and your heart can't understand what's going on. It goes into overdrive, pumping harder to widen your blood vessels, to bring more blood close to your skin. When it's really hot out, your heart may have to circulate four times the usual amount of blood per minute.

Second, you can get rid of heat through evaporation of sweat. (A liquid evaporating from any surface cools it down.) Ordinarily, a teaspoon's worth of sweat, cumulatively evaporating from all over your body, can cool your bloodstream a full 2 degrees.

But when it's humid out, sweat *doesn't* evaporate from your skin, because it has no place to evaporate *to*. The air around you doesn't want it; it's already saturated with water vapor. That's why humidity magnifies the effects of heat waves.

Nighttime is supposed to give your body a break. The cooler air gives your heart a rest, and your body can get some sleep. But nowadays, the rising temperatures rise *most* at night, especially during a heat wave. You never get the chance to cool off.

The unfortunate payoff of all of this follows these lines:

- **Asthma.** Hot air and humid air are both triggers for asthma attacks. To make matters worse, hot days are usually windless days. If the air is polluted, it just sits there, stagnant, and you're breathing it. That's why asthma and other lung problems get worse during heat waves.

  If you're an asthma sufferer, monitor the air quality outside with a free app like Air Matters or AirVisual Air Quality Forecast. During heat waves, the best thing you can do is go inside in air-conditioning. (AC reduces the temperature *and* the humidity.) If you must run errands, do them early in the day.

- **Heat cramps.** When you lose enough water and salt, you start getting muscle spasms, especially if you've been playing sports or working outdoors. Consider these your early-warning system. Stop what you're doing, start drinking water, and get out of the heat.

- **Heat exhaustion.** If you've been in extreme heat for a long time—a couple of days, for example—you might start sweating heavily; feeling your heart race; and feeling faint, dizzy, and tired. You may also get nausea and headache. Key symptom: Your skin is *cold and clammy*.

This is really bad news. Stop what you're doing, start drinking water, get out of the heat, and seek medical treatment fast.

- **Heatstroke.** This is the final stage of heat injury. Your body can't control its own temperature anymore. Your internal temp spikes above 104°. Congratulations: You've got hyperthermia (heatstroke).

  You get confused, dizzy, and mumbly, with a splitting headache. Sometimes, you feel nauseated, your skin is flushed, you're breathing fast, and your heart goes nuts, beating fast to try to cool you down. Key symptom: Skin is *hot and dry*.

  You might not even know what's happening, because—unless you've been exercising—you're not sweating. But unless you get emergency treatment fast, heatstroke can do damage to your brain (because it swells in your skull), heart, kidneys, and muscles.

  Five hundred Americans die every year from heatstroke, and the numbers are going up (shocker). You're more likely to succumb if you're older, homeless, drunk; working out in the sun; unused to hot weather; already vulnerable (heart disease, lung disease, out of shape, obese); or on certain medicines like those for high blood pressure or ADHD, antidepressants, cocaine, or diuretics. If you're not among those categories, you should make it a priority to check in on people who *are* during a heat wave.

  If you suspect that someone is having heatstroke, and there's no prospect of an emergency room visit, treating them is one step: *Cool them down.* Get them into shade or air-conditioning; take off as many of their clothes as possible; spray them with water; and put cold, wet towels on their neck, armpits, and groin.

Heat Wave Terms

Of course, you've read Chapter 8. You've equipped yourself with a NOAA-alert radio or an emergency app, so you'll get heat wave warnings in plenty of time.

You should, however, know the difference between the three kinds of alerts you'll hear about:

- **Excessive heat watch** means that there *might* be extreme heat in the next day or two. Conditions are right.
- **Heat advisory** means that there *will* be extreme heat for the next couple of days, with heat index highs between 100 and 105°.
- **Excessive heat warning** is bad. It's going to be brutally hot, with a heat index over 105, for at least two days.

The Wet-Bulb Temperature

There's the temperature. There's the heat index temperature. And then there is, if you can believe it, yet another "feels like" heat statistic: the wet-bulb temperature.

You're entitled to wonder why we need two apparent-heat measurements. The answer goes like this: The *heat index* takes into account temperature and humidity, but it's measured in the shade.

The *wet-bulb temperature* factors in the temperature *in the sun* and humidity *and* wind, sun angle, clouds, and the amount of physical activity you're doing. When it really matters—if you're an athlete or in the military, for example—wet-bulb measurements are far more important than heat index readings. (You may see wet-bulb readings referred to as WBGT. That's not a radio station; it stands for wet-bulb global temperature.)

Sports team guidelines might forbid practice times longer than two hours when the WGBT is above 87° and limit practice to *one* hour—wearing no protective gear—when it's above 90°.

And above 92° WGBT, nobody should be playing sports outside.

You can look up the current wet-bulb temp either online (www.weather.gov/tsa/wbgt, for example) or in an app like WeatherFX ($1).

The heat index, by the way, is not the same thing as temperature. Remember how humidity makes heat worse, because you can't cool off by sweating? The heat index is a "feels like" number that *incorporates* the humidity, much the way the "wind chill" is a "feels like" temperature for cold weather. At 65% humidity, for example, an air temperature of 96° *feels* like it's 121°. You might want to rethink your marathon.

(You can calculate the heat index yourself, using this simple formula, where T is the temperature and R is the relative humidity:

$$\text{Heat index} = -42.379 + 2.04901523T + 10.14333127R - 0.22475541TR - 6.83783 \times 10^{-3}T^2 - 5.481717 \times 10^{-2}R^2 + 1.22874 \times 10^{-3}T^2R + 8.5282 \times 10^{-4}TR^2 - 1.99 \times 10^{-6}T^2R^2$$

Fun for the whole family!)

The AC Problem

Hotter days and nights might cost you even if you don't get sick or die. During these periods, more people run their air conditioners longer.

The thing is, air conditioners are power beasts. They're the costliest items on your electric bill, accounting for half of it in hot weather. Americans now pay $27 billion for that cool air, and as the world heats up, those prices will climb. The amount of energy we use for cooling our buildings has more than tripled since 1986, and, worldwide, is expected to triple again by 2050. At that point, the world's air conditioners will use as much power as the entire nation of China does today.

Another problem: Air conditioners work by pumping *indoor* hot air out of the building. The net effect of those thousands of air conditioners is, quite literally, to heat up the outdoors by as much as 2 degrees. They contribute to the heat island effect, a measurable spike in temperature in densely populated areas (page 351).

Running the AC is expensive in another sense, too: It's a greenhouse-gas disaster.

First, air conditioners consume huge amounts of electricity, which is still produced largely by coal, gas, and oil-powered power plants.

Second, air conditioners contain refrigerant chemicals called hydrofluorocarbons (HFCs), which are much, much more dangerous greenhouse gases than carbon dioxide. They were, in fact, *1,300 times worse* than CO_2 during their first century in the atmosphere. Next to HFCs, methane looks like a warm-up act.

(Bitterly ironic historical fact: Our AC units *used* to contain *chloro*fluorocarbons, or CFCs. As you may recall, it turned out that CFCs were burning a hole in the ozone layer, a thin layer of atmospheric gas that's our only protection against becoming a species of skin cancer victims from the sun's ultraviolet radiation. In 1987, incredibly, the nations of the world agreed to ban CFCs. We began using HFCs instead, little realizing that they contribute to a *different* pathway to self-extinction. There's an international treaty to phase out HFCs, too—but guess who hasn't signed it? China, India, and the United States.)

HFCs wouldn't be much of a problem if they stayed *in* the air conditioners. Unfortunately, they leak at every stage of their lives: when they're manufactured, when they're installed, and especially when they're thrown away as trash.

Tossing an air conditioner into the landfill is illegal in the United States. What you're *supposed* to do is dispose of your air conditioner "properly." In many cities, the power companies host turn-in days, where they collect old window air conditioners and recycle them safely. Some even pay you, in the form of a discount on your power bill. To look up which incentives are available in your town, visit the government's Energy Star Incentives Database at www.energystar.gov/dime, select your state, and choose "Room Air Conditioning (recycling)."

If you have no luck there, ask your city or town waste department what options they offer for collecting and safely draining your AC unit. And if you strike out on *that* avenue, you can hire an HVAC company to drain the HFCs before you put the thing out on the curb. Just make sure it's equipped to capture the HFC gases, and maybe even supervise the process; shady companies have been known to just release them into the air.

The Urban Heat Island Effect

If you were a satellite 12,500 miles above the earth, with magical eyes sensitive enough to detect temperature variations below, you'd see that *cities* are much hotter than the suburbs and countryside around them.

That's because a city's made of concrete, pavement, and dark roofs. On sunny summer days, those surfaces can get 90 degrees hotter than shady surfaces, which stay about the same temperature as the air.

There's not much shade in the city, either, since pavement has replaced whatever trees were in the area. City buildings also block the wind, meaning that the hot air can stagnate, trapped, baking everyone who's outdoors.

Sunshine isn't the only source of the heat, either. Cars, air conditioners, and factories are all common in cities, and they all pump heat into the air.

Add it all up, and you've got cities that are 4 or 5 degrees hotter than the surrounding countryside during the day, and as much as 22 degrees

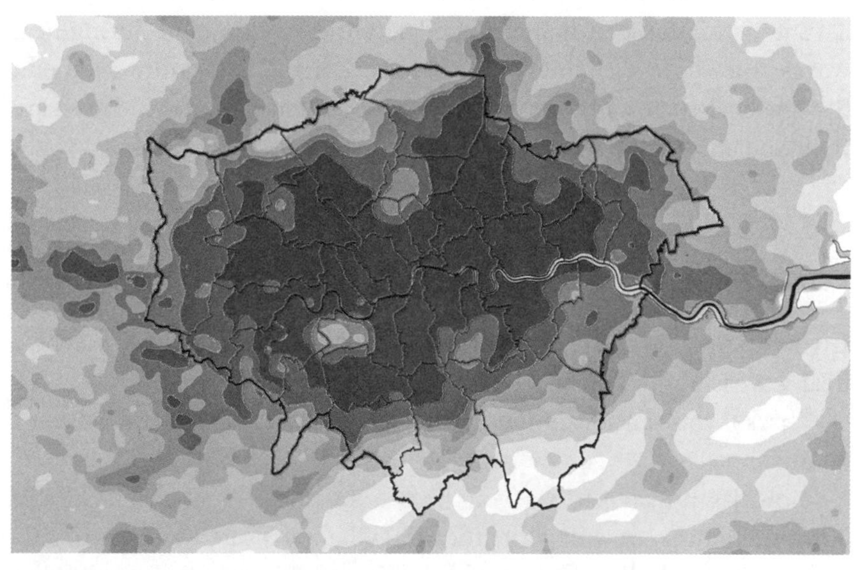

Figure 10-6. This is London, as viewed from space with an infrared camera. The deepest blue spots are about 5 degrees warmer than the lightest ones, even though the weather is identical. That's the urban heat island effect.

hotter at night. Typical temperature readings in Phoenix, for example, show that when grassy areas have cooled to 88° at night, concrete surfaces nearby are still at 113°.

Considering that 80% of Americans live in cities, you could call that a problem.

Like so many other aspects of climate change, the heat island effect winds up having a disproportionate effect on low-income areas and communities of color. "I use geographic information systems, GIS, a lot," says Jalonne White-Newsome, an expert on poverty and climate change who manages climate-resilience grants at the Kresge Foundation. "If you overlay maps of urban heat islands on areas of lower-income households, it's typically the same places."

That's because, she says, poor communities have a smaller tax base, less green space to mitigate the heat, and fewer resources to adapt to these extremes. They're also often in industrial and manufacturing districts, where pollution makes the problem worse.

"So when you talk about the impacts of climate change, you have to look at it with a racial-equity analysis," she says. "That is what it boils down to: Certain folks get hit first and worst, and recover last."

She notes, too, that homeless people wind up in a bind during heat waves: They have no homes, and yet they're made to feel unwelcome in cooling shelters, or are kept out because they have animals. "Those are the populations that people forget about," she says.

Keeping Cool at Home

If you're inside, with the air conditioner humming away, then maybe a heat wave is nothing to fear.

That's *if* you can stay home, *if* you have AC, and *if* the power doesn't go out.

Now, it's possible to rig your house so that it stays cool without air-conditioning, or with very little of it. Chapter 3 offers a range of AC-free cooling tricks like white roofs, plantings, ceiling fans, awnings, open layouts, and heat pumps.

You can prepare for the next summer's heat spikes by checking over your home's air-conditioning, fans, dehumidifiers, and insulation, to make sure they're up-to-date and good to go. If your window air conditioner is old, it's inefficient and expensive to run; if you have any inclination at all to replace it, do it now, before you need it. When an actual heat wave strikes, supplies in stores will be low and prices will skyrocket.

Find out where your neighborhood cooling centers are (search Google for *cincinnati cooling centers*, for example). These are public, air-conditioned spaces like libraries, shelters, churches and temples, community centers, movie theaters, and malls. In many cities, you can get free public transportation to these centers during a heat wave, too. Ask in advance.

But what about when the heat wave itself is upon you? Here's what to expect, and how not to lose your cool when it happens.

Power Outages

The power often goes out during heat waves, right when we need it the most, simply because so many people crank up their air conditioners simultaneously. All those air conditioners put an increased strain on our power plants. Sooner or later, you've got electricity plants pushed beyond their limits.

In most states, when the primary power plants get overloaded that way, the utility fires up "peaker plants," which are small, dirty, low-efficiency power plants that sit around waiting for those overflow situations.

And if the *peaker* plants get overloaded—well, then it's lights out. The utility cuts off people's power to save the plants. They may, for example, institute rolling blackouts, where neighborhoods get to take turns being hot and dark.

Unfortunately, the problem is only getting started. "Based on a 6.3 to 9° temperature increase," says the EPA, "climate change could increase the need for additional generating capacity by roughly 10–20% by 2050. This would cost hundreds of billions of dollars." Translation: more blackouts, higher taxes, or both.

Believe it or not, heat waves in some cities are such a new phenomenon that most homes don't even have air conditioners. When heat waves blasted San Francisco in 2017 (109°), 2019 (105°), and 2020 (106°), residents bolted for appliance stores, only to find all the window air conditioners long since sold out. They huddled in shopping malls and movie theaters or sweated it out with fans at home.

Power outages can be especially devastating to older, sick, and disabled people. If you're in a wheelchair, or you live on the 18th floor of a high-rise apartment building and you have mobility challenges, what are your options if the power goes out and the elevator stops working? You may not be able to leave your apartment.

In any case, you get the point: You should know how to cool yourself even without power. And even if you have AC, you should minimize its use, both to save money and to help prevent a blackout in your region, which will *really* make you feel miserable.

Eliminate Heat Sources

Once the heat wave is upon you, you can take some defensive steps to keep the heat out—handy when you don't have air-conditioning or would like to run it less.

First, eliminate as many heat sources as possible. They include:

- **The sun.** You know what else experiences the greenhouse effect? Your house. About a third of all the summer heat in your home comes from the sun's radiation blasting heat in and getting trapped inside.

  Close the shades or curtains, especially on windows that face south and west. Instantly, you've created a midday temperature drop of an almost 20-degree difference.

  Why south and west? Figure 10-7 reveals the bizarre truth: In the summer, the sun doesn't rise in the east and set in the west, at least not exactly. It rises and sets a little bit *north* of east and west, thanks to the earth's tilt. And then it swings toward the south during the hottest part of the day.

  If you don't have window coverings, get some. Or hang some kind of improvised white drapes, or even fill the windows with reflectors made of foil-covered cardboard. You want to bounce the worst of the sun right back out again.

- **The air outside.** Every time you open a door or a window, you're letting the heat into your inner sanctum. Keep them closed.

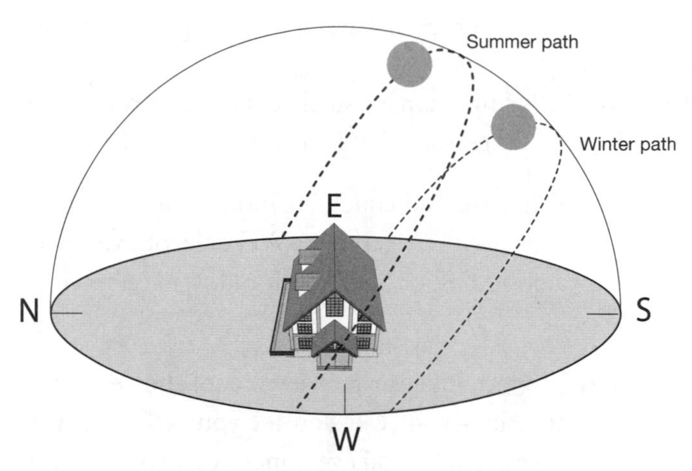

Figure 10-7. In the Northern Hemisphere, the sun doesn't rise exactly in the east and set exactly in the west—except on the equinoxes twice a year.

- **Appliances.** Incandescent light bulbs are heat machines. Keep them turned off. (LED bulbs are fine; they don't produce heat.) Don't run the washer, dryer, or dishwasher until it gets cool at night; they pump heat into your house.

  Same with the oven and the stove: Why would you deliberately make your home hotter? It's sandwiches or salad night, kids! The internet is full of great no-cook, hot-weather recipes.

  And if people start getting sick of variations on salads, sandwiches, and cold soups, you can compromise by using a slow cooker, which radiates far less heat to your living space than ovens and stoves do. It keeps its heat to itself.

Keeping Yourself Cool

With or without air-conditioning, there's no need to suffer.

- **Let the air flow.** Open all the doors of all the rooms inside your home. More air flowing means a cooler interior.
- **Designate a "cool room."** Dirt is a fantastic insulator. That's why a typical basement stays at around 55° all year long, no matter how hot

or cold it is above the ground. A basement is a great place to hang out when it's baking outside.

Also, a north-facing room is usually cooler than rooms whose windows face south, for reasons that Figure 10-7 should make clear.

- **Cool packs.** Mattresses and couch cushions tend to trap body heat. You can fight back with refrigerated cold packs or even plastic water bottles, hot-water bottles, or buckwheat pillows you've chilled in the freezer.

- **Cool shower.** A short, lukewarm or even cool shower can drop your body temperature in seconds. If you let yourself air-dry, including your hair, you'll keep cooling off even once you're out. There's nothing to stop you from taking a few of these showers a day.

- **Cool sheets.** Chill your bedsheets in the freezer. They'll feel amazing.

- **Cool neck.** If you apply your antiperspirant to the back of your neck, the skin there won't get all sweaty and gross, and your hair won't stick there to trap heat. It's perfectly okay to take this step in the privacy of a closed bathroom, so people don't think you've lost your mind.

- **Fans.** Fans are amazing. They use only a tiny fraction of the power required for air conditioners, and pass the savings on to you.

  A ceiling fan, for example, costs about a penny an hour to run. Compare that with 14 cents an hour for a window air conditioner, or 36 cents for central air. That's 30 watts of power for the ceiling fan, versus 1,200 watts for a window AC unit or 3,500 watts for central air.

  Desk fans and standing fans use even less power, and cost even less to run.

  Even in the absence of air-conditioning, a fan makes you feel cooler by whisking away your perspiration, just the way nature intended, and also by pushing your own body heat away from you. It can easily feel 10 degrees cooler next to a fan.

  Not good enough? Then fill a shallow pan with ice water and set up the fan to blow across it. Instant MacGyver air conditioner!

  But fans are also fantastic if you *do* have AC. With a fan on, you can raise your air-conditioning thermostat by 10 or 12 degrees and feel exactly as cool.

Here's the catch, however: *Feeling* cooler is not the same as *being* cooler. When the temperature indoors is above 95°, a fan can't prevent you from getting heat illnesses.

- **Use the ventilation system's fan.** Every air conditioner has a Fan On setting, which blows air around without actually cooling it. You'll find this control on your window AC unit or, if you have central air, on the thermostat. Running the fan on the On setting (meaning, "blow all the time") instead of Auto ("blow only when the AC is running") costs far less than actually running the AC. But the moving air feels cooler to you, and the cooler air from the lower floors can now neutralize some of the stifling air upstairs.
- **Swamp coolers.** In dry areas, like the Southwest of the United States, you can save a lot of money by using a machine called an evaporative cooler—or, more entertainingly, a swamp cooler.

Figure 10-8. An evaporative cooler, or swamp cooler, is like an inside-out dehumidifier. Instead of converting water vapor to liquid (and creating heat in the process), it converts water to vapor (creating cooling).

Some homes have big, whole-house units the size of air conditioners, but you can also buy a personal-sized, $25 swamp cooler about the size of a toaster.

You plug it in and fill it with water. A fan blows air across wet pads; as the water evaporates, it gets cooler.

In other words, this device is a mechanized version of sweat evaporating off your skin. What you feel near a swamp cooler is moving air that's been cooled by water's transformation into vapor.

These coolers are far less expensive to buy—and to run—than air-conditioners. They work great in places like Nevada, Utah, and Arizona, where the air could, frankly, use a little humidifying.

Alas, these machines don't produce any cooling in humid places, for exactly the same reason that *you sweating* doesn't produce any cooling in humid places: The water simply doesn't want to evaporate.

Finally, for heaven's sake, adjust your schedule. The hottest part of the day is roughly from 10 a.m. to 5 p.m. Anything active or hot that you can shift *outside* of those hours, you should shift. "Just changing when you exercise, when you cook, when you do laundry—it's all these little things that can reduce your vulnerability during heat events," says Jalonne White-Newsome.

Sleeping in the Heat

Trying to sleep at *night* during a heat wave without air-conditioning can be truly miserable. And then the next day, enduring the heat wave is twice as hard, because you're tired and cranky.

Fortunately, the worlds of science and commerce are always available to help:

- **Sleep low.** Heat rises. The hottest parts of your house are the highest. If you're having trouble sleeping because it's too hot, drag a mattress to the basement or whatever lower floor you've got. Lower is always cooler.

- **Fix your sheets.** Cotton, percale, and linen are great materials for sheets, because they're breathable. But you can also buy special hybrid-fabric sheets that are expressly designed for cooling: They're extremely lightweight, breathable, and moisture wicking, and designed to absorb heat from your body. They're called things like Slumber Cloud Stratus, Comfort Spaces Coolmax, Sheex Micro-Balance, and Sleep Number True Temp.

- **Dampen your sheets.** If your sheets are cotton, try this classic trick: Get them wet, then wring them out until they're only damp. Their arrival on your body marks the official end of your "it's so hot I can't sleep" phase.
- **Do the ceiling fan thing.** A ceiling fan is a classic, age-old trick for preventing your bedroom from becoming sticky, hot, and stagnant. If you've got one, use it—but be sure that it's spinning counterclockwise. (The other way is for winter; it blows warm air *down* on you.) Prices start at about $150, installed.

Once the Air Cools Outside

When the sun finally sets, and the air outside becomes *cooler* than it is inside, get all the hot air out:

- **Open the windows.** If you have double-hung windows, raise the lower window and lower the top one. Now the hot air goes out the top opening, and cool air comes in the bottom one.

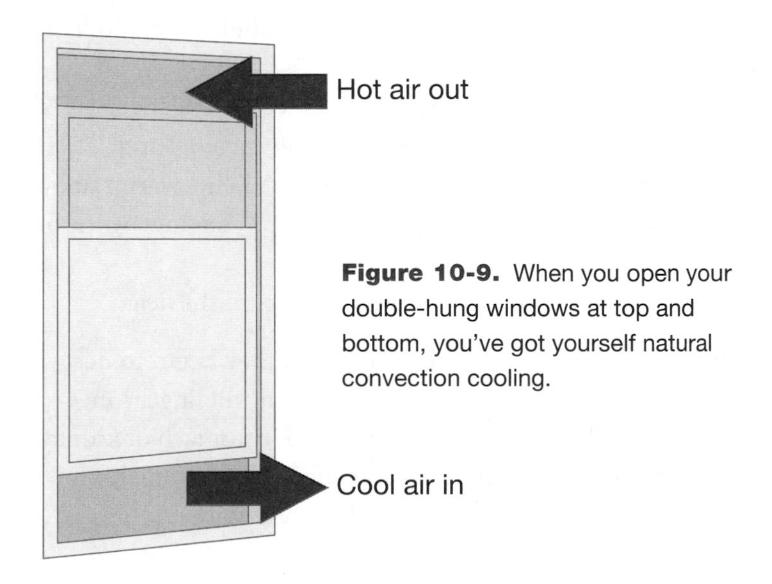

Figure 10-9. When you open your double-hung windows at top and bottom, you've got yourself natural convection cooling.

- **Set up a cross breeze.** That's where the outside air is encouraged to blow *through* your house, shoving out all that stagnant hot air. To do that, open windows on opposite sides of the house.

  If you don't have that kind of layout, but you do have a room with two windows in it, set up a fan to blow out one window; the resulting low pressure sucks cool air in the other one.

  And if you have a room with only one window and a door, well, open them both.

  You can create *vertical* cross breezes, too. Hot air rises, so big fans aimed out the windows on your upper floors do an amazing job of forcing hot air out of the home. Cooler air automatically gets sucked into the lower floors, where you're watching Netflix or playing board games.

- **Use exhaust fans.** Your bathroom probably has an exhaust fan in the ceiling, and your kitchen may have a hood or a vent over the range. They exist to blow out inside air, so why not? Turn them on. They can only help.

Medicine and Heat

Your medicines aren't little capsules of magic. They're complex chemical mixtures, and chemical reactions are incredibly susceptible to temperature.

The big drug companies generally guarantee the effectiveness of their drugs only when the storage temperature is between 68° and 77°. They're being conservative, but it is true that once a medicine warms above 86°, the molecules can actually change. A medicine may not work or may even make you sicker.

Here's what happens to some common drugs in the heat:

- **Antibiotics.** When these drugs get warm, they begin to decay. They lose their potency; as a result, your infection will linger longer.

  Warmed antibiotics can also damage your stomach or kidneys. For example, tetracycline, a broad-spectrum antibiotic, breaks down into toxic chemicals that can give you kidney failure.

- **Aspirin** breaks down into vinegar (acetic acid) and salicylic acid, which can upset your stomach—and do nothing for your headache.
- **Hydrocortisone** cream begins to separate, at which point it's useless.
- **Test strips**, like the ones in pregnancy tests and blood sugar tests, can give bad readings in high humidity.
- **Hormone medicines**, including thyroid meds and birth control pills, are usually based on proteins. And protein, as you know if you've ever made meat or eggs, cooks in the heat.

  Even if your thyroid medicine loses only a few percentage points of potency, you're still getting, in effect, a smaller dose. You may feel really terrible, and you won't be able to figure out why.
- **Lifesaving drugs**—insulin, heart meds, seizure drugs—break down in the heat and lose their effectiveness. Never let them get warm, even briefly.

Some meds increase your photosensitivity, too, meaning that you're more likely to get burns and rashes in the sun. They include antibiotics, antifungals, antihistamines, cholesterol drugs, diuretics, ibuprofen (like Advil), and oral contraceptives.

Sometimes, you'll *see* changes to your meds. The pills may be stuck together, gooey, or a different color, or they may have a smell when you open the bottle. Those are warning signs that say, "Replace us as soon as possible."

The more worrisome situation is when you *can't* detect any changes to your medicines. You have no idea that these chemical changes are taking place.

The only solution is to keep the meds cool and dry. Never leave them in a parked car, especially not in the trunk. (Don't check them in luggage when you fly, either, for a different reason: because the high-altitude air can reduce the medicines' potency from *cold.*)

If you must venture into the heat with insulin or antibiotics, buy cooling packs for them at the drugstore.

For any meds, if you're leaving a cool home for the day, take with you only what you'll need. Leave the original bottle at home where it's cooler.

Medicine and Your Senses

But medicine, heat, and people—especially older ones—interact in another dangerous way, too. Just ask Jalonne White-Newsome, who studies the intersection of climate change and low-income communities of color.

Whenever she paid summertime visits to her grandparents in Detroit, "It would be like a furnace," she says. "I'd be complaining: 'Why is it so hot in here? You don't have the AC on, you don't have any windows open!'"

What she didn't realize at the time is that many medications affect how people *perceive* temperature. "Their pills caused them to have chills, so they weren't feeling the heat stress that their body was actually realizing," she says. "They'd be sweating, but complaining about how cold they were!"

That effect is especially characteristic of beta-blockers, Parkinson's meds, antidepressants, and diuretics. As a result, White-Newsome says, there have been many cases where seniors were found dead in their homes—the air-conditioning turned off, windows closed, and no airflow. "They literally died in a heat box."

Sometimes, it's the illness itself that messes with your senses. Stroke, diabetes, Parkinson's disease, and Alzheimer's disease can all interfere with your sense of thirst. You might be desperately dehydrated, with a glass of water sitting right next to you, and have no inclination to reach for it.

Now that you know how bad heat is for medicines, you can now see why the bathroom medicine cabinet isn't actually a very good place to store medicine! You take showers in that room, making it humid.

Keeping the bottles on top of the fridge is also a bad idea, because a refrigerator works by pumping heat out of its interior. Your pills are bathed in heat all day long.

Some injectable medicines—usually very important ones, like insulin and antibiotics—have to be refrigerated all the time. Unfortunately, heat waves are frequently accompanied by power outages. That's all the more reason for you to know how long your fridge stays cold when the power's out (hint: four hours) and for you to consider a generator, as described in Chapter 2.

Outside in the Heat

Heat poses the greatest danger to people who have to be outdoors in it, like road workers, farmers, and utility repair technicians.

Heat is especially dangerous to anyone who's very old, very young, or has had brushes with heart disease. A significant fraction of heatstroke victims every year are homeless people, who have very few resources to protect themselves.

What's really sinister is that hot weather can work on you slowly and invisibly. You may think you're just feeling cruddy today, but you're actually on your way to heatstroke. Tune in to how you're feeling when it's hot out.

Your best bet is not to exert yourself outside, if you can help it. If you want to work out, for example, visit a gym or swim.

Or if your home isn't cool, spend the hottest part of the day somewhere that has air-conditioning, like a mall, a movie, a senior center, a library, or a friend's house. Or, for the price of a single fare, you can even ride the bus for a few hours, settling back with a good book or YouTube on your phone.

Working Outside

But what if you must be outside in the heat? What if it's your job?

OSHA, the Occupational Safety and Health Administration, has some stern words for your employer, which you're welcome to pass along:

- Schedule frequent rest breaks in cool or shady places, especially for people who have to wear protective gear or focus on operating dangerous equipment. If it's over 95°, OSHA requires at least a ten-minute rest break every two hours.

- Use a buddy system, so workers can watch each other for signs of heat illness.

  During the hottest parts of the day, switch to slower-paced, less-demanding work. Save the heavy lifting for early morning or evening, when it's cooler. If necessary, split your shift in halves so that you work before and after the hottest hours.

- Rotate employees through less physically demanding jobs.
- Bring on more workers, so each individual employee is exposed to fewer hours in the heat.
- If it's bad, cut the shifts short or stop work altogether.

Drinking More Water

When you get hot, your body sweats, which means that it squirts water out your pores. The more you sweat, the more water you lose. That's why people get dehydrated so quickly on hot days and why, when you're out in the heat, you have to drink a *lot* more than you usually do, just to stay normally hydrated.

The golden rule: Drink enough so that your pee comes out clear. The oranger it gets, the more dehydrated you are.

The silver rule: If you're thirsty, you're already dehydrated.

Even being slightly dehydrated can bring you headaches, irritability, weakness at sports and work, and the inability to concentrate. Dry eyes and blurry vision aren't uncommon. If you're very young or very old, dehydration can send you to the hospital.

And if you're dehydrated often, you can get urinary tract infections, kidney failure, and kidney stones.

Of course, sweat isn't *just* water. It also contains electrolytes—minerals like sodium, potassium, calcium, and magnesium—that your muscle and nerve cells need for everyday operation. Most of it is sodium, which is why sweat tastes salty.

Lose enough of your electrolytes, and you get cramping, dizziness, and headaches.

There's even such a thing as water poisoning (also called water intoxication or water toxemia). That's when you drink so much water that your electrolytes get out of whack. Water intoxication has killed people, but only in two situations: water-drinking contests (yes, that's a thing), or being outside in baking hot weather when you're exerting yourself all day.

Climate Change and Kidney Stones

It's strange but true: Kidney stones are a side effect of climate change that nobody talks about, maybe because they're terrified.

A kidney stone is an ugly, jagged little crystal, made of salts and minerals, that grows inside your kidneys. If it's big enough to get stuck in the tube (the ureter) that connects to your bladder, the pain is beyond description. You'll remember it for the rest of your life.

About 19% of all men, and 9% of all women, will one day get a kidney stone. That's half a million emergency room visits every year—and climbing.

There's a direct correlation between climate change and kidney stones. When the temperature goes up, so does the number of kidney stone patients in the emergency room. That's because heat leads to dehydration, and dehydration leads to kidney stones. More people get stones in the summer and in hot climates. Kidney stones became a shockingly common problem among American soldiers in Iraq and Afghanistan—because of heat and dehydration.

In our new, hotter world, more people are dehydrated—and more people get kidney stones.

You do not want a kidney stone. Drink a lot of water. Aim for a glass every hour. (That's *water*. Caffeine and alcohol contribute to dehydration.)

No medicine can prevent kidney stones (at least not the usual calcium-oxalate type), but there is *one* thing you can do: *Drink lemon water.* A lot of it. Every day. Carry around a water bottle with you and drink all day long.

The citrate in the lemon juice changes the chemistry of your kidney and makes it hard for stone crystals to develop. If you keep up with your lemon juice drinking, you can cut your chances of getting a kidney stone by *half.*

Store this concern away in your emergency memory: If you find yourself drinking a lot of water in hot weather, find something to replenish your electrolytes.

Athletes and runners guzzle sports drinks like Gatorade, which contain precisely those minerals. You can also make up your own facsimile by dumping half a teaspoon of salt and eight teaspoons of sugar into a quart of water.

Eating a salty snack and washing it down with water is a worthy second choice tactic.

The Heat Wave Survival Guide

No matter what you're doing outside, this classic behavioral safety guide applies:

- **Take breaks.** Heat + humidity + exertion = heat illness. You have control over one of those factors. Take frequent breaks.

- **Eat for the heat.** Digesting food heats up your body like a furnace. To avoid feeling that extra metabolic heat, eat smaller, more frequent meals.

  What you eat can help replenish your fluids, too. Juicy stuff like strawberries, melon, cold soup, Popsicles, Jell-O, diced fruit cups, applesauce, and Italian ices are all on the menu.

- **Wear the right clothes.** Dark colors absorb heat and make you hotter. White and light-colored clothes *reflect* heat and make you cooler. And this part you could probably figure out on your own: Lightweight clothes are cooler, because they let your sweat reach the air and vice versa.

- **Wear sunscreen.** Sunscreen's a great idea for lots of solid reasons, including not getting skin cancer. But sunburn also impairs perspiration functioning—so the more sunburned you are, the hotter you'll feel.

None of this is necessary if your day consists of moving from your air-conditioned house to your air-conditioned car. This advice is for people who exert themselves outside for an hour or more.

Kids and Heat

Children are especially vulnerable to heat waves, for all kinds of reasons. First, kids run around a lot. Second, they produce more heat than adults. Third, their little bodies can't get *rid* of that heat as well, because they don't sweat as much.

And fourth, they're less likely to speak up when they start feeling the effects of heat. They might keep right on being active when they should be resting.

The Hot Car Problem

For years, safety experts cited 40 as the average number of American kids who die in hot cars each year. But as the planet warms, that number is going up; more than 50 children died in hot cars in 2018 and 2019. Most of them were under two years old (the kids, not the cars).

Half of these children are left in the car accidentally; 20% are left there on purpose, in the "I just have to run into the store" scenario. And 25% of the time, the children crawl into cars by themselves.

Trouble is, car interiors get hot incredibly fast—20 degrees every ten minutes—thanks to, of all things, the greenhouse effect. The sun's visible light passes easily through the windows, where it's absorbed by the car's seats, carpet, and dashboard. They store the heat and then radiate it as long wavelengths—infrared heat—that *doesn't* pass well through glass. So the heat is trapped, and the car gets *really* hot. It can easily reach over 200° if the car's interior is dark colored.

Sadly, once a kid's internal temperature reaches 107°, that's the end.

Here's an idea: Whenever you put your kid in the car seat, put your wallet, purse, or phone back there, too. You're a lot less likely to stride away and forget that you've left something valuable behind.

Finally, over half of America's children already don't drink enough water. A quarter of them don't drink any water *at all.*

When the going gets hot, therefore, you have a job to do, parents.

- Make sure your kids are drinking enough.
- Steer them toward wetter forms of fun, like running through sprinklers or swimming. When they're not in the water, they should be in the shade whenever possible.

  Related safety note: About 350 kids under five drown every year in swimming pools. *Watch them.* If you have to leave the pool, put somebody else on kid-watching duty.

  Slather the kids with sunscreen. Slather them again every two hours or whenever they come out of the water. Yes, even if it's cloudy.
- If it's over 90° and humid, don't let your kids play outside for more than 30 minutes at a time.

Pets and Heat

Dogs and cats, too, suffer in the heat. They have fur coats that they can't remove, and they don't sweat; their sole body-cooling mechanism is panting, which exchanges hot air from their lungs with cooler air outside.

That information explains three quirks of dogs and cats. First, fans don't cool them off very well. Second, humidity is brutal on pets; their heat-exchange system doesn't work. Third, putting a muzzle on your dog in hot weather interferes with her ability to pant.

The telltale signs of heatstroke in a pet are heavy panting, intense thirst, restlessness or clumsiness, thick saliva, lethargy, lack of appetite, a dark tongue, fast heartbeat, throwing up, and bloody diarrhea.

The treatment: Cool the animal down as soon as possible. All the same techniques that work with people also work with dogs: air-conditioning; spraying with water; lowering into cool water (but not ice water, which would cool the animal *too* fast); providing water to drink (but not forcing him to drink); wet towels on the stomach, chest, groin, and paws. And then get the dog to a vet as fast as possible; that's the only way to make sure there aren't invisible problems like shock, kidney failure, organ damage, and clotting.

As with humans, being old, sick, or overweight makes a pet more susceptible. Animals with flat faces—bulldogs, pugs, Persian cats—get overheated more easily than other breeds, because it's harder for them to pant.

You even have to rethink taking your dog for a walk. Paved surfaces, like roads and sidewalks, absorb and store heat; when the air is 87°, the sidewalk can get up to 140. That's enough to burn and scar a dog's foot pads—and meanwhile, her body is low to the ground, in perfect position to bake in the radiated heat of the pavement and increase the risk of heatstroke. If you must walk the dog on pavement during the hot part of the day, test it with your hand first.

Otherwise, do your walking early and late in the day, keep it short, and don't take your dog on runs. And:

- Provide plentiful water, indoors and out. Put out *two* bowls, in case one gets spilled. Some pets like ice cubes in the water, too; give it a try.

- If you're outside, find shade for the dog.

- A doghouse can be hotter inside than out.

- Take your dog swimming, if he's into that sort of thing.

- Beware windows that don't have screens. Pets can fall out of them.

- A haircut can keep your dog cooler, but don't *shave* your dog down to the skin; the last inch of hair protects him from sunburn. (Don't cut or shave cats at all. They're more efficient than dogs at regulating their own temperature, and their fur is part of the system.)

- Brushing a dog's or cat's fur helps in two ways. First, it removes dead hair from the undercoat (a lower layer of fur), so that air can help cool the skin. Second, it prevents that fur from getting matted and damp, which can lead to skin disease.

Hundreds of dogs die in hot, closed cars every year, too. Animal advocates stress that it's *never* okay to leave an animal alone in a car, even if you take steps like:

- **Running the air conditioner.** A frantic dog can bump the controls, turning off the AC or even switching it to heat.
- **Leaving the window cracked** just isn't enough to cool the car. A car with windows opened 1.5 inches is only 2 or 3 degrees cooler inside than a car with closed windows.
- **Returning to the car quickly.** That'd be fine if you really did it. But it's too easy to encounter a long checkout line, a friend who wants to chat, an item you've forgotten, and so on—and suddenly you've been away from the car much longer than you planned.
- **Parking only on cooler days.** Truth is, if the temperature is in the 70s, a car interior can still heat up to 105 in 20 minutes. Depending on your pet's breed, weight, and health, even weather in the *60s* is too dangerous for car-leaving.
- **Leaving water.** That helps. But dogs don't release heat by sweating; they release heat exclusively by panting. If the car is hot and stuffy, the dog can still overheat.

If you *see* a dog alone in a car on a hot day, let the police or humane society know; they'll want to know the license number and car description. If there are businesses nearby, ask if they'll make an announcement to find the car's owner. (Plenty of people don't realize the danger of leaving animals in cars.)

Now, the "no dogs alone in cars, ever" rule isn't without controversy, even among true dog lovers. Many dog owners are careful. They park in the shade, leave the AC running (in cars where the dog can't reach the controls), and return quickly and responsibly to the car. They insist that the "no dogs in parked cars" movement is overblown and overreaching. They note that leaving the dog in a shaded, air-conditioned car is better for her welfare than leaving her alone at home or tying her outside the store, where she might get stolen or attacked.

Trouble is, leaving a dog in a parked car can be a disaster for you for

Figure 10-10. On a Tesla, you're far less likely to get your window smashed by a well-meaning dog lover, thanks to Dog Mode.

entirely different reasons: In most U.S. jurisdictions, you can be arrested and fined for cruelty to animals. In 11 states and a growing number of cities, it's legal for passersby to smash your window to rescue a child or an animal left inside.

Tesla's cars have a novel solution called Dog Mode. When you turn it on as you leave the car, the car maintains whatever temperature setting you've specified—70°, for example—and its dashboard screen says, in gigantic type, "My owner will be back soon. Don't worry! The AC is on and it's 70°F."

It's a safe bet that fewer people will alert authorities or smash your windows if they realize that the dog is comfortable and you're a caring owner.

You may not own a Tesla, but maybe Dog Mode is the solution to the problem. There's nothing to stop you from making a sign of your own to put in your car when you have to leave your dog alone inside—briefly, responsibly, with water at hand and air-conditioning on.

Chapter 11

Preparing for Drought

AS NATURAL DISASTERS GO, DROUGHT IS BORING.

It's what scientists call a *creeping* natural hazard. You can pinpoint the day a hurricane or a wildfire hits, but droughts can take months to start and end. No app is going to give you a notification that says "Drought alert 4:34 p.m. tomorrow."

Drought isn't very dramatic, either. Photos of brown, motionless fields can't compete with terrifying shots of hurricanes, floods, and wildfires.

Droughts aren't even very measurable. For example, how do you measure the *costs* of a drought? It's not easy, because droughts create massive secondary effects like these:

- **Restricted food supply.** You can't grow food in a drought, either for us or for our meat animals. In rich countries, drought makes food prices go up. In poor countries, it can mean famine and mass migration.

- **More wildfires.** Dried-up vegetation makes an all-you-can-eat buffet for wildfires. It's no coincidence that California's most devastating wildfires all erupted at the end of a six-year drought, or that Australia's nightmarish 2019–2020 fires followed one of the worst droughts in the continent's history.

- **Blockaded cargo.** Much of our food, fuel, and products moves around on river barges. One key reason: River barge shipping is 95% cheaper than trucks and 54% cheaper than trains.

  But the rivers have to be at least nine feet deep, or the barges can't get through. During the 2013 drought, the Mississippi River's water

Figure 11-1. Historic low water levels in the Mississippi in 2012 required barges to park off to one side. The low water level revealed sandbars on the opposite side.

level became so low that barge traffic nearly ground to a muddy halt. "I have been in this industry for 48 years, and I have not seen it this bad," a shipping executive told the *New York Times*.

Fortunately, the U.S. Army Corps of Engineers stepped in with an expensive list of emergency procedures that allowed the barges to squeak through. The protocol included blasting and dredging the river to deepen it, releasing water from reservoirs, and insisting on lighter cargo loads.

- **Unemployment, crime, disruption.** In areas where agriculture is the primary occupation, drought means farmers losing their jobs, which begins a domino effect of social problems. "Unemployment translates into higher crime, education issues, pressure on children, malnourishment, and all kinds of things," says Amir AghaKouchak, professor of hydrology and climatology at University of California at Irvine.

- **Civil unrest.** Wars over water in hot countries go back millennia—and more drought means more drought-triggered conflict.

  One of many factors that led to bloody civil war in Syria, for example, was the brutal 2006–2009 drought—the worst ever measured there. Over 1.5 million people, desperate for food, water, and a livelihood, fled the rural regions and headed to the cities. The resulting social stresses led to an uprising against Syrian president Bashar al-Assad, and the rest is civil war history.

But even if they're nebulous and not very flashy, droughts are a natural disaster all the same, and among the most expensive. Between 2000 and 2017, the United States experienced 14 major droughts, costing over $120 billion.

How Drought Comes About

Some drought is natural. "Drought happens. It's a natural part of nearly every ecosystem on the planet," points out Kelly Smith, assistant director of the National Drought Mitigation Center.

But an increasing number of droughts are linked directly to climate-change phenomena. Warmer temperatures reduce water levels in rivers; less snowpack means there's less melting water.

Climate change can lead to more drought even if the rainfall amounts don't change. "Consider a place that has a certain risk of drought naturally," says Ben Strauss, CEO of Climate Central. "Every so often, you'll get three years without much rain. If that happens in today's world, it's *hotter* during those three years, so you get more evaporation, and the drought is worse. It develops more quickly, goes deeper. You don't even need to change precipitation patterns to have a worse drought."

You might suppose that heat leads to drought, because heat makes more water evaporate. True. But Amir AghaKouchak points out that, weirdly enough, *drought* also causes *heat.*

To understand why, remember the scientific principle that *evaporation cools a surface.* That's why perspiring cools you off. The same effect applies when moisture evaporates from soil on the earth's surface: The soil cools down.

In a drought, though, there's not as much water to evaporate, so the ground doesn't cool and the temperature rises. "It's called land/atmosphere interaction, and it intensifies warming of drought," says AghaKouchak. "A drought today is generally warmer than droughts 50 years ago." Twice as warm, in fact.

Since 2000, about 20% of the United States has been in drought at any given moment. But recent droughts are far worse, affecting greater areas and lasting longer.

In 2005, for example, the worst drought in 20 years struck the Midwest. And then came 2012, when a massive drought covered 81% of the United States, costing the Midwest $35 billion; it was the worst American drought since the 1950s. More than half the counties in America were declared federal disaster areas. "There is virtually no other explanation other than climate change," wrote NASA scientist James Hansen.

Finally, there was California's monster six-year drought (2011–2017), the worst in that region in *1,200 years*. It killed more than 100 million trees, devastated the salmon population (because they couldn't navigate the depleted rivers to spawn), and resulted in mandatory water restrictions for the first time in the state's history. The governor required each California city to cut its water use by 25%, which translated into citizen restrictions on filling pools, watering lawns, washing cars, taking showers, and even serving water in restaurants.

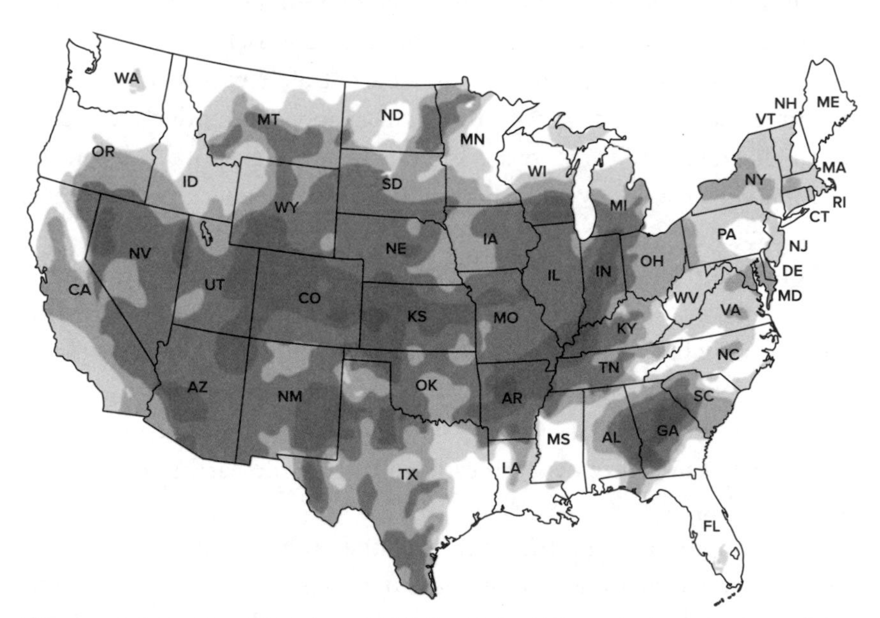

Figure 11-2. The 2012 drought hit the United States in the worst possible places: the central states, where much of the nation's food grows. (Deeper blue areas show more intense drought.)

The Argument for Saving Water

Drought preparation guidance is surprisingly hard to come by. The reason is fairly simple: Planning for drought isn't considered a project for mere mortals like you. It's the job of public health experts; industrial reps; farmers; governments; and people who run municipal water systems, factories, and commercial buildings. They, after all, are the ones who monitor aquifer and reservoir levels, who have the funds to implement substantial changes, and whose job it is to make sure that everyone has the water they need.

If the public officials say anything at all about steps *you* can take to prepare, it's usually all about *saving water*. Most of it's pretty obvious:

- Never let water in your house just *run*. In the shower, turn off the water while you're soaping up. Don't run the water while you're brushing your teeth or shaving, either. Don't run the faucet, waiting for the water to get cold; instead, keep a pitcher of water in the fridge.
- Don't flush the toilet after every pee. Don't flush away materials like Kleenex, baby wipes, food scraps, and squashed bugs.

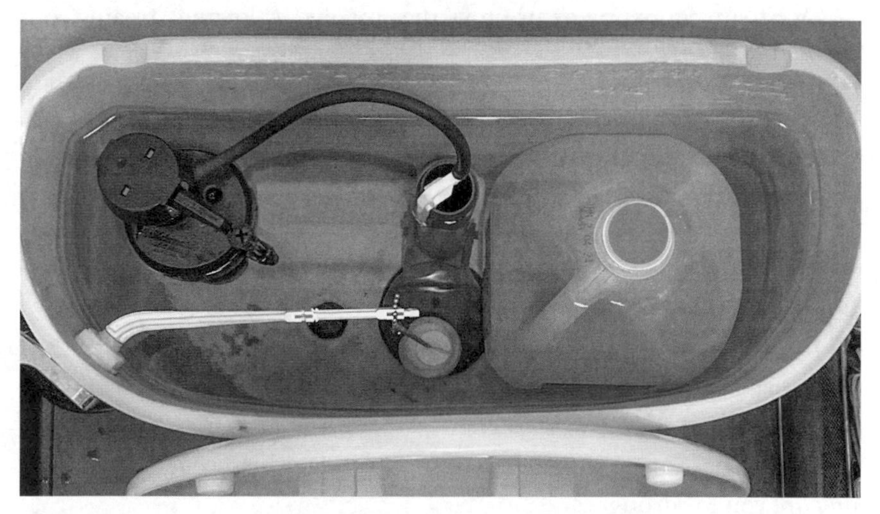

Figure 11-3. A gallon jug saves you a gallon of water per flush in older toilets.

- ♦ Repair dripping faucets; check pipes for leaks. Install low-flow faucets, showerheads, and toilets. If you have an old toilet, put a gallon milk jug, filled with water, into the tank so that each flush uses less water.
- ♦ Limit watering your lawn. You're probably watering too much anyway; one inch of water a week is usually plenty, and after a heavy rain, you can skip watering for two weeks. If you must water, do it early in the morning or late at night, so the water doesn't just evaporate away.
- ♦ Set the mower so it cuts the grass higher.
- ♦ Better yet, replace your *grass* lawn with one of the no-maintenance yard plantings described on page 160. And use mulch around trees, shrubs, and gardens to keep moisture in the ground.
- ♦ Run the dishwasher or washing machine only when it's full. Don't rinse plates before you put them in the dishwasher. Use the light cycle on the dishwasher.
- ♦ Avoid pouring any water down the drain. Capture it in a bucket and save it for watering your garden or indoor plants, washing your car, cleaning tools or your grill, and so on. That includes shower water and faucet water while you're waiting for it to get hot.
- ♦ Don't run fountains or use toys that involve a flowing stream from the hose.

Those are all excellent suggestions. If more people followed them, then less belt-tightening would be required, region wide, when a drought comes.

But keep them in perspective: 70% of all the water in the world is used for agriculture. You won't make much of a dent by taking shorter showers.

In other words, those aren't what you'd call *preparation* steps. They don't make life easier for you during a drought; they actually make it harder.

So the question is: How does conserving water help *you*? How does it *prepare* you for drought?

Part of the answer is staring you in the face from California. Following its devastating drought, the state passed a new water-restriction law in 2019 that limits each resident to 55 gallons of water a day. In 2030, that daily limit will drop to 50 gallons.

Fifty gallons might seem generous—just how thirsty can a person be?—but the average American uses *88* gallons a day. That's not just drinking water, of course. It's the water we use in washing machines (30 to 40 gallons per cycle), showers (17 gallons for eight minutes), toilet flushes (about 1.5 gallons each), and lawns (280 gallons a week for a 550-square-foot lawn).

Our cities spend billions of dollars a year to produce clean, chlorinated, fluoridated, drinkable water. In a perfect world, we wouldn't use that delicious, pristine water for flushing toilets and watering lawns. That seems like a massive waste of water treatment and transport.

But that's exactly what we do. In fact, we take water for granted. Over the decades, we've cultivated a sense that water is essentially unlimited and free.

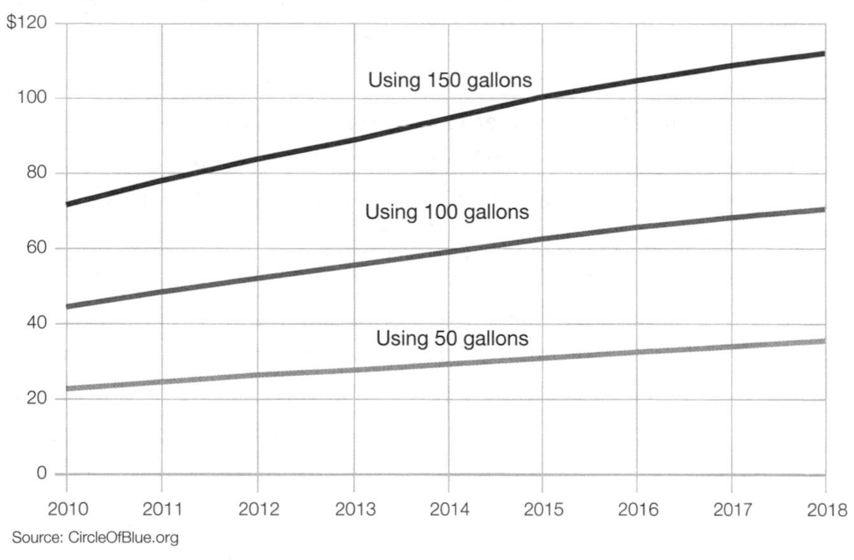

Figure 11-4. The price of water in the United States is rising steadily. Shown here: the price of water for a family of four using 150 gallons per person daily.

In fact, though, clean water is valuable and becoming more so every year, and water prices are finally beginning to show it: 2019 marked the eighth consecutive year of U.S. water-price increases; since 2012, the price of water in the United States has gone up 31%. Incredibly, as of 2017, one in ten American households could no longer *afford* their water bills—and that number will triple by 2022.

Those numbers put California's law into perspective. It's intended to introduce the notion that water can't be taken for granted, that it has value just like electricity or gasoline.

Idaho, Oregon, and Montana also implemented various forms of water restrictions. California-style limits on watering lawns, washing cars, and filling pools are also in place in Australia, India, and several Middle Eastern countries.

In Cape Town, South Africa, water shortages reached desperate levels in 2018. "No one should be showering more than twice a week now," said the premier of the province, where water was rationed to 11 gallons a day per person. "I regard oily hair in a drought to be as much of a status symbol as a dusty car."

Sooner or later, one of those two trends—more expensive water or laws that limit your daily use—will catch up with you.

Saving water, in other words, means saving money, more every year. And if your state imposes water restrictions, you'll be able to worry less about your total: If you're watering your lawn half as much, for example, you may not have to yell at other household members for spending too much time in the shower.

Eventually, most droughts end, but they also come back. It makes sense to begin treating water as something that's *always* valuable and costly, not just during droughts. Because, as John Steinbeck wrote in *East of Eden*: "And it never failed that during the dry years the people forgot about the rich years, and during the wet years they lost all memory of the dry years."

Your Own Water Supply

During a drought, you can expect the water supply to diminish in both quantity and quality. Your drinking water gets more contaminated as the plants that usually filter pollutants and sediment out of our rivers die, insects attack trees that once helped to filter our groundwater, and there's less dilution of agricultural fertilizer runoff.

But shortages of clean, drinkable water aren't unique to droughts. Hurricanes and floods routinely swamp local water-treatment plants, resulting in sewage in the drinking water (Chapter 9). Wildfires may shut down the treatment plants altogether and may lead to ash and other nastiness washing down burned-out slopes into the water supply.

But if you put together a backup water system as described in Chapter 3, you'll have something on hand to drink, wash, and flush with.

Preparing for Drought's Side Effects

Preparing for drought also means being ready for the new risks it brings to everyday life.

Power Outages

It doesn't take a genius to understand why droughts affect hydroelectricity—power created by water that flows through gates in dams. When water levels are low, power production drops off.

What's less obvious is why droughts also threaten all the *other* power plants, like coal, oil, natural gas, and nuclear.

All of them work by generating steam to turn turbines—steam that must then be cooled back into water to be used again. If you can believe it, 41% of *all freshwater in the United States* is used to cool power plants.

But in a drought, there's not enough water for power-plant cooling. The plants have two ugly choices. They can shut down, causing a blackout, or they can seek governmental approval to dump the 100°F water into the rivers and lakes, which is devastating to fish and other aquatic life.

In any case, the discussion in Chapter 3 about equipping yourself with a generator—even a tiny, cheap one capable of recharging your phone and a couple of lights—applies equally well in times of drought.

Mosquito Breeding

During droughts, outbreaks of mosquito-borne diseases are rampant. Droughts play a bigger role in West Nile virus outbreaks, for example, than any other weather factor, including temperature. During California's six-year drought, West Nile infections hit the highest number of cases ever recorded there.

At least three factors are at play:

- **More mosquito breeding.** Stagnant water is a mosquito baby factory. And in a drought, a lot of water stops moving—rivers, creeks—creating thousands of standing pools that are just right for mosquito parenting.

  To make matters worse, humans in a drought often put out rain barrels and other collection containers—because, hey, free water. Trouble is, we've just granted the grateful mosquito population even more convenient breeding zones.

- **More biting.** Mosquitoes actually *bite* more during droughts. They're thirstier.

- **More disease transmission.** In a drought, sources of water are smaller and less numerous. Wildlife is forced to crowd together to get at whatever pools of water still exist, including birds and mosquitoes. When they're crammed into close proximity, the West Nile virus has more opportunity to spread from birds to bugs.

During a drought, therefore, the advice in Chapter 14 is twice as important. Drain your puddles, spray yourself, and avoid walking and playing at dusk and dawn.

Unexpected Critters

During a drought, animals can't find anything to drink at their usual watering holes; can't find food growing plentifully as it usually does; and may discover that their homes are dried up, collapsed, or burned. So they go looking. You, human, can expect to find wild animals in unexpected places.

You might encounter raccoons eating your pets' food or your garden's goodies. You may spot bears rummaging through your garbage, licking the grease off your grill, and raiding your bird feeders. Car collisions with animals are more likely, as deer, elk, possums, skunks, turkeys, and bighorn sheep have to forage closer to roads.

Preparing for a drought, therefore, involves bringing garbage, pet food, and animal feed indoors. Make sure your trash is regularly collected, and, to be safe, wash out the cans so they don't still smell like garbage and food. Good idea to bring your barbecue inside, too—into the garage is fine, if you have one.

If bears are around, you may have to rethink those bird feeders, at least until winter.

Finally, pick up the fallen or rotten fruits and vegetables in your yard and garden. For skunks, raccoons, and bears, they're literal low-hanging fruit.

Health Effects

A general drying-up of water around you can have some unexpected effects on your health, too.

- **Lung irritation.** Dry landscapes mean dustier ones, which irritates the lungs of people with bronchitis, bacterial pneumonia, and asthma. Smoke and ash particulates from nearby wildfires don't help, either. As ponds dry up, furthermore, they tend to concentrate the contaminants in the water, including cyanobacteria (blue-green algae), which blooms like crazy. Those bacteria can become airborne, adding yet another threat to people with lung conditions.

  That's why, if you've got asthma or another lung condition, you'll be encouraged to stay indoors or even to wear a face mask when you go out.

- **Hygiene shortfalls.** Drought periods coincide with outbreaks of diarrhea and skin and eye infections—because people don't wash their hands as much.

  Your water conservation in a drought shouldn't go so far that you cut corners on hygiene, like washing your hands, your fruits, and your vegetables. You can minimize the amount of water you're using, sure, but don't skip the washing altogether.

- **Swimming hazards.** Two things happen at your favorite swimming and boating spots. First, the water level might be much lower than you're used to, meaning that your boat, or your head, could strike the bottom.

  Second, shallower, stiller water is often *warmer* water, which means more conducive to the growth of sick-making germs like *Naegleria fowleri*. If in doubt, swim in chlorinated pools, not the ol' muddy swimming hole.

  On the other hand, here's glass-half-full moment: Water at the *beach* can be cleaner and healthier for swimming during a drought, for the simple reason that there's not as much runoff from the land to carry chemicals, fertilizer, and other contaminants into the surf.

Preparing for the Rain

These days, preparing for drought also means preparing for what often happens next: too *much* rain.

"Drought can make soil hard and not as absorbent, so rain doesn't soak in as well as it normally would," says Kelly Smith of the National Drought Mitigation Center. "There may also be less vegetation, so heavy rain after drought may cause worse flooding."

This alternation of droughts *and* heavier rains is a one-two punch that does us no favors. After months of drought, a torrential rain suddenly deposits thousands of gallons an hour onto dried, hard earth. Drought and flooding are like toxic siblings.

If you're a farmer, this sort of thing can keep you up at night for months on end.

If you're not, your main concern should be flash flooding of your home, yard, and garden. Most of the preparation steps have a single principle in common: Keep huge masses of rain off the ground.

- **Check gutters and drains.** You probably already have a water-elimination system: gutters and downspouts. They're supposed to keep rainwater off your house (to prevent rot and mold in your roof, shingles, fascia, soffits, and masonry), off your windows (to prevent condensation and water damage to sills and inside walls), and away from your foundation (to prevent erosion, cracking and crumbling, and basement flooding).

  Rain is supposed to roll off your roof, into your gutters, and from there into downspouts. From there, the water may go into downspout extenders (eight-foot extensions that dump the water far from the house), into rain barrels, or into underground cisterns.

  But gutters need regular attention. They get clogged with leaves. They develop leaks from attacks by both storms and squirrels. Clogs and leaks both defeat the purpose of having gutters.

  That's why you're supposed to clean and inspect your gutters at least once a year. If, standing there on your ladder and pouring some water tests, you spot a leak, you can fix it on the spot with gutter-sealing spray or caulk.

- **Collect the water.** If your downspouts are currently set up to dump water onto the ground, well, there's your problem. Route them instead into some big plastic rain barrels beside the building (with lids, to prevent mosquito breeding). Not only does that setup keep the water from saturating the ground and flooding, it also gives you a handy backup store of water (see page 100).

- **Stop watering the water.** If your lawn or garden has an irrigation system, then any rain double-waters it, wasting a lot of water and money. A rain sensor, which costs less than $20, detects rain and prevents your watering system from turning on for no reason.

  If the rain-sensor idea doesn't appeal to you, then you can always turn off your system *manually* whenever a big rain is approaching.

- **Drain the water.** If you've noticed that certain low areas of your yard develop puddles, ponds, or lakes after every rain or during every winter, fix it. Hire a contractor to reshape the contours, dig a trench, or make a swale to let the water drain.

If you're a gardener, Chapter 4 offers more tips for preventing your little farm from getting baked by drought and/or drowned by intense rainfall.

Chapter 12

Preparing for Hurricanes and Tornadoes

HURRICANE, TYPHOON, AND CYCLONE ARE THREE WORDS FOR the same thing: bad news.

If they form in the Atlantic Ocean or the eastern Pacific, they're called hurricanes. The same storms in the western Pacific are called typhoons. In the Indian Ocean, they're known as cyclones.

But they're all violent storms, they all begin near the equator, and they all work the same way: Warm, moist air rises, sucking in new air below, which itself warms and swirls, continuing the cycle. Pretty soon you've got a cylinder of rising, spinning, hot, wet air. Warm ocean water feeds it at the bottom; cooling, condensing water vapor forms clouds and rain at the top.

The death and destruction come from high winds, torrential rain, storm surges, flooding, tornadoes, landslides, and rip currents. The misery comes from losing power, clean water, passable roads, and maybe your car and your home.

How We Measure Hurricanes

To help you gauge your terror, meteorologists have a system, unfondly known as the Saffir-Simpson hurricane wind scale, for classifying the power of these weather bombs:

- **Tropical depression** is a wind spiral whipping around at up to 38 miles an hour.

- **Tropical storm**. Uh-oh. The storm's winds are greater than 39 mph.
- **Category 1.** The winds have reached 74 mph, capable of knocking down trees and power lines. The accompanying storm surge (unnaturally high water levels at the coasts, pushed inward by the storm) can be up to five feet tall.
- **Category 2.** This hurricane's winds have surpassed 96 mph, with storm surges up to 8 feet.
- **Category 3** means winds exceeding 111 mph. Storm surges reach up to 12 feet over usual sea level.
- **Category 4.** The winds are now greater than 130 mph, with storm surges up to 18 feet. Hurricane Harvey (Texas and Louisiana, 2017) was one of these. So was Hurricane Laura (2020).
- **Category 5** is anything blowing harder than 157 mph. Storm surges tower higher than 19 feet. An hour of Category 5 delivers the same force as five Hiroshima atomic bombs. At its peak, before making landfall, Katrina was a Cat 5 hurricane. Its storm surge was 28 feet tall as it hit the land.

How We Name Hurricanes

Before the 1950s, people referred to hurricanes by their latitude and longitude, which was confusing and clumsy, let alone almost impossible to Google. That's why, starting in the early fifties, the National Hurricane Center started giving hurricanes women's names, which made the storms more memorable and easier to tell apart. Hurricane names went coed in 1978.

Today, the World Meteorological Organization generates and maintains hurricane names. What's weird is that they have only six lists of alphabetical names—which they *repeat* every six years. The hurricanes of 2021 already have names—Ana, Bill, Claudette, Danny—which will also be the names of 2027, 2033, and so on.

The WMO retires a name from these lists only for sensitivity purposes. The name Katrina, for example, is off the lists forever; those *K* storms will now be Katia instead.

2020	2021	2022	2023	2024	2025
Arthur	Ana	Alex	Arlene	Alberto	Andrea
Bertha	Bill	Bonnie	Bret	Beryl	Barry
Cristobal	Claudette	Colin	Cindy	Chris	Chantal
Dolly	Danny	Danielle	Don	Debby	Dorian
Edouard	Elsa	Earl	Emily	Ernesto	Erin
Fay	Fred	Fiona	Franklin	Francine	Fernand
Gonzalo	Grace	Gaston	Gert	Gordon	Gabrielle
Hanna	Henri	Hermine	Harold	Helene	Humberto
Isaias	Ida	Ian	Idalia	Isaac	Imelda
Josephine	Julian	Julia	Jose	Joyce	Jerry
Kyle	Kate	Karl	Katia	Kirk	Karen
Laura	Larry	Lisa	Lee	Leslie	Lorenzo
Marco	Mindy	Martin	Margot	Milton	Melissa
Nana	Nicholas	Nicole	Nigel	Nadine	Nestor
Omar	Odette	Owen	Ophelia	Oscar	Olga
Paulette	Peter	Paula	Philippe	Patty	Pablo
Rene	Rose	Richard	Rina	Rafael	Rebekah
Sally	Sam	Shary	Sean	Sara	Sebastien
Teddy	Teresa	Tobias	Tammy	Tony	Tanya
Vicky	Victor	Virginie	Vince	Valerie	Van
Wilfred	Wanda	Walter	Whitney	William	Wendy

In 2020, there were so many hurricanes that the WMO ran out of names from the list. It called subsequent hurricanes Alpha, Beta, and so on.

Hurricanes and Climate Change

Does climate change bring us more hurricanes? Proving that is difficult, because our ability to count and measure hurricanes has changed radically over time.

Hurricane tallies go back to 1851. But before the satellite era, we had no way to count or measure the hurricanes that lived and died out at sea. As a result, "we're stuck looking at data that only goes back to around 1980 or so. It's a much shorter record," says Tom Knutson, climate-modeling director for NOAA.

Within that short time line, the *number* of hurricanes/cyclones/typhoons has been remarkably steady: about 80 a year, worldwide.

What does seem to be changing, though, is the *intensity* of the storms. For every 1.8 degrees Fahrenheit of planetary warming, we get about 30% more Category 4 and 5 storms. At this rate, by 2100, we can look forward to 240% more hurricanes reaching Category 5.

In fact, some NOAA scientists suspect that hurricanes may become so potent that they'll leave the five-category Saffir–Simpson rating system in the dust. Soon, we'll have to define a Category 6.

To make matters messier, hurricanes are also getting to their maximum ferocity faster, slowing down more when they hit land, reaching new areas farther from the equator, and dumping much more water.

You've probably heard about some of the most recent examples of hurricane devastation in the United States, but here's a refresher:

- **Andrew** (1992). Category 5; $27 billion in damage (1992 dollars); 65 lives lost; worst hit: Florida, Louisiana, Bahamas. At the time, Andrew was the most damaging hurricane ever to hit the United States. It drove the flood-insurance companies either out of Florida or out of business, with industry repercussions that last to this day.

- **Katrina** (2005). Category 3 at landfall; $125 billion in damage; 1,833 lives lost; worst hit: Florida, Louisiana, Mississippi.

  New Orleans sits an average of six feet below sea level, protected by an enormous system of earthen walls, from 13 to 18 feet tall, called levees. Katrina's storm surge sent Gulf waters pouring over those levees and breaking them down in 50 places. The combined waters of the Gulf, the Mississippi River, and Lake Ponchartrain poured into the great New Orleans bowl. The floodwaters destroyed 200,000 homes and left 80% of the city underwater—an area the size of seven Manhattans. Or, rather, under a toxic brew of water, sewage, chemicals, and gas.

- **Ike** (2008). Category 5; $38 billion in damage (2008 dollars); 214 lives lost; worst hit: Texas, Florida, Texas, Louisiana, Arkansas, Bahamas, Haiti, and other Caribbean islands.

- **Sandy** (2012). Category 3 at landfall in Cuba; tropical storm at U.S. landfall; $70 billion in damage (2012 dollars); 233 lives lost; worst hit: New York and New Jersey, plus Canada, as well as Haiti and other Caribbean islands. At almost 1,000 miles across when it hit the United States, Sandy was the biggest hurricane ever measured and the fifth costliest.

- **Harvey** (2017). Category 4; $125 billion in damage (2017 dollars), tied with Katrina as the costliest ever. Harvey was also the wettest U.S. hurricane, dumping *60 inches* of rain on Houston. According to computer models, that level of inundation should happen only one every 9,000 years.
- **Irma** (2017). Category 4/5; $77 billion in damage; 134 lives lost; worst hit: Florida, Bahamas, Caribbean islands. The most intense hurricane to hit the continental United States since Katrina.
- **Maria** (2017). Category 5; $92 billion in damage; 3,059 lives lost; worst hit: Puerto Rico, Dominica. The worst natural disaster ever recorded on those islands. Puerto Rico suffered the longest power blackout in United States history and a humanitarian crisis that lasted for months.
- **Laura** (2020). Category 4 at landfall; $10.1 billion in damage; 77 lives lost; worst hit: Louisiana. Laura's 150-mile-an-hour winds and 9-foot storm surge made it one of the most powerful storms ever to hit the United States—even stronger than Katrina. Laura blew apart buildings, flattened homes, snapped trees, and tossed utility poles like toothpicks, especially (as usual) in communities of color and low-income areas. Fortunately, thanks to efficient evacuation and a landfall centered on a relatively unpopulated marshland, Laura's devastation wasn't as apocalyptic as predicted.

The costs of hurricane damage have been hitting record levels in recent decades. Part of the reason, however, is that we're *living and building* more where hurricanes like to land. A certain hurricane will cause far more costly damage in 2020 than it would have in, say, 1920, simply because we've built so many homes, buildings, factories, and oil refineries in its path.

That doesn't mean that the dollars-of-damage statistic is irrelevant, however. If anything, it only underscores how vulnerable we've become to hurricanes—and how important it is to get ready for the next one.

How to Prepare

If you live in a hurricane-susceptible region—that is, within 150 miles of a coast—you have an ace in the hole: advance warning. First, you know that there's a hurricane *season* (June 1 to November 30); September is usually the peak. Second, if you're not completely oblivious to news and conversation around you, you'll keep hearing that something big and rainy will be arriving in a few days.

But that doesn't mean you can wait until the last minute to think about how you're going to manage.

The following pages guide you through the preparation you can do, both now and when the storm is imminent. Some of them cost money; all of them require some effort.

In each case, however, the payoff can be immense, as University of Utah sociology professor Sarah Grineski can tell you. In 2012, she had conducted a study that involved interviewing Houston residents. Among other questions, she asked what steps they had taken to minimize (mit-

Figure 12-1. These are the regions of the United States where hurricane preparation is a wise idea. Deeper blue areas are more susceptible.

igate) their flood risk, like installing hurricane shutters, reinforcing the roof, and moving major appliances upstairs. Some people had taken several steps; some hadn't taken any.

Five years later, Hurricane Harvey hit Houston. Grineski and her colleagues realized that they had a rare research opportunity: They could track down some of the same people and find out how their preparation steps had helped them weather the storm.

As you'd guess, the homeowners who had prepared experienced less damage and financial hardship. But there was a surprise, too: "If you had done more mitigation actions at your home," Grineski says, "you had significantly fewer physical health symptoms after Harvey, fewer post-traumatic stress symptoms, and fewer adverse-event experiences" like being separated from family or pets; having to live on the street; or going without food, water, or medicine. Better yet, she says, your recovery was much faster.

And why would those steps protect your *mental* health? "You took actions to protect your home, so you feel better," Grinesky theorizes. "You feel like you have more control over the situation, which has positive benefits on your mental health."

Preparations Today

A hurricane shoots with two barrels: wind and water. That's why every word about flood preparation in Chapter 8 applies to this chapter, too.

Remember: Your homeowner's insurance doesn't cover flooding. If you live near the coast, it probably doesn't cover wind damage, either. Those are added-cost policies. And they don't kick in until 30 days after you sign the contract, so the time to get it is *now*, while skies are clear. That's why you should review Chapter 6.

You should also review Chapter 3, which offers guidance on how to build your home, or reinforce it, to withstand that wind and water.

Finally, well before hurricane season, you should review Chapter 8. It covers the smartest and least-expensive preparations to make in advance of any natural disaster. In particular, read the sections about installing an early-warning app; setting up a water source for "camping at home" after the storm; buying enough shelf-stable food to tide you through a couple of weeks; working out a plan for communicating or meeting up if cell towers are down; studying evacuation routes, taking elevation (that is, flooding) into account; building a go bag for each family member; and learning how to shut off your home's systems, like gas, electric, and water.

But some to-dos are unique to hurricanes. People are always surprised, after a hurricane or tornado, at the crazy wind-blown projectiles that turn up: garden gnomes in palm trees, LP records sliced into tree trunks, sailboats in treetops. In other words, it's not just the winds you're worried about; it's the stuff that gets picked up and shot at your home.

You can tackle the prep steps over the months—it doesn't have to be all at once—but tackle them you should. Here we go:

- **Upgrade the garage door.** A garage door is a wide, weak, thin, lightweight, hinged covering for the biggest opening in your house. And it's connected to the house only by its feeble metal tracks.

  If a gale pushes or pulls hard enough, the door rips right off the track, and your garage door is gone. Now the wind can explode into the house, doing vicious damage and even blowing off the roof.

  Yes, that really happens: Once the garage door gives way, the wind slams into the garage, increasing the air pressure and basically blow-

Figure 12-2 Garage doors are an easy entrance point for hurricane winds.

ing your home up like a balloon. The bigger the garage, the greater the pressure buildup (that's for you, two- or three-car–garage owners).

Impressively enough, *80%* of all residential hurricane wind damage begins with garage door failure.

If your garage door doesn't have a sticker that indicates its pressure rating or wind-speed rating, it may not be hurricane ready. In that case, you have three options.

The least expensive option is a temporary garage door reinforcement kit. It's a vertical aluminum post that bolsters any garage door that's in good shape. When the weather's good, you (or someone you hire) can install its mounting brackets above and below the garage door. Then, when the storm approaches, you snap the metal post into those brackets to strengthen the door. You remove the post after the storm is over. These brace kits cost about $200 per door (at, for example, Lowe's or securedoor.com). For a wide door, you'll need two or even three of these posts.

Figure 12-3. You can reinforce your garage door during stormy times by snapping a wind-load post into place.

If having to set up and take down those posts with every storm sounds like a pain, you may prefer plan B: replacing your current garage doors with heavier, stronger, hurricane-rated doors (about $1,800 for a two-car garage).

And if it's too late for all that, here's an almost free garage door hack worthy of MacGyver: Stand a two-by-four beam up against the inside of the closed garage door. Carefully back up your car until it pins the beam against the door.

You've just made a do-it-yourself wind-load post for next to nothing.

♦ **Prep your windows.** You always see people on TV nailing plywood sheets over their windows before a storm. That's certainly better than leaving your windows unprotected. But the time to buy and cut these panels to fit your windows is *now*; you don't want to be cutting and sawing as all hell is breaking loose.

There is, however, a far superior way to prepare windows: Install either hurricane shutters (metal covers that unroll or unfold over the window as necessary, either manually or electrically) or impact windows (plastic-bonded, double-paned glass). Both will lower the cost of your insurance, and both are described in detail in Chapter 3.

♦ **Prep your doors.** A hurricane considers the exterior doors of your home to be a weak spot and a convenient entry point. Each one should have three hinges or more and a deadbolt at least an inch long.

♦ **Trim the trees.** You're looking for limbs and branches that, in high enough winds, could snap off and smash your house.

♦ **Reconsider gravel.** If your landscaping incorporates gravel or rocks, think about replacing it with shredded bark. In 100-mph winds, bark chips do a lot less damage than gravel.

♦ **Clear the gutters.** Hurricanes bring a lot of rain. You want your house to have a fighting chance of directing it away from your walls and windows.

♦ **Seal the openings.** The walls of your house have all kinds of holes: openings that the builders made to accommodate vents, outdoor power outlets, garden-hose faucets, and anywhere else a cable or a pipe has to pass through a wall. Each one offers wind and water an opportunity to penetrate and damage the house.

Your job is to seal around each one of them with urethane-based caulk. Draw the caulk gun's tip around the seam where the faucet or outlet plate meets the wall.

- **Buy your materials.** When a storm hits the fan, stores like Home Depot and Lowe's are mobbed. You want to have your plywood, plastic sheeting, hammer, and nails on hand ahead of time for prep and repairs.

- **Choose a safe room.** When an evacuation order comes down, evacuate, if you can. But what if there isn't an evacuation notice? Or you have no transportation, or it's too late, or there's some other reason you're stuck inside?

  For those reasons, you should figure out where in your home to hunker down when a hurricane makes every 150-mph effort to get inside.

  Branches and debris may explode through your windows; trees may even force themselves through your walls. That's why the ideal safe room is an interior room without glass windows or doors. It should be the lowest room in the house that won't flood. If you have a basement that doesn't flood, that's a great option for a safe room (or in which to *build* one), because its underground location means instant protection from howling winds and flying debris.

  You can *create* a safe room, too. You can build one yourself—see www.fema.gov/safe-rooms—or buy a prefab, ten-foot-square model for about $3,000.

As the Storm Approaches

As the sky gets dark and you start monitoring your emergency app or radio, it's important to know your terms.

- **A hurricane advisory** means "There may be some non–life-threatening hazards resulting from this bad weather."

- **A hurricane watch** means "A hurricane is *possible* in the next 48 hours. Check your emergency kits and get your home ready."
- **A hurricane warning** is the most dire. It means "A hurricane is *coming your way* in the next 36 hours. Get set and be ready to evacuate."

As soon as you get word of a hurricane watch or warning, begin these preparations:

- **Beat the crowds.** You can count on one thing every time a major storm approaches: The public will panic, descend like locusts on stores, and strip the shelves bare of water, packaged food, flashlights, and batteries. If you've followed the prep steps above, you won't care, because you'll already be equipped.

  But if not, go early, while there's still something to buy.

  Get some cash from an ATM while you're out. Assume that the power will fail citywide, and most places won't be able to process credit cards.

- **Prep the property.** Loose stuff like yard or patio furniture, bikes, trash cans, and your grill don't belong outside in a superstorm. First, you've paid good money for them; presumably, you'd rather not have them blow away. Second, in the monster winds, they can become missiles. Tie them down or bring them into the garage or the house.

- **Prep your windows.** Here's a fun hurricane fact: Taping a big *X* on your windows, the way people do on TV, is a dangerous mistake.

  First of all, it's pointless; do you really think a piece of tape will protect the glass against a 140-mph flying mailbox? Second, the *X* tape ensures that when the window does shatter, the glass shards flying inward at your family will be much bigger than they would have been.

  If you have hurricane shutters, now is the time to close them. If you don't have them (or *impact glass*, which is reinforced like a car windshield), plywood panels are the way to go. They should be *exterior* plywood—cheap and not pretty, but fused with waterproof glue that resists coming apart when wet—at least ⅝ of an inch thick. It's usually called CDX or marine-grade plywood.

- **Prep your fridge.** While you still have power, crank the fridge and freezer as cold as they'll go. The colder they are now, the longer your food will last when you lose electricity.

- **Gas up the car.** Gasoline is another essential that often runs out as word of an impending hurricane spreads. The sooner you can fill up your tank, the better shape you'll be in should you have to evacuate. And if you fill up a couple of additional gas jugs, you'll be okay longer if the gas stations remain closed.

- **Load the car for evacuation.** The evacuation order may not come, but you want to be ready. Put your go bags in the trunk. Stash some food and water in the car, too; if an evacuation order comes down, you may be sitting in motionless traffic for *hours.* If there's room, you may as well add some sleeping bags, a couple of tools, and flares and jumper cables. You're departing for the unknown at a risky time, and you never know what's going to happen.

- **Fill some jugs.** Contaminated water is one of the most common—and disastrous—effects of superstorms. Your municipal water-treatment plant gets swamped, and raw sewage spills into the water that feeds the city. You're going to need drinkable water.

  Chapter 8 identifies some other great sources of drinkable water that are already in your home, like your hot water heater and your toilet tanks.

- **Fill the tubs and sinks, too.** Most people use this water for cleaning things and flushing toilets, not drinking—but in a pinch, you can drink it. If you line the tub with a tarp or a shower curtain, your water stash can sit there, leak free, for weeks.

- **Test the flashlights.** Do you have spare batteries?

- **Bring a hatchet to the attic.** If floodwaters drive you up to the attic and floods the entrance, you'll need some tool to hack through the ceiling to escape to the roof.

- **Unplug appliances**. First, you want to protect them from power surges in the coming days of iffy electricity. Second, you want to protect yourself and everyone else from electrocution if the place floods.

Leave the fridge plugged in, though. Once again, it's more important to keep the food cold as long as possible. You may have to live off it for days.

The Red Cross suggests leaving one lamp plugged in and turned on—so you'll know when the power has come back.

♦ **Charge your phones.** You'll want them full before the power goes out.

♦ **Deal with special requirements.** If someone in your gang is old, is disabled, or otherwise can't keep up, call a hospital, public health department, or the police to get advice.

♦ **Check over the go bag.** If you've followed the advice in Chapter 8, then your bug-out bag has a piece of paper taped to it that reminds you of the *last minute* things to stuff in there, like current prescription meds, your purse or wallet, a phone charger, and important documents.

Make sure the batteries in your go bag still work, the water's in there, and that none of the food has expired.

♦ **Communicate.** As the storm bears down, let friends and relatives know where you are.

The flood-prep steps in Chapter 9 apply here, too. Start monitoring an NOAA radio or app; move the cars to higher ground; turn off your propane and gas; carry valuable stuff upstairs; sequester the pets.

Since you probably live on the coast, you might be on a boat. You have one job: *Get off it.* Whether it's docked or out at sea, a boat is a deadly place to be in a superstorm.

If you have a travel trailer or a boat—on or off its trailer—well, you have some extra homework. You need to know where to move them before a storm where they'll be safer.

If you have a mobile home, check the condition of its tie-downs, which anchor it to the ground; now is their big moment. (You do *have* tie-downs, right?) Be aware, though, that even a properly anchored mobile home is no match for a major hurricane; its long, wide sides act like a gigantic sail. Attachments like porches, carports, and awnings aren't anywhere near as strong as the home itself. It might be worth asking an inspector what you can do to secure them—and to make sure that they don't decrease the actual home's chances to survive the storm.

Finally, you'd better be familiar with how your boat, trailer, or mobile home is treated (if at all) in your insurance policies.

The Evacuation Moment

After every hurricane, news reports inevitably feature a few residents who have ignored the order to evacuate.

During Hurricane Katrina, for example, tens of thousands of people "just sat there and crossed their fingers and hoped for the best," says Jed Horne, author of *Breach of Faith: Hurricane Katrina and the Near Death of a Great American City*. "They figured, as in so many storms before, that this one would simply veer away at the last minute—they have a wonderful habit of doing that—and the city would be spared."

Local police were reluctant to force people to leave their homes, but they weren't above using persuasion tactics. "Do me a favor. Make life easier on us," New Orleans SWAT team member Dwayne Scheuermann would tell stubborn residents. "Take a permanent marker, write your social security number along one arm and one leg. So when we find your body, we can find out who you are."

The stay-behinders later required expensive and dangerous rescues—if they survived.

You might watch these reports, shaking your head and muttering, "What kind of idiot ignores the evacuation orders?"

Fifty-seven percent of those who ignored the evacuation order before Hurricane Katrina did so because they misjudged the devastation of the storm. Part of that was a longstanding New Orleans cultural tradition of riding out storms, stemming from the days when the city *couldn't* be evacuated, before the highways were built.

"People would just settle in and ride it out," says Horne. "And not only was that the mark of resignation, but it was a good time. You'd pull a grill out. Everything's going to rot anyway, so you'd unload the freezer, cook up all the chicken, have a big party in the street."

Some thought their homes were strong enough, or far enough away from the coast, to withstand the hurricane. Or they didn't want to leave their pets.

But 39% of the stay-behinders *couldn't* leave. They had no car, no money, or no place to *go*.

If you have a choice, don't ignore evacuation orders when a hurricane approaches. Here's the thing: Your governor probably *hates* ordering an evacuation. It's a political nightmare. If the storm turns out to be as bad as everyone expects, well, the order won't be hailed as a great stroke of genius. But if it's *not* as bad as expected, everybody will blame the state leaders for causing a world of disruption for nothing. Ordering evacuation is a no-win situation for a politician.

In other words, if you do get an evacuation order, you *know* it's serious. You know that meteorologists and emergency personnel have pushed your leaders way out of their comfort zones, far enough to overcome their reluctance to take this drastic step.

If you get word to evacuate:

- **Don't dally.** Start *now.* The roads out of town will become increasingly choked; you want to be at the *front* of that long, slow line. The longer you wait, the greater the chances that you'll get trapped by high winds or floods, too. Get going *now.*

  The Insurance Information Institute advises that if you have a pet, you should leave *before* the evacuation order comes down. That's because evacuation notices often include orders to leave your pets behind. But if you're already on the road, what can anyone do?

- **Shut off gas, power, and water** if you have time.

- **Lock up the house—doors and windows.** You don't know when you'll be back, and you don't want your place to be a sitting duck for looters in the storm's aftermath.

- **Don't count on shortcut roads.** They may be closed during an evacuation. Don't drive around barricades, either. Every year, some people think they know better than emergency personnel—and wind up stuck or underwater.

- **Check the routes out of town** on your state's office of emergency management website, if you have internet.
- **Check the apps**—the Red Cross or FEMA apps—to find the available shelters.
- **Tell somebody.** Let *somebody* know that you're leaving and where you intend to go, preferably somebody out of town.
- **Avoid the public shelters** unless you have absolutely no other place to go. Not only do shelters tend to be hot, wet, understaffed, under-equipped, and miserable during a disaster, but your presence makes those conditions even worse for people who truly have no alternative.

  If you're with someone who needs medical attention or requires oxygen, look up "special-needs shelters" in your Red Cross or FEMA app or on the Web, using Google (*special needs shelters houston*, for example). These shelters aren't hospitals—all they're allowed to do is *monitor* someone's medical condition—but they're better equipped than, say, the school gym.

  Pets usually aren't allowed in public shelters, by the way.
- **Keep your receipts.** Your insurance will reimburse you for whatever you spend on living expenses while you're out of your home—but only if you have receipts as proof.

During the Storm

Sometimes, instead of ordering you to evacuate, officials order you to stay home. That's likely if traveling is too dangerous. That's also likely in the case of a tornado, which comes and goes too fast to justify evacuating an entire area.

If you're at home, the main thing is: Keep away from *glass*. Windows, doors, and skylights are weaker than the surrounding walls and roof. They can burst without warning, bombarding everyone inside with shards. Go to your safe room, the basement, if it isn't flooding, or even a

Why Hurricanes Target the Poor

Some writers have referred to Hurricane Katrina as the "great equalizer," because the storm destroyed the homes, cars, and businesses of rich and poor, black and white. But anyone who takes more than a quick peek into the situation discovers that the damages, deaths, and recovery delays were far worse in poorer neighborhoods. Same thing for Hurricane Harvey. Same for Sandy.

"This is nothing new," says Jalonne White-Newsome, who studies climate change's impact on lower-income citizens. "It's been going on for decades and decades."

Question is: Why? Surely a weather event can't be racist, targeting which houses to blast apart.

No, but not having much money exposes you to the worst of the storm in all kinds of ways. You can't afford flood insurance. If you don't have a car, you can't easily evacuate—and even if you did, you wouldn't be able to afford a motel room to wait out the storm.

Your home may be poorly constructed and therefore among the first to go in hurricane winds. And your neighborhood may be in an industrial part of town, where chemicals and petroleum spill during a disaster and get into your water supply.

Then there's the recovery period. After a superstorm, your workplace, or the roads to get there, may have been destroyed. FEMA offers disaster relief funds to help you rebuild—but after Katrina, Black homeowners received an average of $8,000 less in government rebuilding money than white homeowners did.

And where are you supposed to stay while you're rebuilding? It's not like you have a second home or can afford a hotel.

Finally, there's the repeat customer phenomenon. If you can afford it, you might move out of the hurricane zone once you've had enough trauma. But if you're poor, you're stuck there, condemned to live through one superstorm after another. The poor get poorer; the rich leave town.

closet. Bring food, water, and your phone or NOAA radio with you. And the kids and pets, of course.

If it's really wild and woolly, and you're terrified, you can lie down in your bathtub and pull a mattress over you. Now you're protected on all sides.

Don't go outside, either. Don't go out to "see what the wind feels like"; what you'll learn instead is what shrapnel feels like.

There will, eventually, be a period of calm in the storm, and you might be tempted to have a look around. But you might be sensing the *eye* of the storm—the relatively still center of it—and within seconds, the far side may hit you and your house, every bit as powerfully as the leading edge did.

Stay tuned to TV, radio, or a NOAA radio or NOAA-compatible app, so you'll know when the storm truly has passed and when it's safe to go back home.

You really, really don't want to be driving around during the hurricane. The winds can make it hard to control the car and could even flip you over. And, of course, you know how foolhardy it is to drive when roads are flooded (Chapter 9).

But if you're somehow caught in your car, here are your options, from best to worst:

- **Best: Get indoors.** A car is a dangerous place to be in high winds. Get inside a strong building at the first opportunity: a truck stop, restaurant, convenience store—whatever's closest. Find the safest room: an interior room without windows or glass doors. A basement is great, if it's not flooding.

- **Okay: Hunker down.** If you're stuck in your car—in traffic, for example, or broken down or pinned—and there's no building nearby, duck below window level. Cover your head with a towel, blanket, or, worst case, your hands. Keep your seat belt on.

- **Last resort:** Find a spot that's noticeably lower than the roadway, like a ditch. Lie there and cover your head with your hands until the wind subsides or you're rescued. With luck, the debris will whip right over your head, sparing you from injury.

High-Rises and Hurricanes

If you're in a high-rise apartment or office building when a hurricane hits—a likely scenario in many East Coast cities—you get some special advice all your own.

Bring in *everything* from your balcony before the storm hits: chairs, tables, plants, wind chimes. Otherwise, you'll lose them to the winds—or, worse, they'll become projectiles that smash your windows.

During the storm, the power will probably be out, either because the storm takes it out or because the building managers turn off the power for safety. Without power, there won't be any air-conditioning, so it will be stifling hot—but resist the temptation to open the windows! Keep them all closed, locked, and covered. Outside, insanely powerful winds are whipping up and down the building, creating tremendous suction forces. If you open your windows, the wind could start ripping the walls apart.

In general, the best place to be is on the lowest unflooded floor, because the higher you go, the more dangerous the winds. Besides, when the power is out, you'll probably be grateful that you don't have to navigate 35 flights of stairs to seek food, water, and help.

A tall building can sway alarmingly in storm winds. If you're terrified being in your apartment or office, wait out the storm in the stairwell, which is the sturdiest part of a high-rise. It's made of reinforced concrete, and it's got no windows.

Before you flee, check on your neighbors who are old, frail, or reliant on medical machines that won't work when the power's off.

Finally, although there is such a thing as condo insurance (an HO-6 policy), which covers all your belongings, your walls, and your floors, there's no legal requirement to have it, so most people don't bother. But if you live in hurricane country, insurance shopping might be worth the effort.

- **Worst:** Here and there, websites suggest parking under a bridge overpass, so that you and your car are somewhat protected from the fusillade of debris and branches. *Don't do it.*

  First, those spots actually *channel* the winds and debris, basically focusing them on you and your car. Wind speeds *accelerate* under an overpass.

  The second problem: The overpass itself can collapse.

  The third problem: You're blocking the roadway. You might be preventing other drivers from escaping or preventing emergency vehicles from getting to people.

  Ohio's government website summarizes all of this nicely. "Remember: Overpasses offer NO PROTECTION."

Figure 12-4. Parking under an overpass during extreme weather is a bad idea. This photo, taken during a 2019 Oklahoma storm, shows why: You're blocking the road and possibly blocking people in.

After the Hurricane

Amazingly enough, more people get injured or killed *after* a hurricane than *during* it. That's because the world after a superstorm is an incredibly unstable place. Branches and building pieces are teetering or dangling. Trees can fall at any time. Power lines are down but still live with 13,800 volts. Water is flowing with toxic juices of every kind.

Chapter 8 describes everything you need to stay safe, keep healthy, and get the insurance money flowing. It covers all your concerns after a superstorm: electrocution, mold, contaminated drinking water, loose critters, mosquito infestations, spoiled food, cleanup, and post-traumatic stress.

And Chapter 6 covers the delightful task of getting the insurance ball rolling if your home was damaged or destroyed.

Tornadoes

You might assume that a tornado is basically a hurricane without an ocean. After all, they're both violent storms, both furiously rotating columns of air, both weapons made of wind that can do tremendous damage to our puny human world.

But the two storm types are different in a few key aspects:

- **Life span.** Hurricanes are slow to form, slow to move, and slow to dissipate. Tornadoes' lives are fluky and fast. They form spontaneously and die quickly—usually within minutes.
- **Size.** Hurricanes can be hundreds of miles across; tornadoes average about half a mile wide.
- **Power.** Tornadoes produce the strongest winds on earth—up to 300 mph, much higher than hurricanes. A tornado can flatten huge buildings; drive blades of grass into telephone poles; shoot broken glass like bullets; and lift passenger cars, railroad cars, and cows into the air.
- **Formation.** Hurricanes form offshore and draw their moisture up from the ocean. Tornadoes form over land and draw their wetness down from thunderstorms overhead.

Figure 12-5. Tornadoes are fast, fierce, and fleeting. You don't want to be in the path of one.

How Tornadoes Form

Tornadoes are frustratingly hard to study and explain. All kinds of conditions have to be just right for them to form—and even then, they sometimes appear and sometimes don't.

Until the 1990s, when Doppler weather radar came along, no technology could spot and count tornadoes. Even today, many of them are discovered and counted by human spotters, because radar doesn't reach everywhere.

We still can't directly measure the wind speeds of tornadoes, either, because they rarely do us the favor of passing by our weather instruments. Instead, we gauge their power by looking at the degree of devastation they cause. As with earthquakes (the Richter scale) and hurricanes (the Saffir-Simpson scale), tornadoes get their own intensity ruler: the Enhanced Fujita scale. It ranges from EF-0 (winds up to 85 mph, "light damage") to EF-5 (over 200 mph, "incredible damage").

Meteorologists don't even know exactly how tornadoes form. But the current theory involves a collision of warm, humid, slow-moving air at ground level—with cold, dry, fast-moving air high above the earth. The clash of air masses generates an immense thunderstorm called a supercell. At that point, the speed difference between the warm and cold winds produces a *horizontal* spinning tube of air, like a tornado lying down.

Eventually, rising warm air lifts this thing into a vertical position. Once it stretches from the clouds to the ground, it's a tornado.

Tornadoes and Climate Change

You might imagine that climate change means more tornadoes, because there's more wet, warm air at ground level. And indeed, the U.S. 2019 tornado season was the worst in a decade; at one point, 500 tornadoes struck within 30 days.

But you might also imagine that climate change means *fewer* tornadoes, because there's less *cold* air up high. If that upper air is warmer, there are fewer forces to begin that horizontal rolling pin of air.

So which is it: more tornadoes or fewer?

The answer is neither. The *number* of tornadoes hasn't seemed to increase over the last few decades, and scientists haven't yet proved or disproved a link between climate change and the number of tornadoes. Of course, it's incredibly difficult to *count* tornadoes, given their relatively small size, their brief life span, and the huge number of variables that affect them (time, season, major weather patterns).

What does seem to be happening, though, is that:

- **The tornadoes are becoming more powerful.** As with hurricanes, it's not the quantity that's going up, it's the quality. And that quality is "more devastating"; tornadoes are becoming about 5.5% more destructive every year.

- **The tornado zones are shifting.** The United States gets more tornadoes than any other country: about 1,200 a year. They've been known to form in all 50 states and in every season. But traditionally, *most* of them arise in Tornado Alley (Figure 12-6), and mostly between March and early June.

  In recent decades, however, they've been creeping eastward, into the Midwest and Southeast. The traditional Tornado Alley has seen a decline in tornado numbers, but the states to the east have been getting significantly more.

"The leading theory is a reduction in soil moisture and precipitation across the Great Plains, due to climate change," says Vittorio Gensini, coauthor of the study that revealed the shift.

That drift means some good news for Tornado Alley: somewhat fewer tornadoes. But it means bad news overall, because tornadoes are far deadlier and more destructive in the Midwestern and Southeastern states. Just ask the victims of the 2019 tornado that struck Alabama and Georgia, produced 170-mph winds, and killed 23 people, or the March 2020 tornadoes that killed 25 people in Tennessee.

Gensini calculates that a tornado in Mississippi or Alabama is about eight times more likely to kill somebody than one in Texas or Oklahoma.

The reason: The Southeast is far more densely populated than the Tornado Alley states. Meanwhile, more than 60% of the residents live

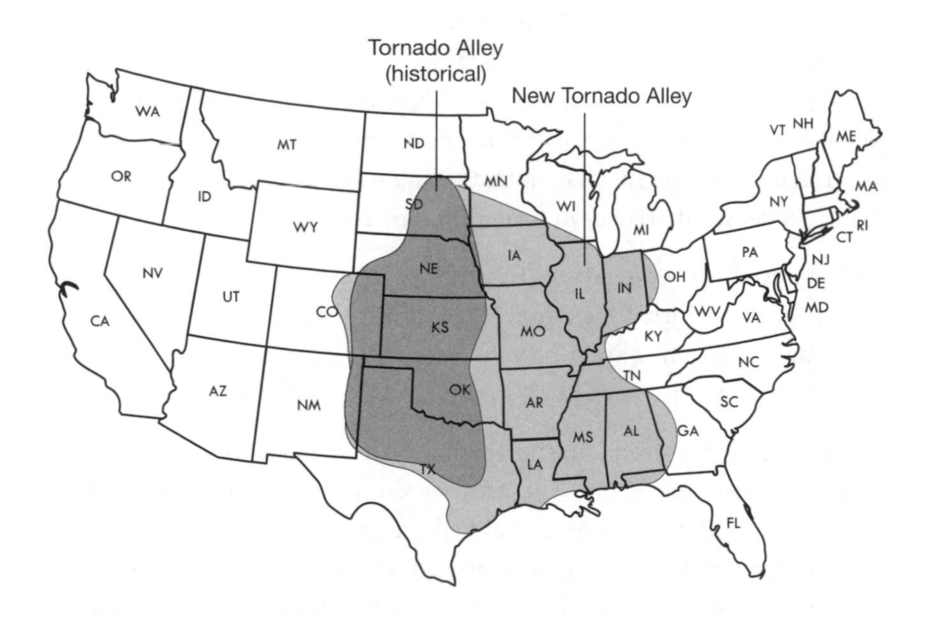

Figure 12-6. The old Tornado Alley, the darker blob, is giving way to an even larger, Eastern tornado zone, covering more states—and more populous ones.

in mobile homes and weakly built buildings. "No mobile home is a match for even an EF-0 tornado," Gensini says.

But tornadoes in the Southeast are also more deadly because people can't see them coming. They tend to strike at night, when people aren't paying attention, and the wooded areas of these states conceal the oncoming tornadoes much better than wide-open plains do.

How to Prepare

As with hurricanes and thunderstorms, you might receive either of two levels of alert: a tornado *watch*, meaning that conditions are right for tornado formation, and a tornado *warning*, meaning that somebody's actually spotted one.

The preparation steps are almost identical to the ones you'd take to protect yourself against hurricanes. Follow, therefore, the guidance on the previous pages that have to do with winds: Windproof your home, windows, and garage doors. Identify safe places to cower, away from windows and glass doors. Bring outdoor stuff inside. Install apps that can play NOAA broadcasts. Set up food and water for a couple of days.

And it's worth repeating: A mobile home is an incredibly dangerous place to be. A tornado can toss the whole thing like a football or blow it apart like a party popper. Get into a solid building.

The biggest differences in tornado prep are:

- **It's not about evacuation.** In general, cities and towns don't evacuate for tornadoes. Evacuation takes hours, and a typical tornado's life span is only ten minutes. It's therefore more about finding that windowless inner room of your house to shelter in place.

- **There may be sirens.** In the Tornado Alley states, many towns have installed tornado sirens. These are very loud horns, mounted on posts up high, audible from almost anywhere in town. You may not be able to hear them when you're indoors, but that's not the point; they're intended to alert people who are *outdoors* that they need to get *indoors*, because a tornado has been spotted, or storm-strength winds or golf ball–sized hail is approaching.

In general, the horn plays a long, loud, three-minute blast, or a series of them. That sound means "Get inside and start listening to the radio or the news." There won't be any "all clear" blast afterward.

Remember, though, that tornadoes spring from thunderstorms, which are noisy. Even if you are outdoors, you may not hear the tornado sirens over the wind and rain. More reliable clues are your early-warning apps, NOAA radio broadcasts, news reports, threatening skies, and the fact that everyone around you is tearing around, battening down their homes and talking about tornadoes.

- **Hail.** Hailstones are balls of solid ice. They form inside the updrafts within thunderstorms—they've been known to become bigger than softballs—and then fall to the earth.

  As you might expect of balls of solid ice falling from tremendous heights, they cause huge amounts of damage. They shatter car windshields, crash planes, and kill animals. They're also frequent companions to tornadoes, becoming ice balls whipped sideways with enough force to shatter windows and rip up the siding on your house.

  Fortunately, if you've taken steps to prepare for tornadoes, you're already protected against hailstorms. You already know to stay indoors, keep away from windows in your home that might get shattered, and put your car someplace covered.

Chapter 13

Preparing for Wildfires

EVERY NATURAL DISASTER IS UPSETTING. BUT NOTHING FEELS like the end of the world like a wildfire.

It's dark in the middle of the day. The sky glows red. The flames tower over you, sometimes 150 feet in the air, making you feel terrified and tiny. The sound is deafening, like a freight train, marked by occasional booms as tires or propane tanks explode. And fires move *fast*, consuming everything in their paths at up to 60 miles an hour.

Mike Kreidler, former firefighter and now insurance commissioner for the State of Washington, visited the 2018 Camp Fire while it was raging—the one that wiped out the town of Paradise, California, and killed

Figure 13-1. Wildfires are one of the most terrifying products of the warming climate.

85 people. "Apocalyptic is the only way I could describe it—I've never seen anything like it," he says. "It was a Dresden-style firestorm."

If a wildfire were an animal, we would consider it a fantastically successful product of evolution. Wildfires feed themselves on oxygen, which they suck up from ground level, and dry plant material, which they create as they go with their intense heat. Wildfires reproduce with amazing efficiency by sending hot, floating embers to spawn new fires miles away. They're capable of creating their own winds, their own weather.

And these aren't tornadoes, which blow out in minutes, or hurricanes, which last a couple of days; a wildfire can last *months.*

You may live in one of the richest and most technologically advanced countries on earth. But when it comes to wildfires, you're helpless.

"Nothing can turn back a roaring wall of flames shooting hundreds of feet into the air, moving faster than you can run, burning at temperatures that can easily melt your refrigerator into a puddle of aluminum and reduce your house to a pile of ashes in a few hours," says Linda Masterson, wildfire survivor and author of *Surviving Wildfire.*

Firefighters can attempt to herd the general direction of the fire, to defend a few houses, or to put out small fires. But nothing man-made can stop a wildfire. "There was no firefighting to be done," says Woody Culleton, former mayor of Paradise. "It was a perfect storm."

When a fire gets that big, all we can do is get out of the its way and pray for rain, humidity, or a change in temperature and wind direction.

Yes, wildfires destroy our homes and buildings with terrifying speed and impassivity. They kill people and animals. But they also kill trees and plants, thereby removing the root systems that used to hold the earth in place. The result can be massive erosion—and, with the next big rain, mudslides, of the kind that buried miles of Interstate 70 (and 30 cars) in Colorado in 1994, or the Santa Barbara mudslides of 2018 that killed 23 people and caused $177 million in damage.

And then there's the smoke. Wildfires produce massive volumes of carbon dioxide, which only fuels the climate-change cycle. Wildfire smoke contains dangerous and even deadly compounds. After a wildfire, the ash gets washed into reservoirs, contaminating the drinking water. The smoke also makes people sick in invisible ways by releasing toxins into the air.

It's easy to consider a wildfire a devastating juggernaut and to think

Figure 13-2. Following wildfires in 2017, rains in 2018 triggered massive mudslides in Santa Barbara, California.

that you're powerless to prepare. "People think, 'I can't stop it, therefore I'll do nothing,'" says Masterson. "'I won't make it so my house can survive. I won't make it so I can collect on my insurance. I'll just be Scarlett in *Gone with the Wind.*'"

No, you won't be able to put the fire out. But that doesn't mean you can't prepare.

How Wildfires Happen

A fire requires three ingredients: oxygen, fuel, and heat. If you can take away any one of those things, you can put out the fire.

For example, covering a small fire with a blanket puts it out, because you've eliminated the oxygen. Clearing the vegetation away from one side of a fire puts it out, because you've eliminated the fuel. Spraying the fire with water puts it out, because you've eliminated the heat.

Unfortunately, once a fire gets big, you can't do any of that stuff. They don't make blankets big enough to cover a county in California.

Once it's going, the ferocity of the fire depends on weather (hot, dry, and windy preferred), the fuel (dry grass, leaves, trees, houses), and the terrain. As it moves, a wildfire preheats the surrounding plants, drying them out just in time to serve as fuel for its advance.

Wind is a fire's best friend. It feeds oxygen to the fire, making it bigger and hotter; it helps to dry out the fuel in the next region to be burned; and it pushes the fire across the land. That's why, when you read about some horrifying out-of-control wildfire, you so often hear the firefighters pin their hopes on shifts in the wind.

Wildfires make things worse for us by generating their *own* winds, which can blow ten times as fast as regular wind—strong enough to fling burning logs long distances.

Fire moves uphill four to five times as fast than it does downhill. That's partly because the flames have a better angle to heat up the next stretch of vegetation and partly because wind itself tends to gust *up* slopes.

Wildfires Catching On

You pick a metric—deaths caused, land burned, cost of damage—and you'll find the wildfire numbers climbing sharply. In the United States, we're getting huge fires (more than 1,000 acres) about five times more often than we did in the seventies. And they're burning ten times as much area—about 10 million acres a year.

In 2018, for the first time, wildfires surpassed hurricanes as the costliest natural disaster in the United States. That year, 8.8 million acres burned—the land footprint of 74 of the 75 biggest U.S. cities combined.

Wildfires seem happy to strike almost anywhere in the United States. In 1998, it was Florida. In 2015, Oregon and Washington. In 2016, Tennessee. In 2017, Oklahoma, Kansas, and Texas.

And California—poor California! Here's its recent wildfire track record:

- **2017:** About 71,500 wildfires burned 10 million acres, killed 23 people, and took out 8,700 buildings. The insurance claims totaled $12 billion.
- **2018:** The Mendocino Complex Fire was the biggest fire in California history by a huge margin. It burned down 460,123 acres. Together

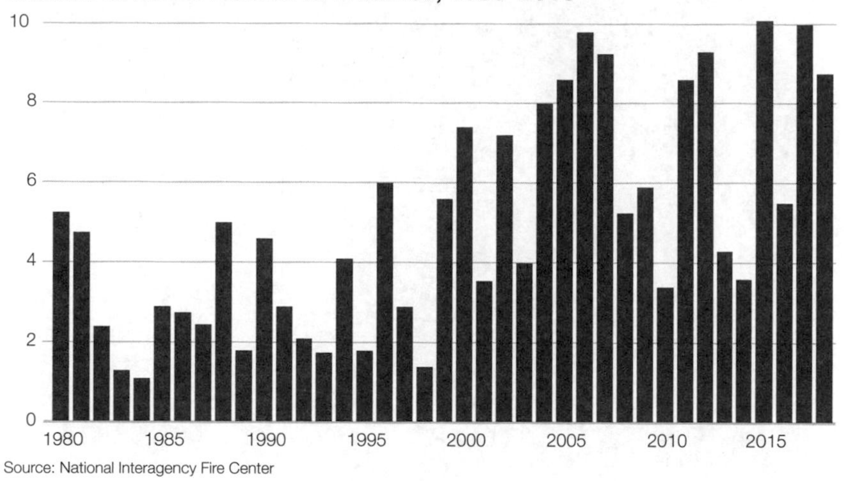

Figure 13-3. U.S. wildfires have steadily been consuming more American land in the last 40 years.

with the nearby Carr Fire, it destroyed 8,900 homes and 800 cars. That was July.

Then, in November, the Camp Fire, at the time the deadliest and most destructive fire ever recorded in California, left 85 dead, 153,000 acres burned, and 18,800 buildings wiped out. The town of Paradise was 95% destroyed.

- **2019:** The late October wildfires in California drove 200,000 people from their homes; the governor declared a state of emergency. The Kincade Fire burned more than 76,000 acres, twice the size of the city of San Francisco. And in LA, the Getty Fire caused the evacuation of 7,000 homes.

- **2020:** An extremely dry winter (including the first rain-free February in San Francisco since 1864), a vast tree die-off, a resurgent drought, and heat waves spelled trouble for California in 2020. And sure enough: As wildfire season began in August, a freak series of 11,000 lightning strikes within 72 hours helped launch the worst California fire season in recorded history.

  Within days, more than 7,000 fires raged, burning up more land

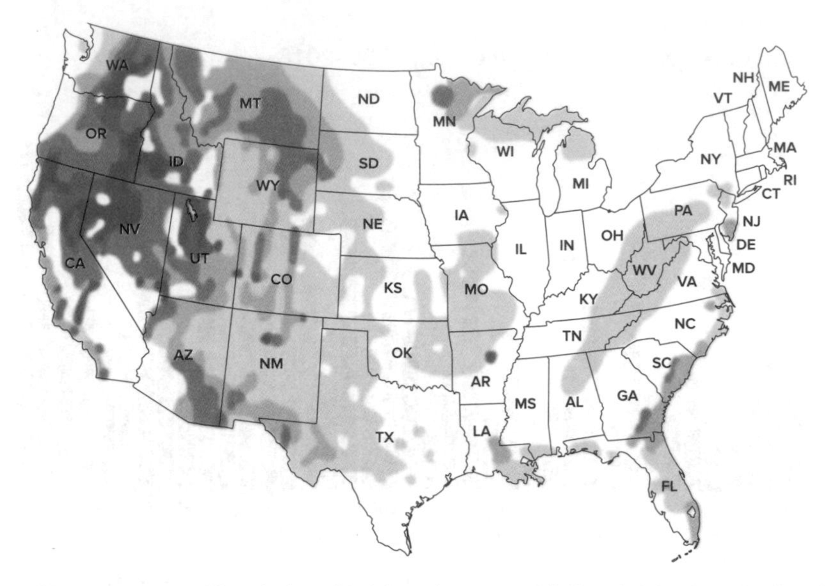

Figure 13-4. The darker the blue, the more wildfire risk in the middle of the century.

than all of California's 2019 fires combined. In the San Francisco Bay Area, three of them became the biggest megafires in state history. Over 100,000 people fled their homes into what, even at midday, looked like an orange-brown twilight. The air quality in San Francisco was the worst on the planet—four times worse than downtown Beijing's.

Dozens of towns were destroyed or evacuated—and not just in California; Oregon and Washington State, too, were experiencing the largest fires in their history. (In ordinary years, Oregon governor Kate Brown tweeted, "an average of 500,000 acres burn in an entire year. We've seen nearly double that in 3 days.")

At one point, the state told 10% of the entire Oregon population to prepare for evacuation—500,000 people. The air became so opaque that Alaska Airlines flights out of Portland and Spokane airports were suspended.

By late October, over 6.3 million acres—an area the size of Connecticut, Delaware, and Rhode Island combined—had burned across 12 Western states, destroying thousands of buildings and killing at least 43 people. Even ecosystems that don't usually encounter wildfires ignited, including forests of Joshua trees and redwoods.

The fires sent a vast stream of smoke, gases, and ash into the sky, hundreds of miles wide; satellite images showed the massive plume crossing the entire United States, where it created yellow haze in New York City and was detected as far away as Europe.

The COVID crisis made the fires' impact even more dire. Residents across three states were told to evacuate—and yet were advised to avoid congregating in shelters. And California's firefighting forces, usually supplemented by inmates from state prisons, were smaller, since so many inmates had been released to protect them from the coronavirus. The firefighters' efforts were further hampered by the necessity to live in hotels and empty fairgrounds, rather than base camps closer to the fires, and distancing rules that dictated fewer firefighters in each truck and greater spacing on the front lines.

But the United States doesn't have a monopoly on monster wildfires. In British Columbia, Canada, 3.1 million acres burned in 2017. It was the largest burned area in a single season in recorded history—until more wildfires beat that record the following year.

In 2015, Indonesian fires reduced 6.4 million acres to ash and burned for much of the year. The record amounts of toxic smog (the "South Asia haze") killed 100,000 people.

In 2018, wildfires swept through Athens, Greece, killing 80 people and burning 1,500 homes. Brazil, Spain, Latvia, Norway, Sweden, Finland, Portugal, the United Kingdom, Siberia, and even the Arctic Circle have all burned in the last two years.

And then there's Australia, where nobody will ever forget the 2019–2020 fire season. Sparked by a three-year drought, the worst in decades, *and* a heat wave that broke all national records *and* abnormally powerful winds, fires raged in all six Australian states and burned several towns to the ground. About 46 million acres burned—an area 24 times as big as the land burned in the 2018 California wildfires.

The fire was so huge that its smoke rose nine miles into the sky and created its own weather systems, which produced lightning, which in turn started new fires. The smoke in Sydney was 10 times thicker than is safe to breathe—so thick that people's home smoke alarms went off and ferry service was suspended because the boat pilots couldn't see. The smoke cloud eventually grew to the size of the continental United States and created hazy skies in South America, half a planet away.

A billion animals died in the Australian fires, including a third of all koalas. Experts assume that a number of species were driven to extinction, because so many creatures exist only in Australia.

And yet, as bad as the wildfire problem has become, it's not as bad as it's going to be. By 2050, the regions of the United States burned each year by wildfires is expected to expand from two to six times what it is today.

Wildfires and Climate Change

There have always been wildfires—and, in the kind of patterns that Nature has in mind, that's a *good* thing. They're part of a natural cycle. They burn away thick tree canopies and deadwood on the forest floors so that sunshine can reach the ground for new plants to sprout. They burn away diseased plants and infestations of bugs. They consume decaying plants and logs, returning their nutrients to the soil. Some plants' seeds, in fact, *require* wildfire heat to split them open.

But something has changed. Wildfires today are bigger, more frequent, and longer lasting than at any time in recorded history. Fire experts still remember when a 30,000-acre fire was considered huge. Today, *300,000-acre* fires burn regularly.

The reasons are all related to climate change:

- **More fuel.** The warmer climate evaporates more moisture, which dries up forests and vegetation. Eventually, millions of acres become a massive powder keg waiting for a lit cigarette, a power line spark, or a bolt of lightning to begin a blaze. It's no surprise that California's horrific run of wildfires coincides with six of the hottest years ever recorded there—2015 to 2020.

- **Less snowmelt.** In the warmer world, spring begins earlier, so the forests are dry for a longer stretch each summer. What used to be a summer-to-autumn wildfire season is quickly becoming a year-round phenomenon.

- **Less rainfall.** Rainfall patterns are changing in the new climate, too. Just our luck: There's *less* rainfall in the places most likely to get wildfires. California's rainy season is rapidly shrinking to a December-to-February stretch. The rest of the year is getting less and less rain, essentially becoming wildfire season.

  Less rain equals more droughts, too, which are welcoming committees for wildfire.

- **Shifting winds.** The Santa Ana winds are responsible for 80% of the damage in California's wildfires. These are annual winds that blow *westward*, from the hot, dry desert regions of Nevada and Utah, across Southern California and out to sea. They blow hard—40 to 100 mph—and they're responsible for California's weird, hot, dry autumn weather. But they're also contributors to wildfires, because they're hot and dry, and they're wind.

  In the hotter world, these winds are becoming even drier, making them even more susceptible to fanning the flames. As a result, meteorological models predict a 70% increase in wildfires by 2050.

Figure 13-5. The Santa Ana winds blow smoke from the Thomas Fire and two others out over the Pacific, burning out of control in Southern California in 2017.

- **Beetles.** The warmer climates have invited massive armies of bark beetles to crawl into northern forests that used to be too cold for them. They've killed hundreds of millions of trees in the United States and Canada—trees that, once dead, dry out and become superb kindling.
- **More lightning.** Among the wildfires that *aren't* started by human carelessness or arson, lightning is the most common fire starter. Around the globe, lightning strikes the earth 8 million times a day—but by 2100, that number is expected to increase by 50%, thanks to an increase in water vapor in the air.

In addition to warming the climate, we the people make two more direct contributions to the growing wildfire problem:

- **More aggressive construction.** A huge part of the problem is the growing WUI. That's the goofily named wildland urban interface, which is pronounced, even more goofily, "wooey." It refers to areas where humans build their homes right up against wilderness, like forests and chaparral.

  We've been building our homes deeper and deeper into wildfire fuel territory, expanding the WUI, for decades. At this point, a third of all homes in the United States sit in these areas.

  That's one reason the *costs* of wildfires keep skyrocketing; for example, nine of the top ten costliest wildfires in U.S. history have all burned since 2003. If a wildfire rages in some unseen forest, nobody adds its damage to the dollar figure of destruction.

  Only recently has it occurred to developers that once-in-a-lifetime wildfires have become an annual threat. All told, 4.5 million U.S. homes sit in areas with "high" or "extreme" risk of wildfire.
- **More aggressive firefighting.** Weirdly, part of the reason we're getting more fires and bigger fires is our century-long habit of *putting out* wildfires when we can.

  When we put out fires at every opportunity, we're stifling Nature's cycles. The brush becomes overgrown; the tinder builds up. "Every time you get one of these big fires, it is the result of 100 years of management decisions where they went and put out lightning strikes, they limited or shut down prescribed fire," Malcolm North, a fire ecologist with the U.S. Forest Service, told NPR. "And those decisions eventually accumulate and bite you in the butt."

Fortunately, we're learning the lesson. Today, most forestry experts are inclined to let natural fires burn (that is, when they don't threaten civilization) and may even start controlled burns ("prescribed fires") themselves. But it will take a couple of tree lifetimes before our better forest management practices begin to affect wildfire frequency.

How They Spread

A building can catch on fire in three different ways:

- **Conduction** is what you probably imagine: flames from the fire lapping against the side of the house, igniting it. But that's actually not what sets most houses on fire.
- **Convection** is how wildfires burn down most houses. Hot air, rising from the fire, creates a column of superheated smoke. Embers flying through the air—sometimes a mile or two in front of the fire—alight in dry leaves or dead needles in your gutters, waft into your attic through the soffit vents, blow in through windows, ignite dried-out roofing tiles, or lodge against the wall and ignite it. Over half of all houses that burn down in wildfires ignite from these embers.
- **Radiation** is heat thrown off by a fire. When it's cold out and you're sitting by a campfire, fire radiation can feel good. But the radiation from a wildfire is far more dangerous. It can set your house on fire from as far as 30 feet away.

How to Prepare Now

Don't go blaming Mother Nature for wildfires. Lightning and lava can start one, but 90% of them are started by *people*, either through maliciousness or stupidity.

Downed wires and improperly maintained power company equipment led to the Camp Fire and 12 other fires in California in 2017 and 2018; it turns out that dry grass, strong winds, and high voltage aren't a great combination. The resulting lawsuits have driven Pacific Gas & Electric into bankruptcy.

By a huge margin, though, the winner of the "dumb ways people start wildfires" derby is campers who don't fully put out their campfires or who have no way to stop campfires that grow out of control. You can make sure you're not one of the responsible parties like this:

- Check with a ranger or park officer to see if you're even allowed to start a campfire, given the current conditions.
- Start your campfire only within a ring of stones or metal.
- Don't start a fire until you've rounded up enough water to put it out if it starts going wild.
- Keep your firewood a safe distance from the firepit.
- After you've put the fire out, it should be so *out*—cold, wet, still, and silent—that you can run your hand through the ashes.

Other popular ways people start wildfires:

- **Burning yard debris (slash piles).** In certain times and places, burning your trimmings is legal; others, not. Either way, wind can quickly carry embers into nearby burnable vegetation, and it's off to the races.

  Call the local fire department before you burn. They'll advise you on current conditions, and maybe suggest that you wait a couple of days.

  Then, before you burn, look around for potential wildfire fuel, either near the slash pile's edges or hanging over the top. You want an open space *above* the fire that's three times the height of the pile and *around* the fire that's at least 10 feet of cleared, wet dirt, pavement, or gravel.

  Watch the pile burn, standing by with a hose. Once the pile is out, water it and shovel dirt over it—more than once. Check it for the next few days to make sure nothing is still burning.

- **Tossing cigarettes.** You can't just toss a cigarette out the car window or drop it on the ground. If you're in the woods, you *have* to squish the butt with your shoe, so that it's completely and utterly out. (And then kindly take it with you.) Logs, stumps, and branches do not make good ashtrays in the wild.

- **Kids**. For heaven's sake, watch your young people when they interact with matches, lighters, firecrackers, or fireworks. Keep them from doing something dangerous.

Preparing Your Home

Chapter 3 contains a concise guide to wildfire-proofing your home. The first step is being aware of your fire risk in the first place. There's no national website where you can type in your address and assess your property's risk, but three sites come close:

- **GeoMac Wildland Fire Support.** This site shows all currently burning fires: maps.nwcg.gov/sa/. But if you click the Layers icon and then Fire History, you can see how many fires have burned here in the past—a likely indicator of future fires.
- **California Fire Risk.** The interactive map at https://ia.cpuc.ca.gov/firemap/ shows detailed fire-risk levels for California, as prepared by the California Public Utilities Commission. It starts out displaying terrain without any street names. To summon a more familiar Google Maps–style display, in the Basemaps control in the panel at the left side of the screen, click OpenStreetMap.
- **SouthernWildfireRisk.** At https://southernwildfirerisk.com/Map/Public/#whats-your-risk, you can dial in the fire risk for any block in the *Southern* U.S. states, from Texas to the East Coast. Even though California's massive wildfires have captured most of the headlines, the South gets more fires every year than any other region.

Chapter 3 explains the concept of defensible space around your home—a 30- or 100-foot no-plant's-land that deprives oncoming fires of fuel. But then you have to *keep* it clear. Keep the lawn mowed, leaves raked, weeds trimmed. No vines on your walls—sorry. No mulch within the first five feet of your house, fence, or deck, either; it's basically tinder. Keep your gutters clean like you're obsessed; everything in them can dry out and become 100% organic, high-efficiency kindling to burn your house down.

Look *under* things, like the porch and the deck; clear out dead leaves,

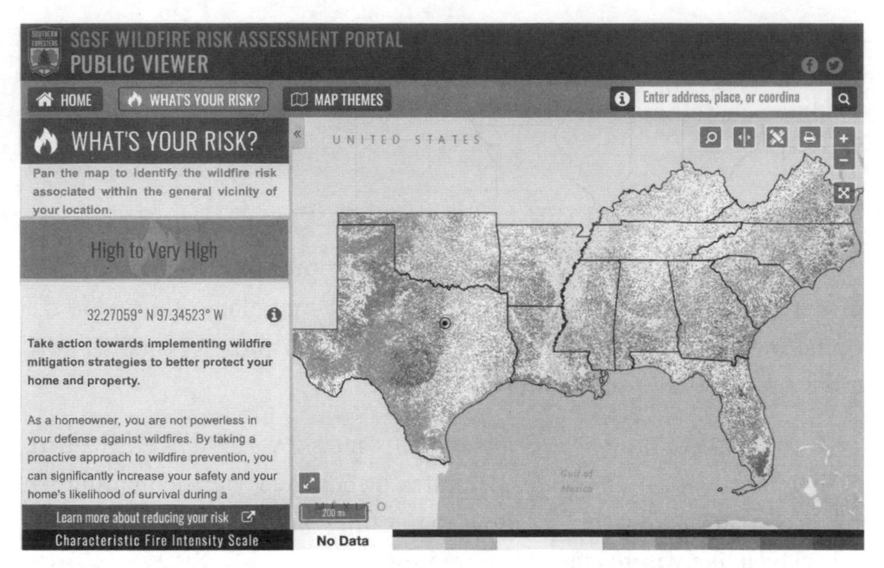

Figure 13-6. At SouthernWildfireRisk.com, you can zoom in to the street level to see your risk of wildfire.

branches, and trash. Look *over* things, too—make sure there's at least 15 feet between the crowns of nearby trees, making it hard for flames to leap from one tree to the next. Check for dead branches over your roof. Trim away any growing thing within 15 feet of your chimney or stovepipe that could serve as a flame bridge to your house. And cut away tree limbs within 10 feet of the ground that could act as ladders for a ground fire.

Your firewood pile should be at least 100 feet away from the house and *uphill* from it. At www.readyforwildfire.org, diagrams indicate safe spacing of trees, which varies if they're on a slope, if they're close to your home, and so on.

Mainly, though, you just have to *think* like a wildfire. What hops could it make from the nearby forest or chaparral to your home? Fires can travel along branches that overhang your roof, across dried wood chip mulch, and through grass clippings that have dried out. Embers love to land in leaves, piles of needles, tall dry grasses. Fire can hop from treetop to treetop if they're close together.

Get rid of it all, and stay vigilant.

Wildfire Gear

Preparing for a wildfire neatly follows the general emergency-prep outline described in Chapter 8. Long before the fire starts, you should create a go bag for each family member, think about escape routes, sign up for early-warning alerts, and equip yourself with apps or a NOAA radio that keeps you apprised of the latest fire news.

But prep for wildfire involves a few extra pieces of gear:

- **A fire extinguisher.** A basic one costs $20. Teach everyone in the family how to use it.

- **Smoke detectors** are required by law in most places, but they're especially important in wildfire country. True, you'll smell and see a wildfire with your own nose and eyes before your smoke detector goes off—but what about when you're sleeping?

  So, of course, it's important to keep the batteries fresh in your smoke detectors. Of all deaths at home in fires, 59% occur in homes with missing or broken smoke alarms.

- **Prepare your air filter.** If your home has central air-conditioning or central heating, you've got an ace in the hole: a powerful whole-house air filter. When the smoke hits the fan, you'll be glad it's there, filtering away ash and particles that you otherwise would have had to breathe.

  You want a filter with the highest MERV value your system can handle. The higher the MERV rating (Minimum Efficiency Reporting Value)—they go from 0 to 16—the finer the particles the filter can scrape out of the air.

  But here's the thing: Not all air handlers have the power to suck air through the thickness of a high-efficiency filter. You can't slap a high-MERV filter onto a medium-MERV furnace; that'll damage your equipment. Look up your air handlers' maximum MERV in its user manual. If that's long gone, Google the model number.

  In any case, before any fire starts raging, you should get some high-MERV filters—at least MERV 8, or whatever is the highest your system can handle—and have them ready, including spares.

  All of this is especially important if anyone in the house is likely to be more affected by the smoke: the very young or old, people with lung or heart conditions, pregnant women.

- **Windex and wipes for the car.** If you have to evacuate or drive through smoke, the *inside* of your windshield can get dark and ashy from the smoke. Have wet wipes or paper towels and water or window cleaner handy.

- **Consider an air purifier.** Chapter 3 describes the different kinds of air purifiers you can buy for your home. Here again, they're mostly important if there's someone in the house who's especially susceptible to the smoke. Keep in mind, though, that most air purifiers don't remove gases from the smoke—a good reason to close your house's cracks as tightly as you can.

  If you think one might be useful, *buy it now.* Once wildfires break out, they'll be sold out of every store.

- **Consider a mask.** If you live in wildfire country, look into buying a couple of N-95 or P-100 respirator masks. They should say NIOSH right on the filter material, meaning that they're approved by the National Institute for Occupational Safety and Health. A box of ten costs about $22 (at least when there's no coronavirus plague going on).

  These masks can be useful if you have to go outside into the smoke, especially if you'll be working or exerting yourself.

  They're not a slam dunk, though. For example, these masks don't work if they're not fitted tightly to your face, which means they don't fit children and they don't work over a beard or sideburns. Note, too, that the masks keep ash and particles out but don't do anything to block fumes and gases.

  The masks come in sizes. You want one that'll fit over your nose and under your chin, and that has two straps for your head, not one. If there's a metal tab over the bridge of your nose, pinch it down.

  These masks can make it hard to breathe, which can lead to a faster heart rate and heat stress; if you're feeling dizzy or having trouble breathing, go indoors and take the mask off.

Depending on your budget, proximity to the WUI, and sense of paranoia, you might also want to equip yourself with some nice-to-have tools like these:

- Fire tools like buckets, shovels, rakes, axes, and saws.
- A ladder tall enough to reach your roof.
- Drapes made of heavy, fire-resistant fabric. You can also get your curtains, drapes, and upholstery sprayed with fire-retardant treatments.
- A hose, or hoses, long enough to reach any spot around your house—and exterior faucets to connect them to. (These should be the frost-free or freeze-proof type.)

Finally, do you know how to open your automatic garage doors manually? Once you've abandoned the house, you'll want them closed—but you may need to open them to get out. And do you know how to shut off the gas (or propane or fuel oil)? That's an extremely important task before you evacuate.

The Clean Room

The EPA suggests designating, as part of your preparation, a *clean room*—the room you'll hole up in when there's a wildfire nearby but you can't evacuate, or have been advised not to.

If you can predict where the fire will be coming from—if, for example, the forest is on only one side of your home—this should be a room farthest from the fire. Not the basement, though, where you'll be trapped if the house collapses.

This room should be big enough to hold everyone in the family, like a bedroom with a bathroom attached. This is the room you'll want to equip with an air purifier; during your stakeout here, you'll close (but not lock) all the doors and windows and run the purifier nonstop.

Insurance

As you know from Chapter 6, the only thing growing faster than the sea level and temperature is the importance of good insurance.

The good news is that homeowner's insurance covers fire and smoke damage. Even water damage caused by firefighters putting *out* a fire is covered. Just this once, the cards seem stacked in your favor.

The bad news, though, is that the insurance companies are quickly getting sick of paying for wildfire damage to homes that everybody knows are in wildfire *country.* It's like, "Dude—you chose to build or move there. Why is that *our* problem?" (That may not be the actual wording in your policy.)

In the wildfire-happy Western states, the insurance companies are eliminating wildfire coverage from home policies, or charging an arm and a leg for it, or refusing to renew people's policies altogether. Chapter 6 has more on this distressing pattern.

Lessons from the Paradise Fire: Fill the Tank

Paradise, California, is what former mayor Woody Culleton describes as "a bedroom community on a dead-end street." And when the most destructive fire in California history (at the time) swept through it in 2018, there was only one way to survive: Join the other 26,000 residents on the single road out of town.

"All the lanes were full of cars going out of town. You were at a dead stop," he recalls. "There were cars that ran out of gas, or had been burnt, on the sides. There was no way to go around anybody. You couldn't go back, you couldn't go forward, you were just stuck—with fire all around."

Culleton is a memorable character in the 2020 Ron Howard documentary *Rebuilding Paradise*, which includes terrifying video taken by residents sitting in that traffic, engulfed in flames.

"There was no place there wasn't fire," says Culleton. "The winds were so strong, embers were blowing sideways. It was 9:00 in the morning, and it looked like midnight. The only light was from the flames. Everybody thought they were gonna die."

Including Culleton's wife. Once she had joined the motionless line of cars, she realized that her car's tank was only a quarter full. She was so sure that she was going to perish that she phoned her son to say goodbye.

In the end, after five hours in traffic, she made it the ten miles out of town. But Culleton's takeaway for anyone who lives in wildfire country—or hurricane country or earthquake country—is this: Keep your gas tank full, or your EV's battery charged—all the time.

"The message that I would share with everybody is: Make sure your gas tank is full," he says. "Whenever your gas tank says half full, consider it empty, and fill it up."

It's not just your life you'll be saving, he says. If your car runs out of gas, you've effectively blocked all the cars behind you. "What about the people four cars back, who are 86 years old and can't get around you, and can't get out and run?"

It's a new way of thinking: "Instead of putting $5 in the tank to get it off Empty, put $5 in the tank to keep it on Full," he says. "Because you never know."

During the Fire

As in other extreme-weather events, the National Weather Service issues both wildfire *watches* (conditions are right for wildfire) and wildfire *warnings* (wildfires currently burning).

If you get an evacuation notice, it may be termed Voluntary or Precautionary (meaning "This is the first sign of danger") or Mandatory or Immediate Threat ("get out now").

But wildfires are different from other disasters in one important regard: The early-warning system may well be *you*.

"Wildfires get reported by human beings. They do not get sensed by some big old wildfire eye in the sky. There are no helicopters flying over wildfire country every day," says Linda Masterson. "People have to call people, and *they* have to call people, and then the county has to figure out who's in the evacuation zone."

She and her husband did get a call, ordering them to evacuate their home—nine hours *after* it had burned to the ground.

If you spot a fire, call 911 to report it. You may well be the first person to call it in, and you might be saving people's lives.

How to Get Warned

Fortune favors the prepared—and in a disaster, fortune favors those who get out first. You don't want to be trapped in traffic as a ten-mile wall of flames bears down.

By far the most essential preparation step is ensuring that you'll *find out* about an evacuation order. See page 250 for details on FEMA's Wireless Emergency Alert network and your local "reverse 911" notification system, which you must sign up for in advance. It's free and relatively simple; do it today, then you can forget about it.

How to Evacuate

If you get a Mandatory or Immediate Threat alert, *get out.* Forget your house, your belongings—just grab your kids, pets, and go bags, and go.

If you feel threatened, no need to wait for the official evac order; just go. Wildfires start fast and move fast. What you see, hear, and smell may provide more recent information than the government's systems.

"The longer you wait, the better the chances you could be trapped," says Masterson. "And trying to get out in the middle of a fire that's blowing up is a heart-stopping experience."

But what if you have a little time? Maybe there's a wildfire in your area, but it's nowhere near your neighborhood.

You can give your home a much better chance of surviving the wildfire by running through as much of this checklist as you can. Naturally, it's quicker if you have family members to share the tasks:

- **Gather the family and the pets.** Outline the plan—which, having read Chapter 8, you've all covered before. Keep everyone calm. Assign everyone jobs.
- **Call your neighbors** to make sure they've gotten word of the evacuation.

♦ **Close up the house—tightly.** The idea is to protect the house as best you can should the fire come near. Close every door, window, and vent on every floor of the house—including dog doors.

Take down your drapes and curtains (unless they've been fire treated). Close your shutters, shades, blinds, and any remaining drapes and curtains (the fireproof ones), to prevent radiant heat from setting the interior on fire.

You should even close all *interior* doors, so that there's no air movement that could spread fire from one part of the house to another. Pull furniture away from the doors and windows. Close the garage doors.

Turn off the AC, the heat, and the fans. You don't want air moving through the house at all.

Someone should do a sweep outside to move everything burnable inside or into the garage: patio furniture, doormats, trash cans, toys, and so on.

♦ **Invite the firefighters.** When a wildfire strikes, firefighters are instructed to do everything they can to defend homes. But there are more buildings than there are firefighters; every year, some homes burn to the ground because they weren't accessible to fire trucks and gear. This is really important: You want to make it easy for them to reach *your* house and protect it.

To begin, turn on every light, inside and outside the house. Make your house visible through thick smoke.

Think about the fire trucks. Do they have room to park? Is there a gate you should open?

Think about the people who'll be trying to defend your house in a dark, smoky, chaotic, dangerous atmosphere. If you have a roof-reachable ladder, haul it out and lean it against a corner of the house. Hook up your garden hoses; make them easy to spot. Fill up buckets with water and leave them around the house, so that a firefighter will have a quick way to douse small flames.

Move your lawn furniture and deck chairs into a cluster downwind from the house, so they won't catch the firefighters' hoses as they maneuver.

♦ **Clear off the deck.** Lots of stuff on a typical patio, porch, or deck is flammable: furniture cushions, woven mats, potted plants, propane tanks on grills. All of that stuff is catnip for embers—and it's right up next to your house. Get it inside or far away.

♦ **Wet everything.** Moisture slows flames. Use a hose or sprinklers to soak your roof, shrubs near the house, and fuel tanks. Don't leave the water running, though; leave enough water pressure for the firefighters to use when they arrive.

If you have a pool or a hot tub, feel free to toss anything valuable into it that won't be ruined by a little water exposure: a box of jewelry, a stone sculpture, a ziplock bag of anything. If your house burns to the ground, this stuff has a shot at being saved.

♦ **Shut off the gas, propane, or fuel oil**. If you have a grill with a propane tank, move it far away from the house or other buildings. If your place does get in the way of the fire, the last thing you need is stuff blowing up.

♦ **Dress for fire.** This might sound grisly, but here it is: Synthetic fibers like nylon, acrylic, and polyester can melt onto your skin, causing severe burns. If cotton and woolen fabrics catch fire, on the other hand, they just turn to ash and fall off.

Note to people with fancy pajamas: Silk is the worst. It catches fire easily and burns fast.

The bottom line: As you escape, put on long, close-fitting clothes like jeans and tight-fitting long-sleeved shirts or tops, and sturdy shoes. No flip-flops. Gloves and bandanas are helpful, too.

♦ **Move the cars.** Get them positioned to zoom away without turning, so you'll be able to go fast if you have to.

♦ **Open the animals' gates.** If you have animals that you can't evacuate—horses, for example—give them a chance to escape.

♦ **Tell somebody out of town**, like a friend or relative, that you're leaving and where you're going.

- **Keep your receipts**. As usual, when the smoke clears, your insurance company will pay you back for whatever you spend during your evacuation.
- **Check InciWeb.** At https://inciweb.nwcg.gov, you'll find a central database of wildfires, prescribed burns, floods, and hurricanes, all over the United States, with information provided by all government agencies. Zoom in to your town, click the icon of the fire in question, and then click Go to Incident to open its information page.

  You'll find photos, videos, maps, and news updates, along with buttons that let you follow updates about the fire on Twitter, Facebook, or Instagram. This info also includes roads that may be closed and shelters that are open.

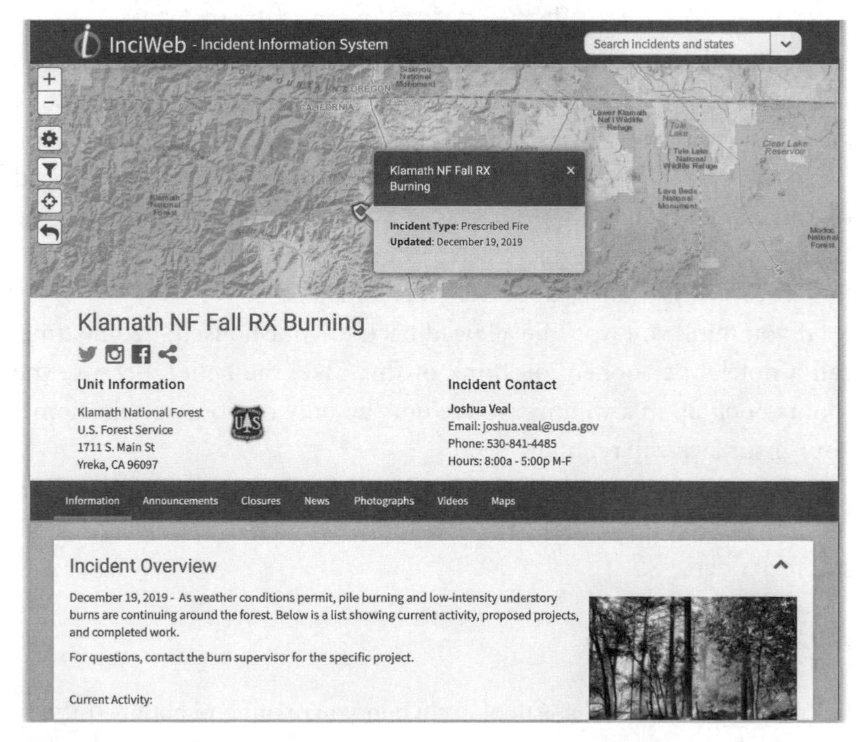

Figure 13-7. Fire experts, watching your particular wildfire, update the government's InciWeb page in real time, so it's likely to be more accurate, and updated more often, than the local news.

- **Finish off your go bags** with the last-minute items: your wallet, ID, latest medicines, water.

Load the car with the go bags, your family, and your pets, and get going.

Where to Go

Okay, so you've evacuated. Now what?

Drive slowly but steadily—headlights on, windows and vents closed. If your gang is in more than one car, the biggest and most rugged vehicle should lead. Be alert for fire and rescue trucks tearing by in the opposite direction.

If you can make your way to the home of a friend or relative out of the fire zone, by all means go.

Otherwise, check the Emergency: Alerts app (the Red Cross app described in Chapter 8) or the FEMA app. Both will provide the addresses of emergency shelters that are operating during the wildfire. If you don't have these apps, you can also check the Red Cross website (redcross.org).

Actually, it's a great idea to stop in at a shelter even if you don't need it. You can let them know you're safe so that rescuers don't waste time trying to find you; share and receive information about the fire; and talk to Red Cross, insurance, and emergency officials.

If you think you won't be allowed back to your house for some time, call a hotel. The sooner you think of this idea, the better, because the rooms book up in a wildfire; you're not the only one out of your home. Your insurance will reimburse you.

If you had to leave animals behind, call the Humane Society, which is often allowed to go back to homes to make rescues even when you're still forbidden.

If You're Trapped

Getting trapped behind a wall of fire when you're out and about—hiking, driving, evacuating—is not a great situation; in a wildfire, you *really* want to be inside a building. Here are the survival techniques for trapped situations:

- **If you're on foot**, drop your stuff and move as far away from vegetation as possible. Never go *toward* a fire, even if that's where your car is, where the trail entrance is, or where the GPS's instructions tell you to go.

  A fantastic option: Hide out "in the black," meaning a spot that's already burned. It won't be comfortable or cool, but at least the fire can't come back to kill you. If you see firefighters, run to them.

  If it's too late to escape the approaching wall of fire, lie facedown in a depression, like a ditch or a streambed, feet toward the fire. Dig a hole for your face to avoid breathing smoke. Call 911. Stay there, no matter how loud and scary it gets. If the fire passes over you, wait to make sure it's not coming back, and then make your way to safety, sticking to blackened, burned areas when possible.

- **If you're in a car**, park as far away from vegetation as possible, and definitely not under trees. Close the windows and vents, lie on the floor, cover yourself with a wool blanket or jacket if you have one, and call 911. Remember: Smoke and gases rise. The lower your face is in the car, the better your chances of surviving. Don't get up if you hear a crash or see flashes. You're safer in your car than out of it; stay there until the fire has passed and the temperature starts to drop.

If you're at *home* and trapped by the fire, call 911. Gather the family, including pets, into the clean room you've designated. Shove wet towels under the doors to keep smoke and embers out.

Fill all the sinks, tubs, and trash cans with water. Close all the doors and windows, but leave them unlocked so rescuers can get to you. Pull down any drapes and curtains that aren't fire-resistant, turn on all the lights, and turn off AC and fans. As noted above, everyone should put on long sleeves—denim, cotton, wool—and sturdy shoes. Stay away from windows. When the fire comes, it's shockingly loud and terrifying, but *don't go outside*; it's four or five times hotter there, and the air isn't safe to breathe. Stay low to the floor, always. Cover yourself with wool blankets or jackets, which don't burn easily.

Linda Masterson notes that it takes about an hour for a house to burn down. Your goal, then, is to wait—long enough for the fire to pass but not so long that the house collapses around you—and then make a run for it. Stay low and make your way to areas that have already burned.

Smoke: A Users' Guide

You know the old saying: Where there's fire, there's smoke.

In a wildfire, smoke is creepy, orange, and surreal. Smoke simulates nighttime even at high noon. Miles away from the actual fire, you can smell and taste it.

Smoke is made up of thousands of chemical compounds, but the stars of the show are carbon dioxide, carbon monoxide, water vapor, dust, ash, hydrocarbons, nitrogen oxides, and tiny amounts of minerals.

Breathing a lot of wildfire smoke isn't good for you. It's not just ash and particles but also gases and fumes. Classic symptoms of "wildfire smoke is getting to you" syndrome are coughing, irritated throat and sinuses, stinging eyes, runny nose, headache, chest pain, or trouble breathing. Most of this goes away when the fires are out, leaving you with no long-term damage.

In general, you *don't* have to worry about getting carbon monoxide poisoning or cancer. The primary risks are to your eyes, nose, sinuses, throat, and lungs.

Sensitive Groups

Things are different if you're a member of a *sensitive group*, a term that pops up often in the government's writing about wildfires and smoke. You're in this special group if you're very old or very young; you have lung problems, such as asthma, allergies, or COPD; you have heart problems; or you're pregnant.

In these cases, the health effects of wildfire smoke can be much more severe and can lead to chronic problems. Every guideline for wildfire safety includes special considerations for sensitive groups.

Avoiding Smoke

The best way to avoid smoke exposure problems is . . . to avoid exposure to the smoke. Mainly by:

- **Staying inside.** The air indoors is at least 50% cleaner than it is outdoors, so close the windows and doors and make yourself comfortable.

In general, you won't be asked to evacuate for smoke conditions—only if your home is in the line of the fire. The exception: sensitive groups. *You* should consider getting out of town until the smoke clears.

- **Relaxing.** Don't exert yourself. When you're exercising or working hard, you tend to breathe deeply—and through your mouth, bypassing all those nice ash-cleaning hairs in your nose. In both cases, you suck more particulates deeper into your lungs.

  Be grateful. You're being *ordered* to sit inside and read or watch movies all day!

- **Running the AC.** If it's hot, run your air conditioner instead of opening the windows.

  And what if you don't have AC? Tough call. You can't open your windows, because you'll let the smoke in—but staying inside invites heat stress. In that case, make every effort to go somewhere that does have AC.

  If you have *central* air or heat, your thermostat has a Fan control. Set it to On (instead of Auto), so that you're recirculating your home's air through its central filters.

  A few air conditioners and homes have a Fresh Air setting, or a way to turn *off* recirculation of the air. Clearly, that's a function you want to turn off. You don't want your own home deliberately pulling smoky air in from the outside.

- **Not making the problem worse.** In a wildfire smoke situation, you've got your home sealed up as much as possible. For that reason, you're supposed to avoid *creating* smoke and fumes within the home—by smoking, using stoves or furnaces that rely on gas, propane, or wood burning, using spray cans, frying or broiling meat, burning candles or incense, or vacuuming. (Vacuums with HEPA filters are okay.)

- **Close the car's windows and vents.** If you must drive, put your car vent on the Recirc setting, so you're not bringing smoky air into the car. But get where you're going and then get out. Don't use the car as a shelter with windows up and the air on Recirc; especially in newer cars, the carbon dioxide buildup from your breath can become dangerous.

Here, for your reference, are a few things that *don't* help, or don't help much:

- **A towel, handkerchief, bandana, or tissue** can keep embers and big ashes out of your mouth, but won't do anything to filter out fine particles or gases. It doesn't matter if they're damp or dry; they just don't fit tightly enough to your face.
- **Surgical masks.** Useless. The seal isn't good enough to keep out particles and gases. As you may have read 60,000 times during the COVID pandemic, these masks are primarily designed to capture germs coming *out* of someone's mouth.
- **Dehumidifiers (and humidifiers).** According to the EPA, humidifiers and dehumidifiers "will not significantly reduce the amount of particles in the air during a smoke event. Nor will they remove gases like carbon monoxide."

Sneak Out When the Smoke Clears

During a wildfire, the smoke levels can change a lot. Smoke may settle in the streets during the night, but then blow out in the morning, for example. You may be able to plan around the smokiest periods, so that you can leave the house and get things done.

To check out the current smoke conditions, visit www.inciweb.com, zoom in, and click the fire near you. Local weather authorities annotate the smoke patterns in real time there.

You should also bookmark www.airnow.gov, a national, multi-governmental-agency air-quality database, updated hourly. At any time—wildfire or not—you can check out the current air quality where you live (or don't live). You can sign up for automatic emails alerting you about air quality problems here, too. (There's an AirNow app, but it seems to have been last updated in the Eisenhower era. You're better off with free, modern apps like Air Matters or AirVisual Air Quality Forecast.)

If you click "Fires: Current Conditions" below the map, you can see all the fires currently raging in the United States at this moment. Click an icon to see the current air quality reading for that area. (The direct link to this Fires page is https://airnow.gov/index.cfm?action=topics.smoke_wildfires.)

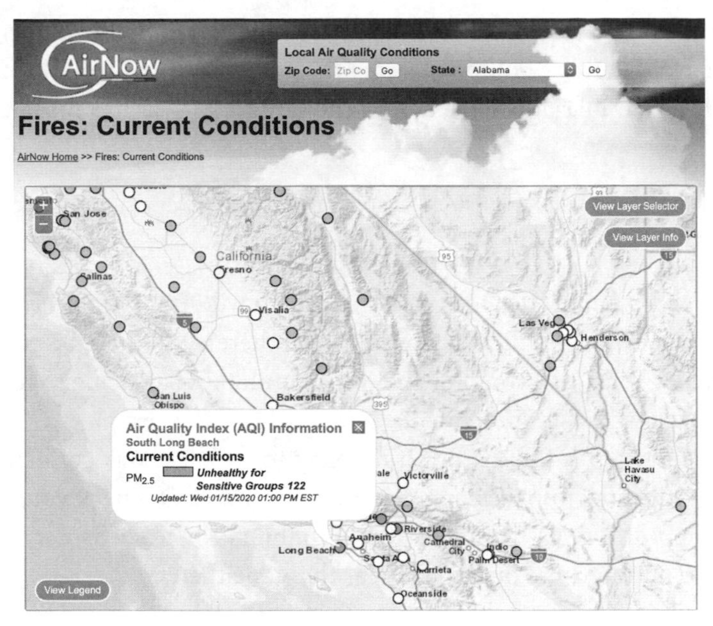

Figure 13-8. At AirNow.gov, you can check the current air quality for your favorite fire. Never say that the government doesn't do anything for you.

Should You Cancel?

But what about the Little League game? Flag football? The soccer match? The company picnic? Should these events proceed as scheduled during wildfire-smoke episodes?

Start by consulting www.airnow.gov to confirm that it is, in fact, still smoky out there. If so, and the event is outdoors—yes, cancel, especially if it's outdoors and *athletic*. People should not be breathing hard in the smoky air.

If it's an *indoor* activity—even an athletic one—proceed as scheduled. Indoors is where you want to be anyway. A school, mall, or gym may well have better AC and filtration than many homes do, too.

Pets and Livestock

Smoke affects nonhuman creatures just as it affects us: irritation of the eyes, nose, and respiratory systems. And the advice for them is exactly the same as it is for you: Avoid smoky areas if possible; drink plenty of water; avoid exercise outdoors; and if there's coughing or trouble breathing, see a veterinarian. (Well, *almost* the same advice as it is for you.)

After the Fire

It's not easy coming back to your home after a wildfire. The *uncertainty* is almost unbearable. Will your house still be there? Will there be anything left inside? What will you do while you rebuild, if you decide to rebuild? How much will insurance cover?

The first step is finding out if it's okay to go back. Ask somebody in charge—the fire or police department, for example. Or check your emergency app or NOAA radio. As the firefighters (or the weather) makes progress putting a fire out, you'll hear that it's either:

- **Contained.** A fire that's "70% contained" means that the fire is still burning, but 70% bounded by barriers: trenches, rivers, burned swaths, or shallow 12-foot-wide trenches dug by firefighters.

Figure 13-9. What's left of a home in Paradise, California, after the 2018 Camp wildfire.

Note, however, that a contained fire can still rage for weeks or months more. It can also hop right over the containment line if it really feels like it.

- **Controlled** is what you really want to hear; it means "almost out." There's nothing left to do but send mop-up crews to put out the remaining hot spots.

When you get back home, make sure that nothing is still smoking, smoldering, or sparking. Check the yard, plants, stumps, roof, gutters, attic. If anything is still a hot spot, call 911 to report it; you really, really don't want to be there when flames flare up again.

The Insurance Process

This is the moment you've been waiting for—the reason you paid such close attention to Chapter 6, learned the ins and outs of your insurance policy, and took pictures of everything in your home.

Call your insurance company and tell them about the fire. They'll send a representative quickly to look over the damage.

The Cleanup

You can begin the cleanup of your burned home in any of three ways:

- **Use a government debris-removal program.** Your county or state may offer a cleanup program for its residents after a wildfire. You have to grant workers entry to your home—but the whole thing is free. In exchange, you agree to turn over your insurance payments for debris removal to the county or state.

  In general, they'll first haul away anything that's dangerous, like asbestos, paint, and batteries. Then they'll come back to remove anything that's been burned, dangerous trees, contaminated soil, and destroyed foundations.

  The nice thing about these government-run programs is that they have an incentive to get the work done fast, so that subsequent rains won't run all the ash and toxic compounds into the water supply.

- ♦ **Hire a contractor.** You can also hire an outside company to do the cleanup. You'll have to pay for this yourself—it's not cheap—but your insurance should cover it.
- ♦ **Do it yourself.** A grubby, sweaty, ugly job—but free.

The Aftermath

Depending on the damage that was done to your home, you'll have to wrestle with a big decision: Rebuild? Or set up shop somewhere else?

Keep in mind, as you ponder, that property values typically plummet by at least half following a wildfire. Who's going to want to buy land that's obviously in a danger zone? It will recover in a few years, but right after the fire is probably the worst time to sell.

If your home was only partly damaged, your big challenge will be getting the smoke smell and soot out of the surfaces and furniture. Masterson notes, for example, that you can't just paint over smoke-damaged walls; the smoke eventually seeps through. You may have good luck first cleaning the walls with TSP (trisodium phosphate) and then using an oil-based stain blocker before painting.

Smoke-saturated rugs, furniture, curtains, and so on might have to be replaced. Don't throw anything away until your insurance adjuster has had a look at it.

The end of Chapter 6 describes the process of working with your insurance company after a disaster; the end of Chapter 8 covers the shaky, upsetting, slow process of recovering psychologically and financially.

All of that is much, *much* easier if you've prepared. Take it from Linda Masterson, who lived through it.

"You can't prevent a wildfire," she says. "If some yahoo throws his cigarette out the window and it hasn't rained in eight weeks . . . or somebody goes camping and doesn't put their campfire out properly . . . or lightning strikes in the middle of the woods in a windstorm . . . there's lots of things that happen that are not in your control. But how well prepared you are, how you know your evacuation routes, how you know your insurance, what you have with you, how well prepared you've gotten your family—*that* stuff is well within your control."

Chapter 14

Preparing for Mosquitoes and Ticks

THE NORTHERN PARTS OF NORTH AMERICA AND EUROPE AREN'T just lucky because they're cool, beautiful, and resource-rich. They're also lucky because ticks and mosquitoes generally prefer hotter territory. Malaria, for example, infects 214 million people a year—and 90% of them live in Africa.

But lately, something is going on.

This planet's hot zone, around its equator, is expanding fast—an average of 5.5 feet a day. And as the warmer belt grows, so grows the territory of every animal, plant, fish, and bug that lives there.

"It's a very well established trend," says Laurence Smith, Brown University earth scientist and author of *The World in 2050*. "These things have been mapped, using climate models to show what their geography is. It's quite startling."

Some of the critters that benefit from the larger, warmer landscape, like starfish and trumpeter swans, may not bother us.

Others, though, spell big trouble. There are the bark beetles that are decimating the forests of the Pacific Northwest, thanks to the shorter winters. There are the crop-eating rootworms that eat more in warmer weather, costing corn farmers $1 billion a year. And then there are the ones that can kill you.

According to the Centers for Disease Control (CDC), illnesses from mosquito and tick bites in the United States more than tripled since 2004. That's a stunning increase, and a hint at the new normal.

"In North America, we haven't had to live our life with that kind of

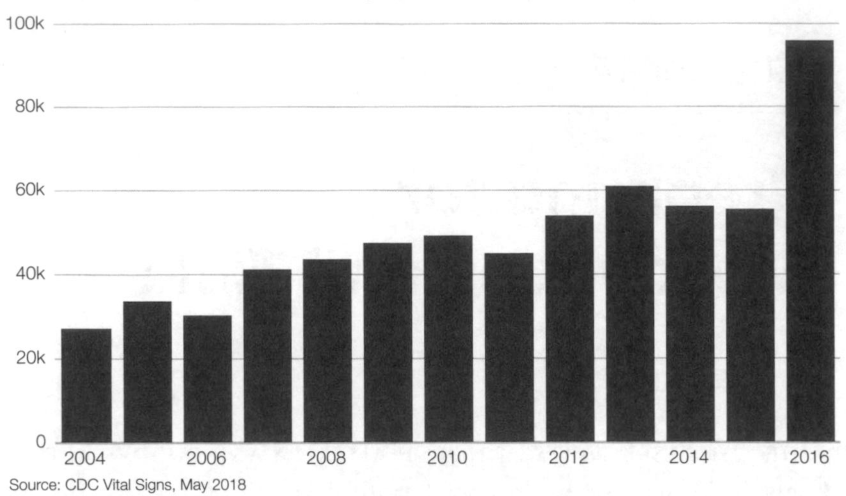

Figure 14-1. The number of Americans infected by bites from infected mosquitoes, ticks, and fleas tripled from 2004 to 2016.

daily thing hanging over us," says Rockefeller University neurobiologist Leslie Vosshall, who's been studying mosquito behavior for 15 years. "But it's going to be here. People are going to be getting sick, with things that they haven't gotten sick with before."

Mosquitoes

The deadliest animal on earth is not the snake (100,000 deaths a year), rabid dogs (35,000), or sharks (6). It's the mosquito. The diseases that it carries kill about 500,000 people every year.

We're used to associating malaria, yellow fever, and dengue fever with faraway climates. Until recently, that's been a good assumption—but not anymore.

"The climate crisis is not a crisis for mosquitoes. They love it," says Vosshall. "It just extends the surface of the earth where they live."

But the expanding habitat for mosquitoes is only part of the problem. In warm weather, mosquitoes actually carry *more* of the viruses that can infect you, so you're more likely to get sick when you get bitten.

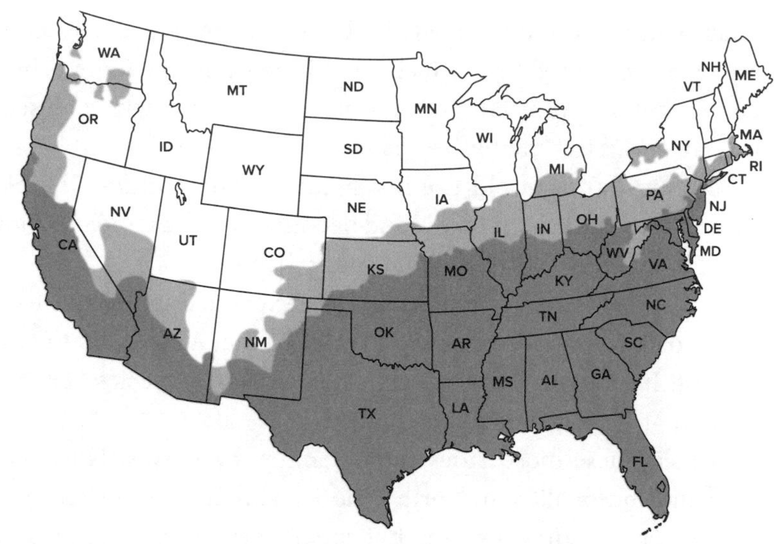

Figure 14-2. This map shows the habitat of *Aedes aegypti*, the most common mosquito breed in the United States as of 2017; darker shades indicate areas where mosquitoes are more likely to live and breed.

Only females mosquitoes bite; the guys don't even have the right mouth parts. To make matters worse, the females live four times longer than the males, whose life span is barely one week.

Those she-bugs aren't drinking your blood because they're hungry; they're drinking it for the proteins that they need to reproduce. (For food, both genders consume nectar and other plant sugars.)

There are more than 3,000 species of mosquitoes, but only a few species bite people. Two species in particular deal out most of the disease:

- ***Aedes aegypti*** carries Zika, yellow fever, dengue fever, chikungunya, and other diseases. It has enjoyed life in the Southern United States for decades—but now it's creeping northward.

  In Europe, *Aedes aegypti* has been found for the first time in Portugal and the Canary Islands. By 2050, these bugs are expected to be thriving in latitudes as far north as Chicago and, in China, Shanghai.

- **Asian tiger mosquito** (*Aedes albopictus*) has white stripes on its body and legs.

  As you'd guess, the Asian tiger mosquito was once a Southeast

Asian specialty. But it gradually built a global empire aboard an improbable vessel: used tires, which are shipped among countries by the tens of millions. (They're put onto new cars and trucks, or used as fuel, crash barriers, rubberized asphalt, and so on.)

To a mosquito, an old tire is a luxury condominium. The black color lets it heat up in the sun—and of course, a tire can hold standing water, so it's ideal for breeding mosquito babies.

Today, the Asian tiger mosquito has expanded its habitat to Europe, North and South America, the Caribbean, Africa, and the Middle East. It's number four on the Global Invasive Species Database's list of the 100 worst invaders.

In Asia, these mosquitoes carry dengue fever, West Nile, yellow fever, and encephalitis. In North America and Europe, they're not yet carrying or spreading diseases, but experts are worried. Viruses constantly mutate; if a strain or two become capable of reproducing in Asian tiger mosquitoes, Vosshall believes that a huge public-health crisis could result.

"Then it's going to be, you know, Zika in Connecticut," she says.

The Catalog of Disease

All right, so the two deadly species are spreading to more places. But the viruses are very different, carried by different insects, with a wide range of devastation. Here's your cheat sheet, presented roughly from most concerning to least:

- **Dengue fever** is a bad one, brought to you by both *Aedes aegypti* and the Asian tiger mosquitoes. Every year, 100 million people in 125 countries get infected and 90,000 die.

  If you develop symptoms, they'll be fever, nausea, vomiting, aches and pains, maybe a rash. They'll go away in a week or so, unless you're among the unlucky one percent who get sicker and sicker until they die.

  There's no treatment for dengue, except to take acetaminophen (Tylenol) to relieve the discomfort. (The CDC notes, however, "Do not take aspirin or ibuprofen!")

  There *is* a dengue vaccine. But if you've never had dengue before, being vaccinated can make your symptoms *worse* if you do get it. Therefore, you're supposed to get this vaccine only if you've already *had* dengue at least once before. Weird.

In the last 50 years, the number of countries where dengue thrives has risen from 10 to 120, and infections have increased some 30 times. That's partly because the planet is warming, partly because the population is growing, and partly because more people are living in cities. That's where mosquitoes can find billions of standing-water breeding pools, and where people and mosquitoes come into frequent contact.

In the United States, dengue outbreaks have struck in Hawaii, Texas, Florida, Puerto Rico, and the Virgin Islands. In Europe, Dengue was unknown until 2010, when the first patient was infected in France. In Europe, a handful of cases were diagnosed in 2018 in Spain and France.

Oxford University researchers have calculated the spread of dengue in the coming decades. Thanks to a general warming, "much of the southeastern USA is predicted to become suitable by 2050," the report concludes. Dengue is also likely to spread faster and wider "in low-risk or currently dengue-free parts of Asia, Europe, North America and Australia."

- **Eastern equine encephalitis (EEE)** may be the most horrifying mosquito-borne disease of all.

  It begins with a couple of weeks of fever, chills, joint and muscle pain, and feeling like hell. Then, if you're really unlucky (and, in most cases, under age 15 or over 50), you get a second wave of symptoms. This is the *encephalitic* (brain swelling) wave, and can bring fever, headache, irritability, restlessness, anorexia, drowsiness, vomiting and diarrhea, convulsions, coma, and death.

  "Even though it has 'equine' in the middle, it has a superhigh kill rate in *people*: about half the people who are infected with Triple E die," Vosshall says. Those who survive often have permanent or progressive mental and physical disabilities—personality disorders, paralysis, and early death. There's no treatment and no vaccine.

  People have been hit by EEE in 21 U.S. states, with the heaviest concentrations in Florida, Massachusetts, New York, and North Carolina. Most of them live or work in swampy areas on the East Coast.

In general, only a handful of Americans contract EEE each year, but the numbers are going up: ten cases in 2010, 15 cases in 2012, and 38 cases in 2019.

- **Yellow fever** may have no symptoms, or it may bring chills, headache, and muscle aches. Those symptoms pass in a few days.

  But in about 15% of patients, that recovery is followed by Round 2: a toxic phase, where your liver and kidneys are attacked, and you exhibit jaundice (yellowish skin—thus the name of the disease), dark urine, abdominal pain, vomiting, bleeding from the mouth, nose, or even eyes. Half of the Round 2 patients die in about a week.

  In all, about 60,000 people die of yellow fever each year. Once again, 90% of them are in Africa and Central and South America.

  The good news is that there's a safe, inexpensive, amazingly effective vaccine for yellow fever. Get one shot, and you're protected for life.

  The bad news is that making the vaccine is slow and difficult, and there's not enough of it. Since 2015, there have been yellow-fever outbreaks in Angola, Uganda, the Democratic Republic of Congo, Nigeria, and Brazil, and residents had to be given partial doses just so there'd be enough to go around.

  The CDC recommends that you get the vaccine if you plan to travel to any of the yellow-fever areas of Africa or South America. You'll probably agree.

  But health experts also fear outbreaks in the southern United States. As our climate warms, cities like Houston, New Orleans, Tampa, and Miami could become yellow-fever hot spots—because *Aedes aegypti* is prevalent, and because those cities are teeming hubs for international travelers.

- **West Nile** virus is the most common mosquito-borne disease in the United States, but it's not the scariest. For 80% of victims, there are no symptoms. The whole thing comes and goes, and they never even know it.

  The other 20% get headache, body aches, joint pains, vomiting, diarrhea, or rash. Most of that goes away, leaving behind only weeks-long weakness. About one in 150 patients, usually older or ill people, develop far more serious illnesses, like encephalitis and meningitis. One in ten of those patients dies.

There's no treatment and no vaccine.

West Nile began life in birds, and comes to you as a gift from mosquitoes who've bitten those birds. The first case was discovered in 1999, near New York City's JFK Airport—a reflection of the fact that planes carry mosquitoes and infected people.

"The first sign that there was a new virus in this hemisphere was that crows started dropping out the sky," Vosshall says. "At the Bronx Zoo, there were all these crows lying on the ground."

Since then, West Nile has infected 3 million Americans, in all 48 contiguous American states. In Europe, the first West Nile infection struck in 1996 in Romania; by 2016, it had spread to 16 countries. In 2018, there were 1,670 cases.

♦ **Zika** is another virus brought to you by *Aedes aegypti* and the Asian tiger mosquito. It generally doesn't make you sick, it goes away in a few days, and it rarely kills anyone.

What struck terror into everyone's hearts when Zika arrived in the United States in 2015, though, was its effect on pregnancies. Your baby can get birth defects like microcephaly—undersized brain and head—accompanied by all kinds of other developmental problems.

In 2015 and 2016, huge outbreaks made headlines in Puerto Rico and the U.S. Virgin Islands, with smaller ones in Florida and Texas. About 5,200 cases were reported, 4,900 of which were in travelers returning from hotter places.

And then, astonishingly, suddenly, Zika occurrences plummeted in the United States. The leading theory is herd immunity: Enough people got unknowingly infected, without symptoms, and developed immunity that the disease had a hard time making further progress.

In any case, be glad. If you don't count cases contracted by travelers abroad, there have been no American Zika cases at all reported since 2017, and Europe has never had any.

♦ **Chikungunya** is yet another fever brought to you by the Terrible Twosome, *Aedes egypti* and the Asian tiger mosquito. This disease brings you a few days of the usual flu-like symptoms: fever, headache, rash—and, notably, joint pain. ("Chikungunya" comes from an African language and means "to become contorted.")

There's no treatment or vaccine, but look at the bright side: Chikungunya generally doesn't kill you, and once you've had it, you usually can't get it again.

Chikungunya was an African, Indian, and Asian disease until 2006, when it first started showing up in the United States. Those cases weren't contracted *in* the United States, though—they were brought home by international travelers. That changed in 2013 and 2014, when people started catching the fever from our own all-American mosquitoes in Florida, Puerto Rico, and the U.S. Virgin Islands.

Since then, a few hundred cases a year pop up in the United States, virtually all in travelers returning from overseas. Europe's Chikungunya cases have ranged from 2 to 330 per year, entirely in travelers returning from India, Asia, and Central Africa.

♦ **Malaria,** carried by the *Anopheles* mosquito—not the two you've been reading about—is the big killer in Africa and South America. Every year, it makes 200 million people sick and kills more than 400,000 of them. Most of them are children.

When you get malaria, you feel like you have a terrible flu, complete with chills and fever. If you're young, old, or weak, and you don't get treatment, you can die.

At one time, malaria was a huge killer in the United States. In the late 1700s, "Carolina was in the spring a paradise, in the summer a hell, and in the autumn a hospital," wrote natural historian Johann David Schoepff in 1788.

Not anymore. Malaria isn't a big worry for Americans and Europeans, even in the climate-change era. Through coordinated programs of spraying, education, improved drainage of standing water, screens, and so on, the United States and the European Union have wiped out malaria. About 1,500 cases of malaria are reported each year in the United States—but they're almost all imported diseases, carried home by travelers from Africa and South America.

As you now know, there are no vaccines for most of these diseases. It may be, in any case, that trying to create vaccines is a fool's errand—because we're discovering new viruses all the time.

The smarter solution, therefore, is to develop ways to stop mosquitoes from biting you in the first place. Presto: One solution for all diseases.

Fortunately, we have a few such tools already.

How to Prepare Your Yard

After that tour of the Museum of Mosquito Misery, you'd be forgiven for wanting to live in an airtight plastic bubble for the rest of your life. In one way, you're on the right track: The best way to avoid getting mosquito-borne diseases is to avoid getting bitten in the first place.

And how do you do that?

- **Drain standing water.** Mosquitoes reproduce only in calm, still water. Take that away, and they go somewhere else to find it.

  Mosquitoes can breed in puddles as small as a bottle cap. "Any container in a backyard that can hold water for more than a week should be removed or dumped," says Jim Fredericks, an executive at the National Pest Management Association.

  So inspect your yard for flowerpots (and saucers), pet bowls, bird feeders, planters, buckets, kiddie pools, garbage cans, tires, puddles, Frisbees, birdbaths, grill covers, leaky pipes, tire swings, wheelbarrows, puddled tarps, and toys left in the yard. "Clean your gutters," says Vosshall. "And no more koi ponds."

  Swimming pools are okay; mosquitoes won't touch chlorinated water. *Moving* water is okay, too, as in a fountain.

  You might wonder how effective it is to clear out standing water from *your* property. The answer is: Very, because mosquitoes can't fly very far! Mosquitoes don't move around more than 100 yards in their entire lifetimes. If you can clear the breeding spots from your property, you can make a huge dent in the local mosquito population.

- **Spray.** You, or someone you hire, can also spray your yard with insecticide. Those chemicals certainly work to kill mosquitoes—and, according to the CDC, "Spraying is safe. You do not need to leave an area when truck spraying for mosquito control takes place. If you prefer to stay inside and close windows and doors when spraying takes place you can, but it is not necessary."

Unfortunately, these insecticides are *not* safe to lots of innocent-bystander bugs—like honeybees, ladybugs, and butterflies—some of which actually help with the mosquito problem. If it's not sprayed correctly, these insecticides can also hurt foliage and pets.

And then there's the resistance problem. In 2016, the CDC discovered that spraying to combat the Zika outbreak wasn't as effective as it should have been. Upon lab testing the mosquitoes, they found that bugs from 30% of the tested municipalities had become partly or completely resistant to the insecticides.

The bottom line: "The risk of pesticide exposure may outweigh the benefits of control," says the EPA.

If you've tried non-spraying techniques and mosquitoes are still a problem, see if your town, city, county, or state has a mosquito-abatement department. (Google it.) If you can convince them to do the spraying, great; they're professionals, and they'll generally do the job inexpensively or even free.

If you have no luck there, then consider a private spraying company—carefully. *Consumer Reports* warns that there are all kinds of outfits, many of which, responding to the outbreaks of mosquito-borne diseases, have sprung up only recently and don't have much experience.

You should ask to see the company's license; the label for the insecticide they plan to use; and their protective gear. Ask what strategies they use to minimize drift of the spray to avoid killing insects you don't want to kill. The better companies also pay repeat visits to see if their spraying worked.

Finally, so that you don't contribute to the resistance problem, avoid automated misting systems that spray the same chemical across the entire yard. And see if the company you're hiring plans to rotate chemicals year by year, which also helps guard against insecticide resistance.

One thing that *doesn't* help, by the way, is fragrant flowers. Sorry, gardening websites.

It's true that mosquitoes are incredibly scent-sensitive. They can find you from 30 feet away, for example, by smelling the carbon dioxide on your breath. (They also sense chemicals in our sweat, the heat from our skin, and other signals.)

But floral scents from lavender, marigolds, catnip, rosemary, basil, scented geraniums, mint, and so on do absolutely nothing to dissuade mosquitoes, no matter how many gardening websites say otherwise. You're falling for an urban legend.

"Lavender and other nice aromatic herbs do not repel mosquitoes," Vosshall says. "The female mosquitoes are busy finding humans to bite and couldn't care less what you do with your garden."

How to Prepare Yourself

With luck, you don't spend your entire life at your home. Sooner or later, you'll want to venture into the world. There's a technique for doing that safely, too:

- **Avoid prime time.** Our friend *Aegypti,* bringer of many illnesses, flies mostly at dusk and dawn. In the middle of the day, they make themselves scarce—so that's the best time to do your work or your hiking.

  The Asian tiger mosquito bites at all times of day, but isn't carrying disease in North America. Yet.

- **Avoid prime locations**. According to Jim Fredericks, mosquitoes' favorite spots are the transitional zones between wooded and open areas.

- **Wear long, loose, light-colored clothing.** Fabric covering your skin is infuriating to mosquitoes. Unless it's very tight, they have a hard time biting through. Wear long sleeves, long pants, shoes (not sandals), and a hat when you're going hiking. Mosquitoes locate you, in part, by your heat signature, and dark colors retain heat. Therefore, go for light colors, which mosquitoes are less likely to notice.

  Obviously, long pants and long sleeves don't feel great when you're hiking in peak summer heat. But that's what DEET is for.

- **Wear DEET or picaridin.** DEET, the active ingredient in bug repellents like Off!, Cutter, Sawyer, and Ultrathon, has been keeping mosquitoes off our bodies since 1944. Picaridin is extremely effective repellent, too. It's been popular in Europe and Australia for years, and began arriving in the United States in 2005.

You can buy these repellents with different percentages of the active ingredient. The percentage has nothing to do with how *well* it works—only how long. For example, 6% DEET lasts two hours; 20%, five hours.

If you're using sunscreen, that goes on first.

Don't spray the repellent onto broken skin, or directly to your face. Instead, spray it onto your hands, and then rub onto your face.

Cover all your exposed skin. No need to spray skin that'll be covered by clothing, although you may want to spray the clothing itself.

DEET is safe for pregnant women and children. Remember, it's not a poison. It doesn't kill bugs; it just smells bad to them. For babies under 2 months old, though, the CDC suggests draping strollers or carriers with mosquito netting.

- **Don't wear botanical oils.** The advice is to wear DEET or picaridin—*not* botanicals or essential oils.

  Health-food stores sell so-called bug repellents containing the oils of citronella, lemongrass, eucalyptus, catnip, cloves, patchouli, peppermint, and geranium. Some of these might have some repellent

Figure 14-3. It's entirely possible to avoid getting a mosquito-borne disease, by applying DEET or picaridin to your skin when you go outside.

ability, others none, but this much is for sure: The EPA has not tested or approved them. The botanicals are under no regulation, and usually don't indicate how often to apply, how much to apply, which species they repel, or what kind of repellence you can get.

"There is a lot of mayhem out there in the field," Laurence Zweibel, insect behavioralist at Vanderbilt University, told *Wirecutter.* "I am very concerned about the lack of regulatory oversight and the ability to disinform or, in some cases, completely misinform consumers."

The other big problem with these oils is that they're *oils*. They evaporate fast. In one study, all but one product had completely evaporated from the skin after 20 minutes. None came close to offering the five-hour protection of 25% DEET or the twelve-hour protection of picaridin sprays.

Some people fear DEET because it's a chemical (never mind that lemongrass and eucalyptus are chemicals, too)—but DEET is safe.

"DEET has a remarkable safety profile after 40 years of use and nearly 8 billion human applications," wrote the authors of a study in *The New England Journal of Medicine* (who also tested seven "natural" repellents against DEET and found them lacking). "This repellent has been subjected to more scientific and toxicological scrutiny than any other repellent substance."

When the problem is as serious as mosquito- or tick-borne viruses, you don't want to mess around with something untested and unregulated that you have to reapply every 20 minutes. Botanicals offer a false sense of security.

♦ **Treat your clothes and shoes** DEET and picaridin repel mosquitoes. *Permethrin,* on the other hand, actually kills them on contact. It bestows muscle spasms, paralysis, and death upon them. You might consider it payback for all the muscle spasms, paralysis, and death they've bestowed upon *us* over the years.

Permethrin is the active ingredient on anti-bug dog collars, in some commercial yard sprays, and plenty of sprays and liquids, but you're most likely to want it as a treatment for your clothes. Those garments become like armor against mosquitoes—and, by the way, flies and ticks. (The military invented permethrin-treated clothing in the early 1990s, because our soldiers hated having to reapply DEET all the time.)

You can buy outdoorsy shirts, pants, and hats already treated with permethrin from LL Bean, InsectShield, ExOfficio, REI, and so on. You'll pay about $50 for a long-sleeved "base layer" shirt, $80 for a button-down, and $100 for pants. The treatment is supposed to keep working for 70 trips through the laundry.

You can also mail your clothes to a company like InsectShield.com. They'll treat it and send it back to you, at a cost of $8 to $10 per item. The treatment is invisible, has no smell, and lasts for 70 washings. (They don't accept underwear.)

Finally, you can treat the clothes—and shoes!—yourself. You just buy permethrin in a spray bottle ($16 for enough to cover 8 garments), spray it onto the clothes, and let them dry. It's just as effective as the pretreated clothes, although it lasts only six washings or so.

This stuff doesn't hurt people, unless they're dumb enough to drink it. But in liquid form, it's toxic to fish and bees, and it can irritate dogs and cats, so choose a good spot to apply it.

Note, too, that you still need DEET on your exposed *skin* even if you're wearing permethrin-treated clothes.

And for what it's worth, *Consumer Reports* testing found that spraying DEET repellent on your clothes is actually more effective than treating clothes with permethrin. (The permethrin-treated clothing manufacturers strongly argue that point.)

How to Prepare for Your Trip

If you plan to visit any of the 21 countries in Africa or 13 countries in South America where mosquito-borne diseases are common, you should take mosquito prep seriously. Remember: Virtually every American and European who contracts Zika, chikungunya, and malaria gets it when traveling to one of those hot zones.

You can look up the latest mosquito-disease hot spots at wwwnc.cdc.gov/travel, and you can get the shot at a local travel clinic. At that point, you should prepare:

- **Get your yellow-fever shot.** If the CDC or your local travel clinic determines that you need the yellow-fever vaccine, arrange to get it at least ten days before you travel. That's how long it takes to kick in.

  They'll give you a special yellow card to prove that you've had the vaccine. Some countries require you to show it before you enter the country.

- **Lodge wisely.** Mosquitoes hate cool air, so choose a hotel with working air conditioning. Confirm that the place has door and window screens, too.

- **Treat your clothes.** This is permethrin's big moment, as described above.

- **Pack wisely.** You can bring along insecticide-treated mosquito nets as a backup. They're cheap and work beautifully. Mosquito coils, which are popular in Asia, Africa, South America, Canada, and Australia, also work. They're basically mosquito-repellent incense spirals. You light the outer end, and they smolder away for eight hours or so.

How to Prepare the World

Intensive scientific efforts to control the mosquito problem are underway.

Some of them involve genetically modifying mosquitoes. For example, a UK company called Oxitec has received EPA permission to release millions of genetically altered male mosquitoes in Florida in 2021 or 2022. (The company will still have to overcome local opposition.) These bugs have been given a gene that makes their offspring die before they reach adulthood—and before they can reproduce. In theory, the mosquito population would crash.

Or at least the population of *that species* would crash. "Even if you get rid of this one species, there are plenty of other mosquito species ready to hand in their résumés," says Vosshall. "Like: 'I like people, I drink blood, and I can spread disease.' And it's like, 'Okay, you got the job! You're in.' "

Another approach, less likely to risk unintended consequences, is the sterile insect technique (SIT). That's where scientists sterilize millions upon millions of male insects, either by irradiating them (tricky) or unleashing a certain bacterium on them (better) that leads to non-hatching eggs. Either way, when these males mate with the females, no babies result, and the population declines.

The beauty of SIT is that no genetic engineering is involved. It's a one-generation process: there's no effect on the rest of the species, no effect on any *other* species, and no opportunity for unintended consequences.

This technique has been enormously safe and successful in fighting off the screwworm fly (United States, Mexico, Central America, Libya), the Mediterranean fruit fly (United States, Central and South America, South Africa, Europe, and Asia), the pink bollworm (United States), and the codling moth (Canada).

Using SIT on mosquitoes is much harder. It involves identifying and segregating millions of males, sterilizing them effectively, and then air-dropping them from planes or drones in the right locations.

But field trials are already underway in ten countries, including the United States. In 2018, one of the largest pilot programs ran its course, run by Verily (one of Alphabet's companies—you know, the Google mother ship).

Figure 14-4. This is the lab where Verily raised 20 million sterile mosquitoes—and reduced the mosquito population of Fresno, California, by 95%.

In a custom-designed, automated mosquito-rearing factory, the company produced 1 million sterile males a week for 20 weeks, and released them in Fresno, California. The results after one season: A 93% drop in the population of female mosquitoes, which are the ones that bite.

Leslie Vosshall, at Rockefeller University, is pursuing yet another approach. It exploits the fact that after feeding on human blood, females do no more biting until after they've laid their eggs a few days later.

Using a modified human diet drug, her team is producing a delicious mosquito cocktail that the females adore. This special nectar has, however, a side effect: "They feel full, like Thanksgiving dinner full. Every female that drinks the diet drug feels like she's had a big blood meal, and she doesn't come after people."

The idea is that you'd put out this nectar in little feeding stations. This approach, too, would be far safer than genetic modification or insecticide spraying. Reduce the biting, and you reduce disease.

Ticks

"The development and survival of ticks, their animal hosts (such as deer), and the bacterium that causes Lyme disease are all strongly influenced by climatic factors," says the U.S. Centers for Disease Control and Prevention (CDC). "Milder winters result in fewer disease-carrying ticks dying during winter. This can increase the overall tick population, which increases the risk of contracting Lyme disease in those areas."

You do *not* want Lyme disease. It comes in stages:

- **First stage.** At the outset, you get what feels like a bad flu: chills, fever, sore throat, exhaustion, muscle aches, and headaches.

  Some people also develop a red rash in a bull's-eye pattern (a red center, then a ring of normal skin, then a ring of red). It doesn't itch or hurt, and goes away in four weeks.

  It takes a while for this classic pattern to form—between 3 and 30 days after the bite. If you have the rash, you have Lyme. Trouble is, only 50% to 70% of people with Lyme *get* that distinctive rash. In other words, you might have Lyme without ever getting the rash.

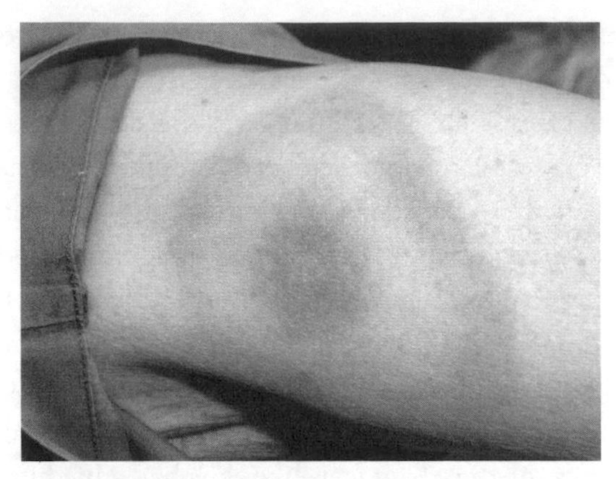

Figure 14-5. The classic bull's-eye-shaped rash means you have Lyme. On the other hand, you may have Lyme without getting this rash.

- **The second stage.** The flu worsens. You may now develop vision problems, numbness, tingling, heart palpitations, and the inability to move one side of your face.
- **The third stage.** If you don't treat Lyme disease with antibiotics right away, then the disease may reemerge weeks, months, or even *years* later. During this round, you may develop arthritis, brain problems (with mood, memory, sleep, conversation, and concentration), heart palpitations, or intense headaches.
- **Post-treatment Lyme.** About 10% of patients *never* get better. They live with the joint pain, exhaustion, and memory loss for the rest of their lives.

Lyme is infuriating because it's so variable and so vague. Doctors call it "the great imitator," because it can look like so many other diseases. Meanwhile, the symptoms vary wildly. Some people develop shooting pains in their arms and legs, or chronic fatigue, memory problems, blurred vision, confusion, stumbling gait, shaking, speech impairments, and so on. Others don't.

Even the blood tests aren't reliable. No test can detect the Lyme bacteria itself in your body; the best we can do is test for the presence of Lyme *antibodies*, the proteins that your body releases in response to the disease. That's an iffy proposition, because you don't know how long ago you had the disease—only that you had it at some point. It could have been months or *years* earlier.

The most common test, and the first one most people get, is called ELISA. The problem is, if you get the test during the first three weeks, your body hasn't really begun manufacturing the antibodies, so the test's accuracy is poor. It might mistakenly report that you don't have the disease.

The test becomes far better—over 85% accurate—once the disease has spread to your neurological system and joints, but by then, obviously, a lot of damage has been done.

"What happens sometimes is you feel sick early on. You go to the doctor, you're tested, and the results say you don't have Lyme disease. Many times the doctors say, 'Come back later if you still feel sick,'" says Mary Beth Pfeiffer, author of *Lyme: The First Epidemic of Climate Change*. "Your immune system may kick in to some extent, so you start to feel better, and you go on with your life. Then it comes back to bite you six months or a year later, when it's much, much more difficult to diagnose."

The second most common test, the Western blot, is also an indirect test—it, too, looks for antibodies, not actual Lyme—but is considered more accurate than the ELISA test.

This time, Pfeiffer says, the problem is that the test is subjective and has to be read with the human eye. "Results came out differently in different labs, and sometimes different in the same lab," she says. Even under the best of circumstances, the test is only 80% accurate.

Lyme is also infuriating to diagnose and treat because often, the tick hasn't given you Lyme disease—or not *just* Lyme disease. About 83% of tick-borne infections are Lyme disease, but ticks can carry as many as 16 different diseases, carried by bacteria, viruses, and parasites.

Here's the list, for your insomniac pleasure. You can assume that "flu-like" means fatigue, fever, chills, nausea, and malaise:

- **Babesiosis.** Either no symptoms or flu-like symptoms. Life-threatening in the very old, very young, and immunocompromised.
- **Rocky Mountain spotted fever.** Flu-like symptoms within two weeks, often with a red, spotted rash. If not treated with antibiotics, can cause permanent cerebral damage, organ failure, necrosis, or death.

- ***Rickettsia parkeri.*** Closely related to Rocky Mountain spotted fever, but less severe. Telltale symptom: a necrotic lesion (black dead flesh) at the site of the bite.

- **Anaplasmosis.** Flu-like symptoms, gastrointestinal problems within two weeks. Unless treated by antibiotics, can cause respiratory problems, bleeding problems, organ failure, or death. Not as deadly as ehrlichiosis or spotted fevers, but dangerous to the elderly.

- **Ehrlichiosis.** Flu-like symptoms within two weeks of a tick bite. Unless treated by antibiotics, can cause organ failure, seizures, and coma. Higher mortality rate in patients with low immune response.

- **Powassan virus.** One to four week incubation. Flu-like symptoms. In more serious cases, it can cause encephalitis and cerebral damage. There is no treatment.

- ***Borrelia miyamotoi.*** Flu-like symptoms; fever comes and goes. Treated with antibiotics.

- ***Borrelia mayonii.*** Causes similar symptoms to other Lyme infections, but with widespread rashes, nausea, and vomiting.

- **Tularemia.** Flu-like symptoms from 3 to 14 days after the bite. Sometimes an ulcer at the site or swelling of the lymph nodes. If untreated with antibiotics, can cause pneumonia, meningitis, bone infections, heart problems, and death.

- **Southern Tick–Associated Rash Illness.** Causes a rash like the Lyme rash, accompanied by flu-like symptoms. Spread by the lone star tick.

- **Heartland virus.** Flu-like symptoms. This one's relatively new, detected only in 2009. Nine cases and two deaths have been reported.

- **Colorado tick fever.** Flu-like symptoms in one to two weeks. Around half of all patients have fevers that go away and then return. Sometimes damages the central nervous system. Rarely life-threatening, but there's no treatment.

- **Tick-borne encephalitis.** Viral disease only in Europe so far. Flu-like symptoms appear between one and two weeks. A third of patients get better for a week, and then experience a second round with more serious complications like encephalitis or meningitis.

- **African tick-bite fever.** Flu-like symptoms may appear after four to ten days, accompanied by muscle pain and a sore at the bite site. Usually improves without treatment, but antibiotics have been effective.

  Those last seven, by the way, have popped up only in the last 20 years.

The point is: One tick can bless you with more than one of them. Good luck to the doctor who has to sort *that* out.

Ticks and Climate Change

On top of everything else we have to worry about, ticks are now making people sick in record numbers on six of the world's seven continents. (If you want confidence that you'll never get Lyme disease, move to Antarctica.)

In the United States, the number of Lyme cases reported in 2017 was around 30,000—more than triple the numbers in the late 1990s.

Trouble is, the CDC can track only what doctors report. If the doctors don't report, or people don't see a doctor, or the disease is misidentified,

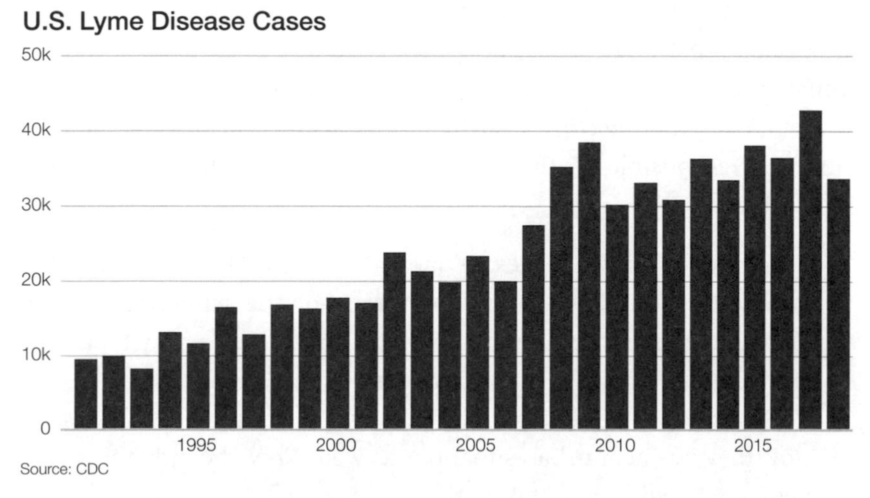

Figure 14-6. Lyme disease cases have more than tripled since 1990.

the CDC won't know about it. For that reason, the agency estimates that the actual number of cases is ten times as high—more than 300,000 cases a year.

As recently as 2010, Japan, Turkey, and South Korea had zero cases of Lyme disease; now, they're seeing 2,000 cases a year. Europe is getting 232,000 cases a year.

To understand why these numbers are exploding, it helps to understand nature's ugly little process.

Lyme disease, named for Old Lyme, the unfortunate Connecticut town where the first outbreaks were documented in the 1970s, is caused by corkscrew-shaped bacteria called *Borrelia burgdorferi*.

But don't blame the deer ticks for *Borrelia*, or even the deer. *Borrelia* lives in mice, birds, and chipmunks.

Baby ticks drink those creatures' blood, slurping in those bacteria along the way. The deer don't carry Lyme disease, but they do serve as convenient all-you-can-drink blood cruises for the ticks; their ports of call are human neighborhoods. The ticks, carried by the deer, bite us, and presto: *Borrelia* has made it, thirdhand, all the way from the mice into our blood.

Neither heat nor rainfall seems to affect the tick population, so tick experts doubt that climate change is the sole factor in the tick expansion. "Climate can certainly play a role," says Thomas Mather, director of the Center for Vector-Borne Disease at the University of Rhode Island. "But the changes that we're seeing have happened at a pace far greater than climate change has occurred."

But if the changing climate isn't directly affecting the ticks' habitat, it's indirectly responsible for the explosion in tick-borne diseases, through effects like these:

- **More deer in more places.** A larger element, Mather says, is the skyrocketing population—and territory—of the white-tailed deer. Warmer winters invite them northward, and as we destroy their natural habitat, we force them into more populated areas.

  "So instead of just being in the deep woods of Maine," Mather says, "now they live near urban situations as well. Now, the only thing that they have to worry about is getting hit by a car."

- **More mice in more places.** Climate change and habitat destruction also explain the spread of mice, the most dominant carrier of the Lyme germ. "The white-footed mouse is able to move and able to live further north than it could before," says Mary Beth Pfeiffer. "It's those little critters—tons and tons of little mice, shrews, and chipmunks—that have very few predators in an unnatural environment, this more urban environment."
- **More people outside.** As winters grow shorter and warmer, people spend more time outside. And that is where the mice, the deer, and the ticks are, too.

These three factors combine to produce a simple result: "More ticks in more places," says Mather. "And they're in more places where there are more people. So instead of just being in the deep woods or wherever, where fewer people are, they're now in suburbs and even semi-urban areas."

The deer tick, the one that carries Lyme, is now in half of all U.S. counties—twice as many as in 2000—in all 50 states. "There's a lot of very good data showing them moving up latitudes," adds Pfeiffer. "They're even moving into Canada, which is a new frontier for Lyme disease."

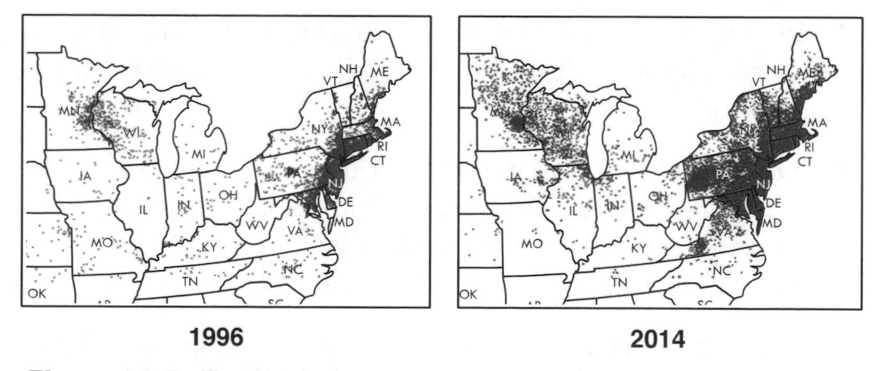

Figure 14-7. The dots indicate reported cases of Lyme disease; the maps show how their territory has grown. Not shown: the 90% of cases that are never reported.

Fun Lyme Facts

Lyme-disease mythology could be a full-semester class. Here are a few fun facts to bring you up to speed, some of which are actually reassuring:

- *Most* tick species don't carry disease, and *most* tick bites don't give you Lyme disease. The tick species you have to worry about is the black-legged tick, also known as the deer tick. Between 30% and 50% of them carry the Lyme bacteria.

- If you do get infected with Lyme, prompt treatment with antibiotics usually makes the symptoms disappear, quickly and for good.

- American dog ticks—the ones you find on your dog—do not carry Lyme disease. "American dog ticks are probably the most common tick across America," says Tom Mather.

  After decades of collecting and analyzing ticks in the wild, he can speak from experience: "Thousands and thousands and thousands of American dog ticks later, not one of them has been positive for Lyme disease," he says.

  (Dog ticks do carry Rocky Mountain spotted fever, though, which is no picnic, either. And dogs can also get *sick* from tick bites.)

- Ticks don't drop onto you from overhead. They don't even have eyes.

- Ticks can't fly or jump. They literally stand up on their hind legs on plants, arms outstretched until somebody passes by close enough to grab onto.

- Today, 95% of all Lyme cases occur in 14 states: Connecticut, Delaware, Maine, Maryland, Massachusetts, Minnesota, New Hampshire, New Jersey, New York, Pennsylvania, Rhode Island, Vermont, Virginia, and Wisconsin.

- You can't get Lyme from another person, even through sex or breast milk.

How Not to Get Lyme Disease

No matter where you live, ticks are something you now have to pay attention to—and prepare for. The good news is that if you follow these steps, your chances of getting Lyme or another disease are very, very small.

- **Treat your skin before you go out.** DEET and picaridin are almost like miracle juices. They're safe, they last for hours on your skin, and they truly work as repellents. And not just for mosquitoes; they also keep *ticks* off you. Put DEET or picaridin on all your exposed skin before you go tromping around in foresty or fieldy places in warm weather.

- **Tick-proof your clothes.** All ticks crawl *upward* once they're aboard. The usual advice, therefore, goes like this: Wear long pants, and tuck them into your socks. Wear a long-sleeved shirt, and tuck it into your pants. It's incredibly difficult for a tick to reach your skin under those conditions.

  The only problem with that advice, says Mather, is that virtually nobody follows it. "When was the last time you saw somebody wearing long pants in the summer and tucking their pants up into their socks?" he says. "Why do we want to continue to tell people to do stuff they're not going to do?"

  To be clear, *he* wears long pants and sleeves when he goes out tick collecting every summer weekend, and *he* tucks them in. But because it's such a goofy-looking, hot way to dress, he rarely sees anybody else following that guideline.

  Instead, he says, it's more effective to treat your shoes and clothes with permethrin, as described earlier.

  Keep in mind that ticks like to crawl up from your shoes. It's wise, therefore, to treat the *inside* of your pant legs, too. Turn the lower eight inches of them inside-out and spray away.

- **The duct-tape trick.** If you spot ticks on your clothes or your dog before they've had a chance to bite, lift them off with a piece of duct tape.

- **Inspect yourself afterward.** Here's a blessing: You generally can't get infected unless the tick has remained attached to your skin for at least 24 hours (if it's a nymph) or 48 hours (if it's an adult).

(Gross-out trigger warning: If you don't remove it, a tick may remain attached for *days*, growing up to *ten times* its original size as it fills up with blood.)

In other words, you should do a tick check every day that you return from outside in warm weather. If you remove the tick the day you got it, it's extremely unlikely that you'll get Lyme disease.

But if you have a luxurious 24 hours to find and remove the tick, why does anyone get Lyme?

Mainly because the ticks that infect the most people are the *really* tiny ones, the young ones in the nymph stage. You don't feel the bite, and the ticks themselves are the size of poppy seeds.

The adults are much easier to see. People find them 80% of the time within the first 24 hours.

Another problem: Ticks like to hide out in the groin, armpits, and scalp, where they're harder to spot. That's why Tom Mather's website, tickencounter.org, recommends doing your lower-body inspection when you're on the toilet. In theory, your pants are already pulled down.

Other common tick hangouts: behind your knees, in and around your ears, and in your belly button.

That's why the CDC suggests that you take a shower after each outing. You may wind up washing away those little nymphs without ever even knowing you had them on you.

- **Pull the tick off.** Don't buy into internet mythology: You can't remove a tick by burning it, suffocating it in Vaseline, or coating it in nail polish. In fact, those efforts only make it burrow into your skin more deeply.

  You should remove a tick by grasping its *head*—or the closest part you can get to your skin—with tweezers. Pull it straight out, without twisting or jerking. If the mouth breaks off and stays in your skin, that's no big deal. Wash the spot, and your hands, with soap and water.

  If the tick is alive, drown it in rubbing alcohol.

  And don't squish it! Instead, take a picture of the tick and upload it to tickencounter.org/tickspotters. By providing your location and the date, you'll help enlarge the national database of what ticks are

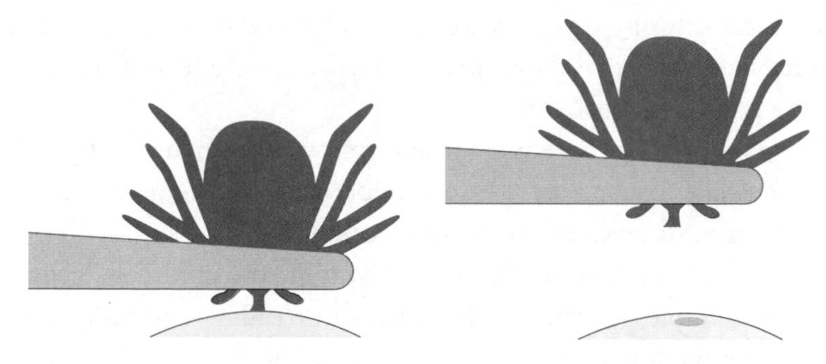

Figure 14-8. There's only one good way to remove a tick: grabbing it by the head and pulling it straight out.

biting where. You'll be helping to strengthen the site's tick-bite warning system.

- **Dry your clothes.** Ticks need humidity. You know where there's very little of that? In a hot clothes dryer. Veteran tick avoiders routinely toss their clothes into the dryer—ten minutes on high heat—to kill any ticks that have crawled aboard. Then, while they wait, standing there naked, they do a body check.

  If the clothes are damp, leave them in the death cycle longer; the clothes should be hot and *dry* when they come out. The ticks will be crispy and dead.

  It's fine to wash the clothes before the dryer treatment. Just remember that washing—even in hot water—doesn't kill ticks.

  The *dryer* does.

Tick-Proof Your Property

Neither heat, nor rain, nor dark of night bothers ticks. What they *hate,* though, is dryness. They need humidity to survive. Once the humidity drops below 82%, they dry out and die within eight hours.

You can exploit that weakness. The more pruning of trees you do, the more sunlight can dry up your property, thus killing the ticks. Rake or blow away all the leaves. Clear away tall grass and bushes around your yard and near your house, too.

Ticks can't fly or jump, so you can lay out gravel or wood chips to create a no-tick's-land that they can't cross. About nine feet of sunny lawn makes a good tick moat, too.

The point is to separate spots where you want human activity to happen safely, like a swing set, garden, or patio, from the ticky areas, like forest, hedges, stone walls, sheds, and wood piles.

Remember, too, that to ticks, deer are public transit. Keep deer out—with, for example, fencing or plants they don't like—and you keep the tick population down.

You can also consider spraying. If your pest-control crew knows what they're doing, they'll use a powder or a high-pressure spray that contains permethrin or bifenthrin. They'll probably recommend spraying twice during the summer—maybe even a third time in the fall, to kill the adult ticks.

Spraying, of course, is controversial. The same insecticide that kills the ticks may also kill pollinators like bees and butterflies. For that reason, spraying as little as possible to get the job done is both smart and economical. Stick to spraying the outer margins of your yard, underneath shady flower beds, and forest trails.

"You can be thoughtful and still protect against ticks," says Mather. "We teach not to spray flowers that are visited by pollinators. Stay away from clover. No reason to spray halfway up a rhododendron that's in flower, because ticks aren't going to be up there anyway. They're going to be underneath the branches."

For the same reason, it's pointless and wasteful to spray your lawn, especially if it gets sunshine each day. Sunny, open-to-the-sky patches of yard are like Death Valley to the ticks, because that's where they dry out and die.

Dogs, Cats, and Ticks

Dogs live down low, at tick level, and most of them don't wear long pants. As a result, they attract far more ticks than you do.

And yes, dogs can get Lyme disease, too, along with many of the other diseases that people get.

There are all kinds of tick-prevention products for dogs and cats, and they work well: medicine-infused edible treats, sprays, powders, collars, shampoos, liquids you apply directly to the pup, and so on. Note that tick collars work only around the head and neck.

Once you're out and about, discourage your pet from wandering through tall grass and weeds.

Then there's the business of inspection. Look over your dog or cat after each outing, just as you look over yourself. If you find a red or irritated patch of skin, slow down and see if a tick is nearby. Starting from the dog's head, comb through the fur with your fingers, feeling for tiny bumps that aren't usually there. Brushing or combing may discover ticks that haven't yet had time to crawl through the fur to the skin. Check especially carefully in the ears, between the toes, under the "armpits," under the tail, inside the groin, and deep in the fur around the neck.

If you find a tick, use exactly the same steps you'd use to remove and kill it that you'd use on yourself.

Now that you know the rules for setting up tick-free zones in your yard, there's no reason you can't use fences (or invisible fences) to keep your pet away from the tick zones, too.

By the way: There's not yet a working Lyme vaccine for humans. There is one, though, for dogs; ask your vet about that.

The Fight Begins

Even now, as climate chaos just gets underway, Lyme and its unpleasant companions have reached epidemic numbers. The CDC's estimate of Americans getting Lyme disease each year—300,000—is 1.5 times more than the number of American women who get diagnosed with breast cancer, and 6 times the number diagnosed with HIV/AIDS.

You'd never know it by the relatively small amount of attention, funding, and media coverage. But that, says Mary Beth Pfeiffer, will have to change.

"We have vastly underestimated the scope and the implications of ticks, and of this epidemic of tick-borne illness," she says. "We need a major infusion of money, similar to what we did for AIDS and HIV. We spent the money, we built laboratories, we funded PhD programs. We need to do better on Lyme disease."

Chapter 15

Preparing for Social Breakdown

BOLIVIA, 2016: FOLLOWING THE WORST DROUGHT IN 25 YEARS, protests and fighting break out over scarce aquifer water.

India, 2016: After a severe drought, riots over limited water supplies erupt.

Nigeria, 2017: A Lake Chad Basin drought drastically reduces farmable land, leading to a long-running clash between farmers and herders. Dozens are killed.

In the end, we're living creatures. Heat, thirst, hunger, and insecurity about our future can turn us on each other. One study after another has established the link between climate gone wild and human conflict. And once the social order falls apart, all kinds of social *systems* can fall apart, too: police, government, services, food chain, communications, media.

Hurricane Katrina, for example, blasted New Orleans back to the 1850s. "The electricity was out across the city, the gas was out across the city, the water was gone," says Pulitzer Prize winner Jed Horne, author of *Breach of Faith: Hurricane Katrina and the Near Death of a Great American City.* "Cell service was pretty much wiped out. All the street lights went out. The sewer system was in a state of complete collapse or paralysis. I mean, it was primitive."

People wandered downtown, flooded out of their homes, unable to contact anyone, cut off from TV, radio, and internet. Desperate for food and water, residents broke store windows and foraged from the abandoned stores.

And where were the police? Sometimes, right alongside the other looters. New Orleans police officers were filmed grabbing stuff from a Walmart and "commandeering" Cadillacs and Corvettes from a car dealership. Meanwhile, 249 officers—17% of the entire police force—simply didn't show up to work.

If it's that easy for our authority system to evaporate, exactly how thick is the wall that separates us from chaos?

Here's a piece of advice from, of all people, Michael Brown ("Hell of a job, Brownie!"): "The American public needs to learn not to rely on the government to save them when a crisis hits. The larger the disaster, the less likely the government will be capable of helping any given individual," he wrote, years after he resigned as director of FEMA during Hurricane Katrina. "The true first responders are individual citizens who take care of themselves."

Climate Migration

In one alarming way, almost every aspect of the climate crisis has the same result: It pushes people off their land.

Wildfires push us out of our homes and towns. Rising seas push us to higher ground. Hurricanes push us inland. Drought and water shortages drive us to places where it's easier to grow food and survive. Almost always, people would much prefer to stay where they are. But when the climate makes it impossible to stay there, the result is *forced climate migration.*

As the hottest spots in the Middle East and Africa become uninhabitable, the people currently living there will have to move; by 2100, as many as 500 million people will be displaced. And it's no help that the countries most vulnerable to climate change are also the ones whose populations are growing fastest. These growing populations, and others, will have to go somewhere, and that's where the problems begin.

"Climate change is pushing groups of people together that have different economic interests in the land," says Erol Yayboke, an expert on refugees and migration at the Center for Strategic and International Studies. "As the population is booming in a place like Nigeria, they're being forced into smaller and smaller territory. Well, you don't need to be a rocket surgeon to figure out how that might cause problems."

Those problems are already sprouting up. The brutal civil war in Syria, for example, was triggered in part by the worst drought in the nation's history, which sent 1.5 million people fleeing to the cities to survive, putting intense social stresses on an authoritarian society.

The United Nations calls climate change a "threat multiplier"—it makes bad situations worse. Calculating exactly how much worse is slippery, considering how many factors contribute to wars. But by one estimate, the changing climate has already influenced as many as 20% of armed conflicts in the last 100 years. At our current rate of warming, we can look forward to a 26% probability of increased conflict risk.

That kind of future is not set in stone. Governments and humanitarian groups can take small steps *now* that will stave off gigantic, expensive conflicts later—like offering resilience support to vulnerable countries.

"If you had more money to have stronger home-building materials—to just use a very simplistic example—then you would theoretically be more resilient to climate change," Yayboke says. "The Bangladeshes of the world, the Hondurases of the world—they need that type of support."

It may be some time before climate migration is fully recognized and measured by governments and the media; it's a new concept, and reporting structures don't really exist yet.

Even the people *doing* the migrating may not identify themselves as climate refugees. On a research expedition where he interviewed refugees from Central America crossing Mexico en route to the United States, Yayboke discovered that the migrants may not say, "We're moving because of the climate." They might say instead, "We're moving in pursuit of a better life." But if such a person had, for example, seven reasons for leaving, Yayboke found that an increasing number of them are related to climate change. "Crop failure, or flooding, or three once-in-a-lifetime storms that happened over the course of five years are oftentimes at the root of these migrations."

The pressure for residents of Mexico and Central America to move to the United States will only increase as more regions of those hotter countries become unlivable. But the need to escape the ravages of extreme weather affects Western countries, too.

The biggest migration in American history, in fact, was climate-related. It was the Dust Bowl of the 1930s, a massive drought that turned America's central farmland states to dust, which literally blew away in "black blizzards" of hot winds. About 2.5 million desperate people were forced to move just to survive.

Cut to the present day, when scientists estimate that a sea-level rise of three feet will require 4 million Americans to move. In fact, these migrations have already begun. Yayboke notes that already, Pacific Islanders, at risk of losing their countries entirely to the sea, are migrating to the mainland United States and Hawaii in enormous numbers.

It's worth contemplating the anarchic situations now, before they occur.

Surviving Disorder

Why do we obey the law? Partly to avoid the consequences of breaking it, of course. You don't want to be arrested.

But most people still follow the rules even when there's nobody around to catch them. You *could* walk out of an understaffed restaurant without paying your bill. You *could* drive through a red light at a deserted intersection. In those situations, far less visible forces compel us to obey the law: a respect for the authorities that created it, for example, or social norms (everybody else obeys, so we do, too).

But when nature goes extreme, both of those forces fall apart. There are no enforcement figures around, *and* your belief that there's an authority evaporates.

"People lost faith in the government" after Hurricane Katrina, says Horne. "You want to be a little bit confident that there is a force for order at some level of government. The Feds did not provide that, and people were extremely skeptical."

Fortunately, once we understand that we're on our own, we don't necessarily fall into animalistic, *Lord of the Flies*–style anarchy. During Katrina, for example, people began to self-organize, says Horne, getting resources together for their neighborhoods and making plans together, even across racial lines.

Even so, when you sense that civility is starting to come apart at the seams, there are steps you can take to stay out of danger and help other people.

Staying Home or Getting Out

When a disaster is unfolding in the streets, the best advice is *not to be there.* Outdoor space is public space, and you can't anticipate what you'll encounter.

In the aftermath of Katrina, for example, there was looting, window-breaking, and carjacking. The traffic lights were out, so cars were crashing into each other at intersections. "I know lots of people who had that fate," says Horne.

And then there were the utterly unexpected consequences. So many roofs blew off houses, notes Horne, that the streets were filled with roofing tacks. If you tried to drive, you got a flat.

If you're enduring the kind of disaster that requires an evacuation—for example, a superstorm, hurricane, or wildfire—the best advice is to get out while the getting's good. Leave before the evacuation order comes—or at the very latest, *immediately* when it comes.

Other weather-chaos situations generally unfold too slowly (heat waves, droughts) or end too quickly (snowstorms, tornadoes) for officials to issue evacuation orders. In those cases, your strategy should be camping at home, as described in Chapter 8.

That means preparing in advance. You need food, water, and heating or cooling sources at home. You should also consider some means to deter unwanted visitors, using powerful lighting or, if you're comfortable with the idea and trained, even weapons, as described at the end of this chapter.

If You're Caught in a Riot

When your city is experiencing civil disruption, there's a third possibility, beyond holing up at home and getting out of town: You're caught out in the streets when the chaos breaks.

That's not an everyday situation, but the survivalist community has nonetheless thought it out, and offers some common-sense advice.

- **Get away.** Your goal is to get away from the mayhem. If the crowd is moving, flow along with them—same speed and direction—while you look for a side street, a doorway, or some other way to exit the blob.

If you're with family, you don't want to get separated. Hold hands or lock arms as you walk. If you've got kids you can carry, carry them.

You don't want to be on the ground, where you can be beaten or trampled. You also don't want to stand out. Don't take sides, don't assist the uprisers, avoid face-to-face encounters. Stay close to walls, head down, away from the center of the throng, and move with the crowd. Take short steps to avoid tripping.

- ♦ **Get inside.** If you see a door, enter it: a shop, a gas station, a bank, anything. You're always better indoors. The space is too small to host a whole riot, so there are fewer people, and at least now you have options—finding a back door, going to another floor, barricading the entrance.
- ♦ **Show the rioters that you're not the problem.** Your looks and your attitude are important. You want to show that you're not the enemy. Look as calm as you can manage.
- ♦ **Show the police that you're not the problem.** At the same time, you want to let the *authorities* know that you're a terrified bystander, not part of the uprising. Cover your head with your arms for protection as you make your way to a wall, away from the crowd. If the authorities address you, be as passive and nonaggressive as you can. That can be a challenge if you're being unfairly targeted by a fired-up police force—all the more reason to abide by the prime directive of public chaos: Get away.

If you're driving, avoid the main roads and public squares, where conflicts usually erupt. If you're forced to drive through a crowd, keep moving forward, honking, to show them that you mean business—but slowly enough that you won't run them over.

Notes on Tear Gas

Tear gas, a favorite crowd-control weapon of police worldwide, burns your skin, makes your eyes and nose run, makes it hard to breathe, disorients you, and effectively blinds you, because you keep closing your eyes to stop them from burning. You cry, you cough, you may vomit.

The skin irritation may last a while, but the other symptoms go away within an hour. That's why police forces use it: It's fast-acting but temporary, and, in theory, nonlethal.

Tear gas, or CS gas as it's known in law enforcement circles, is not actually a gas; it's a crystalline powder. (CS isn't short for some cool chemical name. It stands for its inventors, Ben Corson and Roger Stoughton, who developed it in 1928.) Police toss a canister that contains a small explosive charge that spews the stuff everywhere, usually at the front lines of the riot. *They* are wearing protective masks. *You* are quickly incapacitated. If you know ahead of time that tear gas might happen to you (or pepper spray, or CN gas, or whatever irritants police have on hand)—if, for example, you're attending a protest on some raw-nerve issue that may escalate—you can take some precautions. A scarf or bandana over your face, especially if it's wet, can block the powder on its way into your mouth and nose. Swim goggles are fantastic; sunglasses are a distant second. (You may have seen people rocking the bandana-and-goggles look in photos from the 2019 Hong Kong street protests or the 2020 Black Lives Matter protests.)

Figure 15-1. Protesters wore bandanas and hats to minimize the effects of tear gas during an Egyptian uprising in 2012.

Tie up long hair. Wear a hat and long sleeves; the less skin exposed to the air, the better. Makeup makes tear gas worse, because the powder sticks to it.

Police usually deploy tear gas at the front lines of the crowd, so you're safer near the back. If you're unlucky enough to get hit, and you have any wits about you left at all, try to remember:

- **Close up your face.** Hold your breath (don't take a big gulp first) and, as much as possible, keep your eyes closed. Tear gas hurts and works fast. Keep the stuff out of you!

- **Leave.** Tear gas is a cloud; it hangs around for a couple of hours. The longer you're in it, the worse the effects. Get to someplace breathable. But don't run at the outset—fast-walk instead—so you're not breathing hard.

- **Air is your friend.** If there's a breeze, face it. Let it blow the stuff off of you.

- **Rinse your eyes in cool water.** Take out your contacts, if you have them. Bend over and stick your head horizontally under a stream of cool water. Let it flow from the inner corner of each eye onto the floor or into the sink. Avoid letting the contaminated water fall onto your clothes or skin.

  If you have glasses, take them off and wash in soap and water, so the tear-gas powder on them doesn't continue to irritate your face.

- **Change clothes** as soon as you can. Leave your shoes outside. Hang up the contaminated clothes someplace well ventilated for a couple of days. Later, you can launder the powder out of the clothes with several washings.

- **Wash your hands** with soap and water, so you don't keep getting more of the powder on your face.

- **Take a very long, very soapy shower.** Remember that the stuff is still on you. You can still breathe it; it can still run from your hair into your eyes. Shower in a way that protects your face and lungs.

Rubber Bullets and Flash Bangs

They may be made of rubber, but at close range rubber bullets can break bones or even kill you. At 15 to 50 feet, a rubber bullet can knock you over, make you limp, or give you a nasty bruise. Farther away than that, you can walk it off. Plastic bullets, wooden bullets, "sponge rounds" (hard foam rubber), and beanbag rounds (tiny fabric "pillows" filled with lead pellets) are similar. They're not actually nonlethal—just *less* lethal. All of them can cause permanent damage or death.

Then there are "flash bangs"—stun grenades that explode with a blinding flash and deafening noise, with the goal of leaving you dizzy, disoriented, and temporarily blind. Unless you happen to be wearing shooting-range earmuffs and a self-darkening welding mask, there's nothing you can do to protect yourself.

All the more reason to get out of a rioting mob, avoid the front lines, and make sure the police realize that you're an innocent bystander.

Misinformation

During Hurricane Katrina, the power, water, and phone services weren't the only things that went out. There was no TV or radio, either. News-gathering relied on word of mouth. Predictably, a lot of the reporting was inaccurate.

To this day, you may remember hearing reports of widespread rioting, of gunfire ringing out in the Super Dome, of widespread rapes.

"Epidemic rape and all of that—it didn't happen," says Jed Horne. "A lot of the reports were wildly exaggerated, and stupid, and racist in part."

Horne himself was told that his home in New Orleans's French Quarter had flooded to the nine-foot level. He raced back to save what he could—and found out that the information was false. The French Quarter hadn't flooded.

During Hurricane Sandy, rumors flew on Twitter that all bridges going to and from Manhattan were sealed off, and that the New York Stock Exchange trading floor was flooded three feet deep. People photoshopped sharks swimming in the streets.

No surprise, really. Rumors and bad information "proliferate before, during and after disasters and emergencies," the Department of Homeland Security says.

Online, people deliberately create or pass along fake news about disasters for all kinds of reasons. They may hope to produce a certain result, like closing schools for the day. Or they want to scam the public, hoping to profit from the panic, or push a political or racial agenda. Or maybe they just want to get a lot of "likes" or retweets.

And sometimes, people propagate false or old news unwittingly.

Bogus news can be incredibly dangerous. During 2017's Hurricane Harvey in Texas, for example, a false rumor spread online that if you showed up at a shelter, you'd be asked about your immigration status, which may well have dissuaded some people from seeking help. And during the March 2017 Louisiana floods, a rumor spread that the American Red Cross had forbidden *praying* in its shelters. It was, of course, untrue.

Similarly, some misinformation artists tried to turn the horrific 2020 West Coast wildfires to their political advantage, writing on Twitter that, for example, some of the fires were set by Antifa (violent left-wing) or Proud Boys (violent right-wing) members. The problem wasn't just another reminder how awful people can be; it's that the authorities, already stretched desperately thin trying to control the fires, now had to work overtime to combat the bad information as well. "Conspiracy theories and disinformation take valuable resources away from local fire and police agencies working around the clock to bring these fires under control," said the FBI at the time. As a citizen, your best course of action in these situations is to:

- **Remain skeptical.** Truth breakdowns *happen* during a disaster, every time. Be a cynic, especially if the news affects decisions that you're about to make, like evacuating, traveling, paying money, or going to a shelter.

- **Check it yourself.** Click the link. Does the article say what the headline says? Is the site a legitimate news organization you've heard of, or only one that *sounds* legit ("HurricaneHarveyNewsToday.com")?

  Google the author's name. Is it even a real person?

  Search the rumor on fact-checking sites like Factcheck.org or Snopes.com.

- **Don't be a part of the problem.** Don't like or retweet news until you've confirmed its source.

- **Be part of the solution.** Part of the misinformation problem is the *true*-news vacuum that arises during a disaster. If official news is available, people will amplify it. But in the absence of actual reporting, people make stuff up.

  If the authorities or agencies are smart, they'll get busy on social media, releasing confirmed details and retweeting rumors with hashtags like #mythbuster, #rumor, or #[HurricaneName]Fact and big red Xs across whatever phony photos are circulating.

  Your job, as a citizen, is to find these *actual* updates and repost them, with a link to the *actual* source. Fight bad information with good information.

Scammers and Opportunists

Whenever money flows in society, scammers are nearby, ready to skim some for themselves. When it happens during the chaos of a disaster, it's worse, because you're already devastated and disoriented. You're an easier target when you can least afford the additional heartache.

The most common scam goes like this: Somebody contacts you—by phone, text, paper mail, email, or door-to-door in person—offering help. How wonderful! But sooner or later, the perpetrator mentions that before anything else happens, *you* have to pay *them*.

You pay them—and then you never see them again.

Here are five of the most common post-disaster scams and how to avoid them.

- **The contractor scam.** "Hi, I'm a contractor. Looks like you've suffered some damage! I could get my crew here by day after tomorrow. I'll just need a down payment, and we can get going."

  Start by asking to see their references. Proof of liability. Proof of workers' compensation insurance. Their driver's license. Write that down, along with their car or truck's license plate number.

  Sometimes, they say they're visiting on behalf of your insurance company. They're not. Don't give them any details, like your policy number, personal info, or coverage specifics! They could be harvesting details like that so that they can turn around and pose as *you,* and collect *your* insurance money.

Sometimes they say that "FEMA gave us your name," or "we're contractors sent by FEMA," or "we're FEMA-certified." It's all bogus; that's not how FEMA works, and FEMA doesn't certify contractors.

Don't be fooled by a website, a business card, or even "FEMA Certified Contractor" painted on the side of the person's truck or van. All of those are easy to fake. They're all part of the scam.

You know what you should do? Scam them back. Ask *them* for as many details as possible about *them,* and then report them to the National Center for Disaster Fraud hotline, (866) 720-5721.

No matter who they are, don't pay them a penny until they've left your property and you've asked around. Check with the Better Business Bureau, contact the local contractor license board, and call the person's references. Most states offer online databases that let you look up contractors' license and insurance details.

Get a written estimate for the job, complete with schedule of work, schedule of payments, and scope of the project—and not just from that one contractor. Get several.

Don't sign a contract that has unfilled-in blanks. You don't want anybody to fill those in later with terms you don't care for.

And when the work finally takes place, don't issue the final payment, or sign a certificate of completion, until the work is completely done, and you've confirmed that it meets building codes.

- **The insurance scam.** "Hi, I'm here with the insurance company. I'm here to help assess the damage, so you can get your money sooner. If you'll just pay for the inspection, we'll get the ball rolling."

  If this pitch comes, unbidden, from a phone call, text, or email, it's probably a scammer, no matter what the caller ID seems to show. Hang up, ignore the text, delete the email—and then *you* initiate contact. Contact the insurance company yourself, using the phone number on its website. If you've been flooded, for example, and you have insurance with the National Flood Insurance Program, call (800) 638-6620.

And if someone comes to the door, don't give any money *or information* until you can verify that they're legit.

Do the world a favor: If you suspect you've got a scammer on your hands, report it to the National Insurance Crime Bureau hotline, (800) 835-6422. Or text your suspicions to them, using tip411 as the number and FRAUD as the message.

Let the insurance company know, too, along with the local police.

- **The government scam.** "Hi, I'm with FEMA, and I'm calling to help you get disaster-relief funds."

  Or maybe they show up at your doorstep—wearing a shirt or jacket with the FEMA logo right on it. How could they possibly be scammers?

  Well, don't you worry: It won't be long before they ask you to pay a "processing fee" to get your paperwork in shape. And that's the last you'll ever see of them.

  No genuine FEMA or government representative will ever ask you for financial information or money, and there's no cost or fee to apply for disaster relief from the government. If someone suggests otherwise, they're a faker.

  No genuine FEMA rep will ever call you and ask for bank or personal information, either. Hang up.

  It's important to note that workers and agents *do* go door-to-door in damaged neighborhoods—but they carry government photo I.D. to show you. Federal and state workers are not permitted to ask you for money, or even accept money if you offer it. FEMA officials never charge anything to inspect your home, help you fill out applications, or get you disaster-assistance money.

  If they don't have the photo ID and a FEMA badge, they're impostors. The FEMA logo on their clothing means nothing except that they ordered custom embroidery from an online logo-wear company.

  As for phone calls, emails, and texts: Once again, hang up, ignore, delete. Call FEMA directly at (800) 621-3362, or look up the phone number of whatever agency you're interested in on its website.

  If you think someone's trying to take you for a ride in your time of disaster, let the National Center for Disaster Fraud team know by calling (866) 720-5721, or emailing *disaster@leo.gov.*

- **The charity scam.** Every time there's a big weather disaster—every single time—new websites pop up, advertised through social-media posts, offering to collect charitable donations for the victims. Lots of them are fake. The only people who will be helped by your donation are the scammers.

  That's true even of the charities that encourage you to donate by text message.

  "Oh, I'm not that dumb," you might say. "I'm smart enough to give money only to charities I've heard of, like the Red Cross." Well, that's the thing: Often, these phony sites *say* they give the money to a well-known charity, or are affiliated with one. They're not.

  Here are some of the Con Artist Red Flags:

  They ask you to pay by cash, by wiring money, or by sending a gift card. Only con artists make those requests—because those methods leave you with no way to get your money back. Credit card or check is far safer.

  Or they try to rush you. Or they thank you for a donation you didn't make. Or their charity names are only one word different from the name of a real charity. Or they use "spoofing" software to make caller ID display a local phone number.

  To avoid falling for these cons, donate directly to charities you've heard of—not charities that sprang into existence in the wake of the disaster. If in doubt, look up the charity at a site like CharityNavigator.org, CharityWatch.org, or Guidestar.org. These outfits research each charity and find out where the money goes. If it's legit, the site will let you know.

  Look up the text-message number on a charity's website before you donate, too.

  If you determine that a charity is legit—the Red Cross, let's say—don't forget to specify the particular disaster you're trying to help. If you don't specify, for example, Hurricane Gertrude, then your money may go into a general fund for that charity.

Finally, if you think you've been scammed, report the phony charity to the Federal Trade Commission at ftc.gov/complaint. You may be able to protect future suckers.

- **The GoFundMe scam.** Often, individual disaster victims appeal to the public directly for small donations on crowdfunding sites like GoFundMe or IndieGoGo. *Thousands* of such campaigns arise after every wildfire and hurricane. After Louisiana's 2016 massive floods—unrelated to Hurricane Katrina—6,500 people set up GoFundMe campaigns; citizen donations totaled over $11 million.

  According to GoFundMe, about 0.1% of them are phony.

  You can be most sure that you're donating to a real victim by analyzing the relationship between whoever has posted the campaign and the recipient of its kindness. Good sign: It's an immediate relative. Bad sign: They're unrelated.

  Also check out who's making donations and leaving comments. Good sign: Immediate family and friends are contributing. Bad sign: All the comments came in around the same day, and they're all about the same length, and all seem to have the same writing (and misspelling) style.

  Finally, beware of the copycat campaign. Sometimes, a particular GoFundMe plight gets some media coverage. Quickly thereafter, scammers create identical-looking GoFundMe pages of their own, in hopes of tricking people into donating to *them*. Bad sign: Even after the publicity, the amount donated is surprisingly low.

The Gun Question

You probably have strong feelings about gun ownership one way or another, depending on where you live and how you've been raised.

You've probably heard about the risks of keeping a gun in your home. 13,000 children are accidentally shot each year in the United States, for example. The vast majority of those guns—89%—went off in the owners' homes, when children were playing with them. Gun owners are twice as likely to be murdered, and three times as likely to commit suicide, as non–gun owners.

Becoming the owner of any gun is a major step. Doing it right involves looking into your city and state laws; taking classes to learn handling, shooting, and safety; and figuring out how to store it safely in your home, where your kids won't get at it—like in a gun safe.

"A firearm as your only response to personal security is not the answer," says survival instructor (and gun owner) Tony Nester. "If you don't have the proper training and mind-set, you might be putting yourself and your family at risk."

Instead—or, at the very least, in addition—he recommends some less expensive, less drastic defenses for your home.

- **Lights.** With one trip to Home Depot, you can equip your home with solar-charged lights for dark spots around your house. When the power's out, automatic lighting deprives potential home intruders of decent hiding places. Remember: They don't particularly care about robbing *your* house; they just want to rob *someone's* house. If yours is well lit, they'll look for one that offers better concealment.

- **Tactical flashlight.** Suppose, when the power's out, you hear something in your yard. "Do you really want to meet that person with a shotgun in your hand?" Nester asks. "Maybe it's just some kid screwing around, or a raccoon."

  A *tactical flashlight* is an excellent weapon. These are rugged, aluminum, shockproof, incredibly bright flashlights, of the sort used by police and rescuers. Getting blasted in the face with bright white light

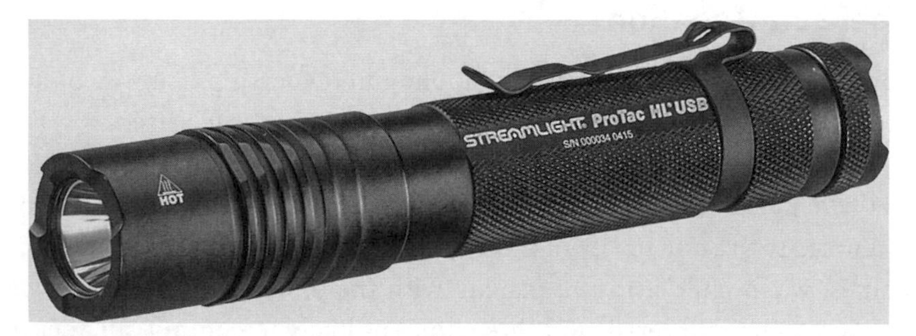

Figure 15-2. This tactical flashlight offers 850 lumens of brightness. You can shine it right into the bad guys' eyes, ruining their night vision, or use the strobe mode to make them close their eyes or turn their heads.

is disorienting and blinding. Often, that's all it takes to send the threat scurrying away.

"Criminals, muggers, and drug addicts hate drawing attention," writes one tactical-flashlight owner on Reddit. "I've actually had really good luck with just a bright light directly in their face and a stern, 'Hey, back the [hell] up!' I've defused quite few sketchy individuals."

The measurement you'd really care about is how many candelas (candlepower) a flashlight gives—that's the measure of a source's light intensity. Most product writeups, though, reveal only a flashlight's *lumens* (total light given off in all directions), which isn't quite as helpful. In any case, for comparison: 80 lumens can temporarily blind someone in the dark; 200 lumens will do the job in daylight.

♦ **Safe room.** Is there a room in your home that you can designate as a safe room? Something with a lockable, solid-core door and, preferably, a window so you can escape? Figure that out now, and stash some pepper spray in there for added defense. If you have kids, keep the spray in a lockbox.

♦ **Seal up.** Look over your home for weak spots. If some of the windows look like they can be jimmied open, you can fortify them, or, less expensively, prop them shut with dowel rods. Battery-powered window alarms, which work even when the power's out, can serve as a further deterrent.

A dead bolt on the front door might be another reassurance.

So might a dog.

Chapter 16

Preparing Your Business

EACH YEAR, ACCOUNTING FIRM KPMG SURVEYS 1,300 CORPORATE CEOs about the business world's future, risks, and growth prospects. The 2019 edition of this report produced an eyebrow-raising finding: The CEOs did not identify cyberattacks, disruptive technologies, or the rise of territorialism to be the number one threat to their companies' growth. For the first time in the survey's history, their greatest concern was climate change.

If you run a small or medium-size business, or if you're the risk or sustainability officer in a larger one, climate change piles a lot on your plate. Your responsibilities include a huge circle of people, places, and things:

- **Your employees** along with their health, their safety, and their perception of your carbon-reduction and resilience efforts.
- **Your customers**, your ability to deliver to them, and how they react to your sustainability policies.
- **Your facilities, equipment, and vehicles**, and their ability to withstand and recover from extreme weather.
- **Your supply chain and raw materials**, and your ability to keep the lights on when *their* facilities get hit.

It's a lot to think about. Maybe that's why 74% of small businesses have no disaster plan in place, and 84% don't have the right insurance.

Maybe, too, that's why business is booming at climate consultancies like Ramboll Group. "Our business has grown by an order of magnitude," says Ramboll principal Sue Kemball-Cook. "Almost a vertical trajectory. People are really paying attention to this now."

The beauty of taking weather-resilience steps now is that they begin paying off immediately. In fact, they pay off even if you never *are* hit by disaster.

Think about it: Your insurance costs drop. You're be more resilient to all *kinds* of disruptions, not just weather. You, as a subcontractor, become much more attractive to prospective government and Fortune 500 clients, who may be required to assess your disruption resistance before hiring you. You make *everyone*—your employees, your customers, and the public—more confident in you.

Finally, as a side effect of analyzing your organization's sustainability weak spots, you may find opportunities to streamline and improve its everyday operations. "Preparing for more extreme weather in the future makes your business better buffered now—better able to absorb shocks," says Kemball-Cook.

And if a weather disaster does befall you, you can recover quickly and inexpensively, and that's everything. Here's a FEMA stat that should curl your toes: Of small businesses that haven't resumed operations five days after a disaster, 90% are out of business within a year.

"When disaster strikes, time is not your friend," says Donna Childs, author of *Prepare for the Worst, Plan for the Best: Disaster Preparedness and Recovery* and CEO of Prisere, a climate-change resilience consultancy for small businesses. "If you're a small business, you either recover quickly or you don't recover at all. You've got to be set up and ready to go."

This chapter is designed to help you create an emergency plan for weather disasters. It's primarily aimed at small and medium-size businesses, which have a much harder time recovering from disasters than bigger companies. The large outfits have a bigger financial cushion, the option of diverting business to other divisions, and, presumably, established emergency procedures.

Even so, the thinking behind the advice here applies to any institution of any size. It's all guided by the wise old saying: "If you fail to prepare, you are preparing to fail."

Sizing Up Your Risk

For any company, the first step in climate resilience is examining your outfit's vulnerability. Not in 2050 or 2100—today.

"Getting a sense of what the climate hazards are *now* is a good place to start," says Kemball-Cook. "Is the business exposed to flooding now? Is it in a FEMA floodplain now? Is it likely to be affected by hurricane storm surges? Is it likely to be affected by wildfire? For industries that are reliant on having a steady supply of water, especially water of a certain quality or temperature, is it in an area that's prone to drought?"

Chapter 2 offers a list of online, interactive maps that reveal the risks of flood, storm surge, sea-level rise, heavy precipitation, and other hazards for any address you supply.

At that point, think about what time period you care about. What's the expected life span of your equipment, your lease, and your investment in your infrastructure? If you think you'll be wrapping up your business at this location five years from now, then climate change may not be urgent enough for you to make any big changes.

But if you've just bought a new property, and you might be here for 20 or 30 years, then the considerations in this chapter could make a huge difference.

What *kind* of hazard makes a difference to your vulnerability calculations, too. "The hazards that cost people the most money are flooding and fire," says Kemball-Cook.

Suppose, for example, that you live on the coast. You're already seeing street flooding when it hasn't even rained ("sunny-day flooding"). You know that the sea level will rise. For you, flooding is the biggest concern.

Now, if your building's ground floor is a parking garage, then flooding might cause only minimal damage. But if your computers and HVAC equipment are on the first floor or the basement, it makes sense to move that stuff upstairs. You might also want to waterproof the floor and wall materials, install pumps, and see if your building materials can be made more resistant to saltwater intrusion.

"Once you've identified what your exposure is," says Kemball-Cook, "there's a long list of adaptation measures you can start to think through."

Your People

When havoc is being wreaked, your first thought should be the safety and welfare of your employees. You care because they're valuable assets, no doubt, but you also care because you *care.*

Having an emergency plan for your business is useless if you haven't communicated it. Once you've thought through the plans described in this chapter, therefore, add a presentation to your next company meeting, or add a brown-bag lunch session and walk through your ideas then.

Building an Emergency Plan

The emergency plan should include:

- **A chain of command.** Who's in charge if the big boss is wounded, missing, or unreachable? Who'll fill in for employees who disappear in a panic and never come back? (Yes, this happens.)

  The command tree should go both ways: If certain employees have gone tearing out of the building, they should know how to get back in touch.

- **A call-down roster.** Create a contact tree for quickly spreading the word in an emergency: Safety and whereabouts checks, disaster news as it's happening, evacuation instructions, rendezvous points, post-disaster plans, and so on. It might work like this: The CEO is responsible for contacting three lieutenants, each of whom contacts two department heads, and so on. Word spreads in parallel, like the branches of a tree.

  As we now know from 9/11, Hurricane Katrina, and recent California wildfires, text messages may still get through even when the cellular network goes down. Therefore, your phone's contact list needs *all* numbers and addresses of the people you're supposed to contact, including their cellphones. If your employees work in shifts, make sure the tree accommodates people who aren't in the building at the time.

  Finally, *test the tree* to make sure you haven't left anybody off the list. Try it both during the workday and after hours (begin with "This is a test," so you don't freak anybody out). As your personnel change over the months and years, of course, you'll need to keep the tree updated.

- **An internal update system.** You need some way for everyone in the company to look up what's happening with your business, from wherever they are. An internal website works fine, or a toll-free number, as long as it's hosted far away from your workplace. This, too, requires testing and training. The website should be bookmarked, or the phone number on speed dial on your peoples' phones.

- **An evacuation system.** Each person should be assigned responsibility for another worker or two, to make sure everybody gets out of the building. Somebody should be the designated bathroom checker, too—you don't want anyone left behind because they were texting in a stall.

  The evacuation plan should also include routes and exits—and *backup* routes and exits. Think, too, about how you'll know if everybody's left the building. How will you be able to account for your workers, your customers, and any visitors?

  Sometimes, after an evacuation, the authorities may be willing to accompany you back inside briefly to grab some essentials. Do you know what that'll be? Are there certain folders, documents, or laptops to grab? Maybe you should flag them with red stickers ahead of time to make them easier to spot in a hurry.

- **A tabletop exercise**, which means thinking through a pretend disaster. In an emergency, who will be responsible for what? Whose job is it to monitor news channels, so you'll all know when an evacuation order has come down? How will you communicate if the cell network is out? What documents or hard drives have to be grabbed from the office before everyone leaves, and who will grab them? Who will shut off the computers, the power, and the gas?

  You, the company's preparer, should do some behind-the-scenes homework as part of this exercise. Figure out what you could use as a backup source of equipment, supplies, and even employees: If you're forced to lay off some of your workforce, how easily could you rehire a team once the business is fully operational?

FEMA offers a free step-by-step "playbook" for conducting this kind of training. It's online at http://j.mp/2tXmdba.

- **Go bags.** As you know from Chapter 8, a go bag is a critical part of any disaster plan. In a workplace, you can pack your employees' bags with gear that's dedicated to getting them out of the building and home: Flashlight, bottle of water, cash, energy bars, phone charger, and a small first aid kit, for example.

  Every worker should have a go bag. You might also hang a few extras in the stairwells or hallways with Velcro, for the benefit of visitors in the building or employees who have stepped away from their desks. You don't want them having to run *into* the office space just to get the stuff they need to run *out.*

- **"In case of emergency" cards.** Every employee should set up the Emergency screen on their phones (page 274) or "In case of emergency" cards in their wallets. This screen or card contains emergency contact details and medical information—critical stuff for first responders who find an employee passed out.

- **First aid training.** For a fee, the Red Cross will come to your workplace and conduct a two-hour first-aid class for your employees.

Don't feel shy about walking your employees through all of this training. Chances are good that they'll appreciate your concern, especially when you show them how to take similar prep steps at home (see Chapter 8).

You need to repeat the training and test its systems periodically, too, because employees come and go, and people forget.

When Disaster Strikes

The preceding chapters offer most of what you and your team need to know about what to do in various extreme-weather emergencies. But in a business context, a few additional tips apply:

- **Flood.** If the water's coming into your building, instruct your employees to get off the elevators at the *second* floor and take the stairs down and out of the building. Why the second floor? So that the elevator motor and its supporting infrastructure don't get ruined by floodwater.

- **Wildfire.** If you're trapped in the building, close all doors, but leave them unlocked, so that firefighters can reach you. Alert everyone in the building; check the bathrooms as you go. Don't use the elevator, because if the power goes, you'll be trapped inside.

  If you can get out of the building safely, call 911; don't assume that somebody else has phoned in the fire. If firefighters are on the way in, let them know if anyone's still inside or unaccounted for.

  If you *can't* get out of the building, make your way to the appointed safe room, which, of course, you've planned in advance. It should be the lowest, most central room—the basement, if you have one. If not, go to the lowest closet or interior hallway. The idea is to put as many doors between you and the fire as possible, and away from windows.

  If there's smoke, breathe shallowly through your nose, close to the floor. Call 911. Seal any openings, vents, or cracks around doors with wet towels or sheets. If you can see flames outside, leave the windows closed. If you *don't* see flames, open the windows from the top and the bottom, and wave something (a shirt, a towel) to get attention from rescuers. If your clothes catch fire, don't run; cover your face and roll.

- **Hurricane.** If an evacuation order has come, get everyone out. No questions, no delays.

  If not, you may be inclined to ask for volunteers who'll stay at the building. Their tasks: Patrol and watch for problems like leaks and structural damage, monitor any equipment that has to remain online, and, if the power goes out, turn off everything electrical or electronic so that they won't get fried when the power comes back on.

  Equip these employees with water, food, flashlights, and walkie-talkies. At the storm's peak, when the wind is dangerous outside, they should take shelter in the predetermined safe room—even a closet.

Emotional Recovery

Recovering from a physical hit to your business involves a lot of visible and logistical elements. But the invisible ones may be the most serious of all.

Putting your business back together is time-consuming, frustrating, and fraught with paperwork and bureaucracy. Clients may be disappearing or canceling contracts. You may be turned down for disaster loans, or your bank may not extend your line of credit. You may be battling the insurance company or the government.

Through all of this, you may also be experiencing post-traumatic stress. That can mean flashbacks, nightmares, trouble sleeping or concentrating, irritability, emotional numbness or depression, and persistent guilt, fear, or shame.

Don't ignore your symptoms, or your employees'. Let them know what kinds of feelings they may experience, reassure them that they're normal and forgivable, and let them take whatever breaks they need. Offer information about getting therapy outside of work. Showing concern for your team's mental health earns you their loyalty at a fragile time for both them and your business.

You may want to invite a counselor to come in to talk to your employees. Your local Red Cross, department of health, or community hospital probably offers counseling services, too.

Even FEMA is there to help. It maintains a free disaster-stress help line: (800) 985-5990, or send a text message to 66746 with the message TalkWithUs to begin a text chat. The counselors in FEMA's Crisis Counseling Program will meet with you at home or at your workplace, one-on-one or in groups. They say that they "treat each individual and group encounter as if it were the only one, keep no formal individual records or case files, and find opportunities to engage survivors, encourage them to talk about their experiences, and teach ways to manage stress."

Keeping the Business Open

Creating a disaster plan for your company is an imagination exercise. Suppose your business is hit by a wildfire, a flood, or a hurricane, or is incapacitated in some other way. What happens next?

Where You'll Go

About half of small businesses *never* reopen after a disaster. In the time it takes for their premises to be repaired, their customers move on and never return. How about you? Could keep your business afloat during the repairing and rebuilding period?

For some business categories, the answer is no. If you have, say, a restaurant, shop, medical office, dental office, or a factory, you're pretty much limited to operating from your original address. Let's hope you've sprung for business-interruption insurance, described later in this chapter.

But plenty of other businesses might stay afloat by moving to a new, temporary location:

- **A home.** If you have a *really* small business, maybe you can set up temporary quarters in somebody's home, or your own.
- **Another location.** If your company has more than one location, make arrangements for collaboration, so that if one site goes down, the other offices can provide space for its people. Try it out some weekend.
- **The B2B buddy system.** If you don't have other locations, apply the same strategy with a fellow business owner: In an emergency, one of you will provide space, equipment, power, and internet to the other. Maybe you'll sell each other's products for a while. It's a business-to-business buddy system.
- **Lease.** You can lease temporary office space. But the *time* to investigate your options, research prices, and introduce yourself to your prospective landlord is *now.* When disaster strikes, you won't be the only one looking for space.

- **Telecommute.** Some or all of your workers may be able to do their jobs remotely. This, too, is something you have to plan ahead of time—if, in fact, the coronavirus pandemic hasn't already given you practice. "If you give your employees laptops, make sure they're not letting their children use them to play video games, because you might get all kinds of viruses that could interfere with your business," says Donna Childs.

  That's not your only security worry. If you make your data available for remote access by your employees, you need to ensure that hackers can't also dial in. You really need to set up a VPN (virtual private network), if you don't already have one in place. Hire an IT person to get that going, if necessary, and teach your people how to use it.

How You'll Run Things

While you're pondering worst-case scenarios, don't forget the nuts and bolts of your operation:

- **Back-office operations.** Will you be able to keep your payroll and human-resources functions running when you can't get into your office?
- **Contractors.** After a disaster, the cellphone of every local builder, plumber, electrician, and debris-removal outfit blows up with calls and texts. The time to research some good ones, and even introduce yourself, is *now*, while the world isn't flying to pieces. The sooner your place is put back together, the sooner your business can resume.
- **Cash flow.** Will you need a loan or an extension of your credit? Here's a great idea: Apply for a line of credit, or a higher advance on your credit card, *now*. That way, you won't get stuck in the heat of battle. Also, if you need to apply for a disaster-relief loan, you're more likely to be approved if you have good credit.
- **Petty cash.** Speaking of cash flow: Your business should have a decent supply of petty cash. If a storm is coming, someone should run to the bank to *get* some. Don't expect ATMs to be running when the power is out.

- **Office phones.** Electronic phones don't work when the power's out, and any phone system might get damaged in a natural disaster.

  It's difficult to set up a landline to roll over automatically to your cellphone, which might be just what you'd want while your main workplace is off-limits.

  As Childs points out, though, it's relatively easy to do the reverse: Set up your *cell phone* to roll over to a landline. One option for a small businessperson, then, is to publish a cell number as your main business number—and set its call forwarding to ring your office landline. That way, in times of trouble, you can simply turn off forwarding. Now incoming calls go straight to the cellphone, and you don't miss any customer (or insurance-company) calls.

  Another option: Sign up for a free Google Voice account (www.google.com/voice). It gives you a magical new phone number. Any time somebody dials it, *all* your phones ring at once: cell, home, work. It's one number to rule them all.

  The beauty of this system is that you can set up rules, like "Don't ring this phone on weekends," or "Go to voicemail after business hours."

  When the weather is calm, customers who dial that number every day ring the office line. But when you can't get to the workplace, you can pop onto any web browser and flip a software switch to make your *cell phone* get those calls instead.

- **Mail.** The local post office can send your mail to a new address while your main outfit is unavailable. Somebody just has to go set that up.

Communicating with the Public

If you hope for your business to recover, the outside world needs to hear from you, too. Your disaster plan should include an outline of how that's going to happen.

- **The master contact list.** In the wake of a disaster, your worst enemies are panic and chaos. They lead to people making mistakes, wasting time, and suffering. That's why you need a master phone/address list for use in an emergency, when there won't be time to go hunting for numbers.

The list should include the usual emergency stuff (fire department, medical assistance, state emergency management agencies); company officials; the most important vendors, clients, suppliers, and contractors; and financial contacts like banks and insurance companies. You need work numbers, cell numbers, *and* home numbers, because you may need to reach these people after hours.

This list needs to be everywhere: Printed and laminated both on-site and off-site; on executives' phones and computers; at work and at home.

- **A communications strategy.** How will you let your customers and clients know what's happening with you, so they won't give up on you and switch to a rival? How will they know about your new, temporary location? How will they know when your original spot is repaired and reopened?

  Update your website frequently, and post updates to your social-media accounts. Update your voicemail greeting often, too, to give people a sense that you're alive and working. You also may find it worth updating your customers by taking out ads, posting flyers, or contacting them directly.

  As soon as you have a reopening date, let your customers know, using your contact list and every channel at your disposal. Let them know again when you *have* reopened. Maybe team up with other businesses near you to take out a big ad in local papers or websites, announcing that you're *all* back in business.

  Consider giving customers a little extra incentive to come back. Consultant Donna Childs notes that when lower Manhattan began coming back to life after 9/11, many restaurants offered free dessert with dinner to attract customers back into the neighborhood.

Insurance

Almost every word of Chapter 6, about insurance for homes and individuals, applies equally if you're a business owner. Chances are good that you, too, are a little vague about exactly what's in your insurance policy. You, too, should keep records and create an inventory of everything your business owns. And as with home insurance, commercial property insurance generally *doesn't cover floods.* You, too, need separate insurance.

Where business insurance is concerned, though, there's more to it than property. If you don't want to be one of those "closed forever" businesses after a disaster, consider these additional policies:

- **Business-interruption insurance.** Here's an ugly little truth: During the time it takes to rebuild your workplace and get ready to reopen, your *expenses* don't take a vacation. You still have to pay your rent, your property taxes, and the paychecks of your employees (if you don't want to lose them!).

  But if you're closed for repairs, with no revenue coming in, how are you supposed to pay all that? Your property and flood insurance covers rebuilding your premises and equipment, but they don't make up for the lost income while you're *waiting* for the rebuilding, losing your mind with stress. They don't pay for your new, temporary quarters during the reconstruction, either, or any other extra expenses to keep the business going.

  That's the beauty of *business interruption insurance*, an inexpensive add-on to an existing property policy. When you've filed a disaster claim, interruption insurance pays whatever you *would* have earned if you were open. You show the insurance company last year's tax return, showing what you earned during the equivalent period. That's how much you'll get paid.

  Business interruption insurance also reimburses you for the costs of your rent or lease payments on your damaged building; up to a year of pay for each of your employees; and whatever taxes you would have had to pay.

  Most business-interruption insurance also includes "extra expense insurance," which covers the costs of setting up a new, temporary location, including hooking up utilities, phone, and internet, and overtime pay.

"In many cases," says Donna Childs, "business interruption insurance is even more important than property insurance. A lot of small businesses don't even know that it exists, and it's really critical."

This insurance generally costs from $500 to $1,500 a year, depending on the size of your company. But it can easily make the difference between reopening your business and closing it forever.

- **Contingent business interruption insurance**—yet another added-cost option on a regular property policy—kicks in when one of your suppliers or partners can't do its job. If you run a landscaping company, for example, and the nursery that supplies your plants gets shut down in a superstorm, this rider kicks in to cover your loss of business.

  Or, conversely, one of your major *customers* might shut down. Maybe, for example, you're a florist, and the downtown hotel that constitutes 45% of your business closes for six months because of a flood. This policy will reimburse you for the lost income.

- **Ordinance coverage.** Suppose your building is a total loss and it has to be rebuilt, but local building codes have changed considerably since, say, 1993 (or whenever it was built). Maybe these days, the building must be two feet above the floodplain, include ember-proof soffits, and meet certain energy-efficiency standards.

  All of this means that your *new* building will cost a lot more than the old one. For that reason, your insurance company will be delighted to sell you *ordinance coverage*, which covers those added expenses when you rebuild.

Once you've made your business resilient, contact your insurance company. You're now in a position to argue that you're a lower risk—and that you should be entitled to lower premiums. "It's the small-business equivalent of a safe driver discount," says Childs. "If I'm doing all the right things to be resilient, I shouldn't be paying the same premium as the guy next to me who's acting recklessly."

While you've got the insurance company on the line, ask them to record your bank's routing instructions (routing number and checking-account number). That way, when your insurance money comes through, it'll go directly into your bank account, same day, same minute.

Otherwise, they'll mail you a check. It could take days to reach you—or, if your workplace was destroyed, a lot longer.

Your Buildings

If anybody knows how to prepare a company for climate change, it's FM Global, a unique hybrid: Part insurance company, part resilience-engineering consultancy.

Its experts first visit your company, looking for ways that smart engineering can make it less susceptible to disaster. Only then do they write you an insurance policy, for whatever part of your risk can't be reduced.

That's an ingenious approach, because both you and they have the same interest: Minimizing your loss and business disruption when disaster strikes.

And what climate hazards are they thinking about when they visit a client? "It all boils down to water," says Lou Gritzo, director of research. "Too much, or not enough."

Flood Exposure

In 2017, Hurricane Harvey hit hundreds of Houston company headquarters, factories, and refineries. "Business as usual" didn't return for many months.

But rivers can flood you, too. Snowmelt can flood you. Heavy rains can flood you. A nearby construction project can break pipes that flood you. Wildfires can burn away foliage that would have protected you from rain floods. With every passing year in the climate-change era, flood insurance becomes a better idea.

Protecting any kind of building usually involves spending money; the amount depends on the permanence of the fix you want to make. Here are three examples, listed in order of best (but most expensive) to weakest (but cheapest):

- **Direct the water away.** If you're in a flood zone (see Chapter 9), the most effective and permanent step is to build structures that keep the water away from your buildings: built-in earthen dams, low masonry walls, concrete structures. This is the most expensive tactic—but not as expensive as the flood damage you'll avoid.
- **Set up temporary barriers.** Plan B is to set up temporary flood barriers when flooding is imminent. That doesn't mean sending your employees outside to stack up sandbags; this isn't 1952. These days, temporary flood barriers come in better-looking, faster to set up, and more durable forms.

  At the inexpensive end, there are plastic "boxwalls"—L-shaped, interlocking panels held in place by the weight of the floodwater itself. They're lightweight enough for a single person to carry, and they stack away when they're not in use.

  A second approach is the inflatable bladder dam: a huge, reinforced vinyl tube that, once filled with water, becomes a gigantic temporary wall.

  You can also get metal barriers that are either removable or that flip up as needed, either manually or automatically.

Figure 16-1. Temporary, removable flood barriers may cost less than permanent water-management construction. The barriers at left slide into permanently installed tracks, panel by panel, in a matter of minutes. The plastic barriers at right use the floodwater's weight against itself.

- **Relocate the crown jewels.** If flood barriers still sound too expensive, at least relocate the circuit breakers, HVAC, and electrical systems above the ground floor, so that floodwaters will have less of a business impact.

Wildfire

"Just within the last few years, wildfire has started to emerge as a bigger threat than just to homes," says Gritzo. "The solutions are still being fine-tuned, but the starting point is a cleared area between the forest and the structures."

That's the defensible-space strategy described in Chapter 3: a no-fuel zone between your buildings and the woodlands.

But for an environmentally aware corporation, the notion of cutting down innocent trees to clear that zone may be problematic. How can you claim to conduct your business sustainably if you start hacking down trees?

You can take some fire-protection steps that don't go quite as far as lumberjacking. First is, as Gritzo calls it, "ultra housekeeping." Clear away any kind of burnable clutter on or around the building itself. Thoroughly inspect the gutters, the roof, and all the openings, vents, eaves, or awnings. Doing that, at least, spares your facilities from wildfires' favorite trick: Sending embers wafting onto combustible materials and starting new fires.

If you're seeking to make your facilities more energy efficient by adding insulation to the outside of your buildings, Gritzo cautions that many insulation types are highly flammable. Following the law of unintended consequences, your well-meaning environmental effort may increase the building's risk of fire damage.

The good news: Not all insulating materials are flammable. The fire-spread rates "are alarming for some of them and basically nil for others," he says. You can install new installation that doesn't increase your fire risk—but you have to research it.

Power Backup

Only 16% of small businesses have backup generators. That's a fairly alarming statistic, since power failures are common to so many kinds of weather disasters.

And don't think that power failures are exclusive to *violent* weather. In early 2019, candy and pet-food maker Mars had to shut down its Australian factory for two days because of a *heat wave.* When citizens in the region all cranked up their air conditioning, electricity prices shot so high that "it was just no longer financially viable for us to have that facility open," according to sustainability officer Lisa Manley.

Even a drought can lead to power outages—not as a result of nature's wrath, but of your local utility's fear of fire and lawsuits. That was the case in the fall of 2019, when California's electric company, PG&E, deliberately cut the power to 2.5 million people, so that sparks from wires and equipment wouldn't fall into the exceptionally dry foliage of the state and start another wildfire. (The company still induces neighborhood blackouts when fire risk is high, although it says it now tries to make them "smaller in scope and shorter in duration." As a company executive puts it: "Turning off power for safety is not how we strive to serve our customers.")

The bottom line: If your business uses power, you're at risk of radical pricing spikes and outages. Preparing means having a generator—Chapter 3 covers various kinds available—or choosing a building with its own backup power.

Water Backup

Your business probably won't get far without clean, running water, either. Your options for creating a backup source are exactly as described in Chapter 3: inline water tanks, plastic storage tanks, or closets full of bottled water.

Your Data and Documents

You can replace a computer, you can order new supplies, you can rent a new building. But if you lose your data, you might just be replacing your career.

The beating heart of almost any company is its records: contact lists, databases, invoices, email, budgets, HR documents, strategy plans, insurance policies, contracts, employee records, lists of leased or bought equipment, and so on.

You desperately need an organized, quickly retrievable backup. When disaster strikes, you won't get a penny from your insurance company until you can produce all the paperwork they ask for—and, of course, you won't be able to get your business going again until you have access to your data.

Off-site Cloud Backup

Back when the biggest threat to your data was a power surge or a hard-drive crash, experts browbeat you with the advice to set up a backup system right there in the office. You'd back up your files onto, say, a tape drive, a Zip disk, or an external hard drive.

Today, that's not good enough. If your place floods, burns, or collapses, having a backup on-site won't help you. You'll lose it along with the original files.

The magic word for good backup systems, therefore, is *off-site.* On-site backups protect your files against human error and drive crashes, but off-site backups protect you against both of those *plus* fires, floods, mudslides, earthquakes, theft, terrorism, sprinklers going off, sabotage by disgruntled employees, volcanoes, sinkholes, and [insert your worst-case scenario here].

Actually, there are two magic words: off-site and *automatic.* That is, you want your company's files backed up, silently and automatically, from across the internet, all the time, to a secured, redundant, heavily guarded data center far away from your office. That way, if your whole building washes away in a flood, you won't care, because all your data will be safe in duplicate data centers in Virginia and Utah. (Okay, you might care a little.)

Plenty of companies offer this kind of internet-based backup service to small businesses. All of them are automatic, encrypted, and easy to set up. All of them work with Macs and Windows PCs; some even auto-backup phones. Some examples:

- **iDrive.com.** $75 a year for unlimited computers, 250 gigabytes max. Other data-capacity plans available.
- **Carbonite.com.** $290 a year for up to 25 computers, with a maximum of 250 gigabytes of data.
- **Backblaze.com.** $60 a year per computer, no data limits.
- **Crashplan.com.** $120 a year per computer, no data limits.

Free Virtual Drives

If your company isn't so much a small business as an *itty-bitty* business, you may be able to get away with a free "virtual hard drive" like Google Drive, iCloud Drive, Microsoft OneDrive, or Dropbox. These services put a "drive" icon on each computer's desktop. When you drag a file onto that icon, it's instantly copied to a data center somewhere, safely duplicated.

This is not an automated thing, though. Your staff has to remember to drag the files at the end of each day.

Also, these services are free only to a point. Dropbox, iCloud, and OneDrive offer you 5 gigabytes free, per person; Google Drive offers 15 gigabytes per person. To get more storage, you'll have to pay a monthly fee. Still: Sometimes, free can do the job.

Physical Drives

Some businesses still make their backups on external hard drives, tape drives, or cartridges. Then, to get around the problem of having the backup in the same building as the machines they're supposed to protect, somebody *carries them home* every night.

That works—as long as the take-homer remembers to do it, and to bring the backups back to work every morning.

Paper Documents

Certain documents—deeds and contracts, for example—are still, for some reason, on paper these days.

Truth is, you should scan all of it. (You can even do that with your phone.) Every important paper file should *also* be a digital file, so that it's backed up and duplicated.

You should then copy the papers, so you have paper copies, too. Finally, you should put the originals into a bank safe-deposit box or a waterproof, fireproof safe. These safes very widely in capacity and ruggedness; for $160, you can buy one that can last 20 minutes of 1,200 degree heat for 20 minutes; $340 buys you a safe that's good for an hour at 1,700 degrees.

Recovering Data from Destroyed Drives

Suppose the news just gets worse and worse. There's been a natural disaster *and* your business was clobbered *and* the fire or water got to your computer *and* you don't have a backup. Now what?

Believe it or not, you're not yet out of options.

You can ship your computer or hard drive to a file-resurrection company like DriveSavers.com. There, for a fee (on the order of $2,500), technicians in a dustless "clean room" can take your drive apart and recover the files from the damaged drive using specialized hardware and software tools.

If they're unable to recover the data, there's no charge, and they ship the drive back to you. Meanwhile, though, they advertised a recovery rate of 90% (until the mid-2000s, when they stopped advertising their recovery rate). They've recovered data from burned and melted drives, from drives on sunken ships, and even from drives with bullet holes.

Your Supply Chain

Maybe your company is sitting pretty in, say, Madison, Wisconsin. You're in a climate haven; you don't worry much about water shortages, or hurricanes, or wildfires.

But what about the people who acquire your raw materials and make your parts? What about the companies that transport your parts and your product? What about the warehouses that store your stuff before it's sold—and the retailers who sell it? *They* are susceptible to climate disasters, too.

One in four companies report that they've already seen a climate-related disruption to their supply chains—but 75% of them say they haven't yet come up with ways to prevent those disruptions.

The results can be costly. In October 2011, the devastating rains of Typhoon Nock-ten flooded 65 of Thailand's 77 provinces. A quarter of the country's rice crop was destroyed, and 815 people died.

The storm flooded 1,400 electronics-component factories—the makers of parts for more than 14,500 car, computer, and camera companies, and even the makers of *parts* for those parts. Western Digital, the world's largest hard-drive maker, lost 45% of its shipments. Nikon and Sony suffered

Figure 16-2. Flooding shut down more than 1,400 factories in Thailand, crippling the computer, car, and camera industries for months.

shortages of their digital cameras and had to delay the launches of their newest models. Toyota, Honda, Nissan, Mazda, Mitsubishi, Ford, and General Motors were caught without essential components. One Honda factory was out of commission for six months; HP lost $2 billion in products. It took more than a year for factory operations to return to normal.

Eventually, Thai factories began moving their fabrication machines up to the second floor of their buildings and setting up subsidiary plants in other regions.

But the changing climate doesn't strike only manufacturing facilities. Our parts and materials also have to *get* to us. In the United States, 99% of all imported goods arrive by ships, at major ports. But port cities are on the coasts, where storm surges and hurricanes do their worst damage.

Nobody wants to profit from the misfortune of others. But if you're the company that thought to line up alternative sources of supplies, you're the one that will keep operating after a disaster.

So what, exactly, can you do to prepare for disruptions in your supply and distribution chains?

If your company is big enough, you can invest in *protecting* your suppliers. Coca-Cola, for example, has funded water-cleanup and water conservation projects in 71 countries, and has joined the World Wildlife Fund in a six-year project to protect the corporation's water supplies around the world.

If your small business doesn't have quite that kind of clout, you have three tactics at your disposal:

- **Identify the alternatives now.** Research some companies, preferably in other parts of the country, that could step in to supply parts and materials if your primary source is disrupted. Vet them; even make contact with them. If you own a store, for example, make a list of companies that could supply your merchandise. Keep shipping routes in mind: Would the supplies still be able to reach you if a certain airport, seaport, or highway is wiped out?

- **Ask your supplier about its own plans.** There's nothing wrong with a little nudging. Ask your current supplier what resilience plans *it* has. Does it have any backups for creating whatever it creates—and for shipping it, if the regular transport company is out of commission?

Asking this kind of question makes the supplier aware that you're interested in this topic, and may even spur them into upgrading their own resilience. Everybody wins.

- **Insure against the disruption.** If you have conditional business interruption insurance, as described earlier in this chapter, you've come a long way in protecting yourself.

When You Are the Supplier

If *you* are the maker of supplies for a larger company, you may think that, from a legal perspective, you're protected when climate change interferes with your operations. After all, almost every supplier contract includes a *force majeure* clause. It says that if you experience an act-of-God-style interruption—earthquake, hurricane, lightning, war, and so on—then you're excused from fulfilling the contract. The customer can't sue you.

If that's your understanding, though, you may be in for a rude surprise. Not all forms of extreme weather that could impair your business are technically weather *events,* and therefore wouldn't trigger the *force majeure* clause. Droughts and rising sea levels, for example, aren't considered *force majeure* events.

Imagine, furthermore, that you're struck by a devastating heat wave. Everyone in the region cranks up their air conditioning, and electricity prices shoot sky-high. That's not covered by the *force majeure* clause, either; you're still obligated to deliver whatever goods or services you make.

And if you sell some product—plants or parts, let's say—that could die or be damaged in extreme heat? The customer can say, "Well, tough; pay for refrigerated shipping." Added costs because of climate change aren't *force majeure*, either.

The lesson here is to write your customer contracts in a way that *does* anticipate some of the new realities of the climate crisis. The old clauses don't.

Disaster Relief

The government may be a massive, seething bureaucracy. But it does try to help you and your business in times of disaster.

SBA Disaster Loans

The Small Business Administration (SBA) is an entire chunk of the U.S. government that's dedicated to helping you, the small business—the lifeblood of the economy.

Ordinarily, the small-business loans you get through the SBA aren't the government's money. Instead, the SBA works with banks to get you long-term, low-interest loans. Since you may be a riskier-than-normal customer, the SBA guarantees repayment of your loan. Very nice of it.

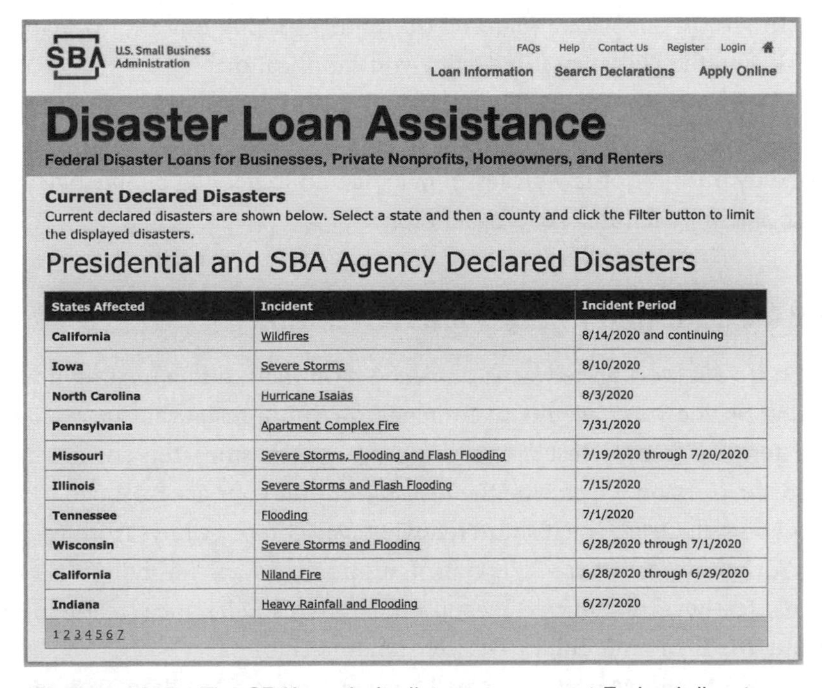

SBA U.S. Small Business Administration

FAQs Help Contact Us Register Login

Loan Information Search Declarations Apply Online

Disaster Loan Assistance

Federal Disaster Loans for Businesses, Private Nonprofits, Homeowners, and Renters

Current Declared Disasters

Current declared disasters are shown below. Select a state and then a county and click the Filter button to limit the displayed disasters.

Presidential and SBA Agency Declared Disasters

States Affected	Incident	Incident Period
California	Wildfires	8/14/2020 and continuing
Iowa	Severe Storms	8/10/2020
North Carolina	Hurricane Isaias	8/3/2020
Pennsylvania	Apartment Complex Fire	7/31/2020
Missouri	Severe Storms, Flooding and Flash Flooding	7/19/2020 through 7/20/2020
Illinois	Severe Storms and Flash Flooding	7/15/2020
Tennessee	Flooding	7/1/2020
Wisconsin	Severe Storms and Flooding	6/28/2020 through 7/1/2020
California	Niland Fire	6/28/2020 through 6/29/2020
Indiana	Heavy Rainfall and Flooding	6/27/2020

1 2 3 4 5 6 7

Figure 16-3. The SBA's website lists every current Federal disaster area in the United States. You may be shocked to see how many places are in trouble at any given moment.

But there's a big exception to the "we don't loan our own money" rule. If you're in a federally declared disaster area, the SBA *does* make direct loans to businesses (and even private nonprofits). They don't have to be *small* businesses, either—any size may apply.

You can borrow up to $2 million to rebuild or replace any buildings, equipment, and merchandise that insurance doesn't cover. The money can also pay for upgrades that'll keep you out of similar trouble the next time, like a drainage system, a safe room, or a storm shelter. The interest rate shows extreme compassion: 2% to 4%. The repayment term can be as long as 30 years.

If you've got a mortgage, the SBA may also be able to help you refinance after the disaster, which could also take off some of the financial pressure.

To learn more about these loans and get your application going, visit the SBA disaster-loan website (disasterloan.sba.gov/ela/), call them at (800) 659-2955, or email disastercustomerservice@sba.gov.

This is, by the way, a loan. It still requires filling out a lot of paperwork, and the SBA may still decline your application.

The SBA can help you fill out the application. But many local business schools or state universities also run free small-business clinics, supervised by business or law professors, dedicated to helping people like you with the impenetrable government forms.

SBA Economic Injury Disaster Loans

Even if your workplace wasn't damaged, you may still be hurting financially. Maybe you have no customers: You run a restaurant or a movie theater, and people just aren't going out much since the superstorm, wildfire, or flood. Or maybe the disaster cut off your access to whatever it is you sell. Or maybe it shuttered whatever company buys from *you.*

You might call these "How am I supposed to earn a living?" situations. The government calls it economic injury, and it's got a treatment in mind: the Economic Injury Disaster Loan (EIDL).

It's an incredible deal for a small business: up to $2 million, maximum interest rate 4%, terms as long as 30 years. (If you also got one of the disaster loans described above, the *total* of the two loans can't exceed $2 million.) It's available only if you *haven't* been able to get a loan somewhere else.

To apply for an EIDL, visit disasterloan.sba.gov/ela/, call (800) 659-2955, or email disastercustomerservice@sba.gov.

Disaster Unemployment Assistance

If you've been thrown out of work because of a disaster, help is at hand. Well, *some* help. The U.S. Department of Labor, working with FEMA, working with your state's unemployment agency, offers another kind of money: disaster unemployment assistance.

(If you, dear reader, are the business *owner,* it would be a much-appreciated kindness if you'd help your own employees with this process. Maybe post instructions on your internal website.)

To get this weekly payment—even if you're self-employed—all of these things must be true:

- You live or work in a federal disaster area.
- You're not collecting *regular* unemployment insurance.
- You no longer have a job or a workplace. Maybe your employer has had to close down—or the workplace is damaged—and you've stopped getting paychecks. Or you can't *get* to your workplace, maybe because the bridge, highway, or public transportation is out. Or you were injured in the disaster, and can't work as a result.

The money can begin to flow the week after the president declares your region to be a disaster area, and carries on for 26 weeks.

And how much will you get? It depends on what your state would ordinarily pay for *regular* unemployment insurance. The most you'll get is your state's maximum payment; in New York, for example, that's $504 a week. The *least* you'll get is half of the state average. The idea isn't to make you rich—just to help you not starve.

To apply, contact the Unemployment Insurance agency of your state—or the state you've landed in after the disaster—which you should be able to find with a quick Google quest.

Deducting Disaster Losses on Your Taxes

When a disaster has dealt damage or destruction to your business property, even the IRS is willing to soften the blow. It's called a casualty loss, and it's a tax deduction for the fair market value of what was destroyed. (If insurance or another source reimburses you for the loss, then sorry—no deduction for you.)

Ordinarily, you have to take the deduction in the year it occurred—but if your business is in a Federal disaster area, you have the option of applying the deduction to the *previous* year. That may mean amending last year's return, of course. But make the effort, because you'll be lowering your tax burden for a year when you actually *had* a tax burden. *This* year—the year of the disaster—the deduction might be wasted, since you may not have much income at all.

Here's a sample scenario. Suppose 2020 was a great year, but a 2021 superstorm destroyed your shop. Now then:

- **If you haven't yet filed your 2020 taxes**—for example, if the storm hit before April 15, 2021—take the casualty deduction when you *do* file your 2020 return. On Form 4684, specify what the disaster was and when, and the address of the destroyed property.
- **If you have already filed your 2020 taxes,** you'll have to file an amended return that includes the casualty loss. Just attach a note explaining that you're taking the deduction for a storm that happened in 2021, but wish to apply the deduction to your 2020 taxes. Again, use Form 4684; specify what the disaster was, date, and the address of the destroyed property.

  You have to file the amended return by the date your *disaster-year* taxes are due, without any extensions. In this example, your deadline would be April 15, 2022.

The IRS cheerfully recommends that you look over its free Publication 547, "Casualties, Disasters, and Thefts," to figure out how much you're allowed to deduct.

The Risks of Inaction

This chapter, so far, has covered the risks that extreme-weather events pose to your operation. But climate change poses another risk to your business that has nothing to do with weather: The risk of doing nothing *about* climate change.

Chapter 17 describes how major companies are carrying out ambitious, sweeping emissions- and energy-reduction goals. That all requires effort, and sometimes even investment—so why do they bother?

Their marketing materials always say that these companies want to do what's right for the planet, and maybe they really do. But let's be real: These are publicly traded companies whose primary motivation is profit. Clearly, they've calculated with precision that adopting cleaner practices is good for business. Here are some of the reasons why.

Your Reputation with Customers

More and more often, your customers care about *your* environmental position. In fact, 81% of consumers today—all ages—say that it's "very" or "extremely" important that corporations should actively improve the environment. It's good business to demonstrate leadership on climate mitigation and adaptation.

If you're a polluter, a denier, or someone who is complacent, they'll rip you apart on social media. They may even boycott you. And they'll switch to a competitor, even if they've known you for years. Half of all consumers say they'd rather buy a no-name product that's produced sustainably than a recognizable brand that isn't. In other words, if your company isn't working on sustainability, you're welcoming rivals to eat your lunch.

Greenwashing

Are you becoming aware that the public cares about your sustainability efforts? Good. But be careful about *greenwashing.* That's where you pretend—or the public *thinks* you're pretending—that you're environmentally friendly, when you're really not. Once you're accused of it, the backlash can be worse than if you'd done nothing at all.

When Chevron ran a series of TV ads in the mid-1980s, for example, showing employees protecting adorable butterflies, bears, and sea turtles, it was simultaneously violating the Clean Air Act and the Clean Water Act. The ads were greenwashing.

Same thing with DuPont, whose 1989 ads showed ocean creatures clapping and cheering to celebrate the company's new double-hulled oil tankers—at the very moment that DuPont was the biggest corporate polluter in the country.

These days, greenwashing is harder to pull off. Just ask Nestlé, the world's biggest food and drink maker, who proudly announced in 2018 that it would phase out all nonrecyclable plastic in its packaging by 2025.

Environmentalists ripped Nestlé apart, pointing out that the company still wasn't *phasing out* single-use plastics. What difference does it make if some plastics are more recyclable? Most countries on earth don't even *offer* consumer recycling programs, so those "more recyclable" packages will still wind up in the landfills and the oceans.

Over and over again, business experts offer the same advice: It's not what you say, it's what you do. Becoming climate-aware isn't just a marketing feature. It has to be part of the company's blueprint. It has to become part of the conversation in every area of the business: design, manufacture, business-to-business, transportation, *and* marketing.

Your Reputation with Employees

What your *own employees* think about your environmental policies matters, too.

In April 2019, a group of 8,700 Amazon workers signed an open letter to their boss, Jeff Bezos, expressing their unhappiness with the company's lack of a climate plan. They protested the company's hosting of oil and gas companies' data on Amazon servers, its purchase of thousands of diesel delivery trucks, and its donations to climate-hostile members of Congress.

The letter was catnip for the media. Amazon's image took a hit. And lo: 17 months later, Amazon pledged to use 100% renewable energy by 2030, to go net zero by 2040, to invest $100 million in reforestation, and to order 100,000 electric delivery trucks.

If you're an employer, you've got nowhere to hide. These days, a successful business is a sustainable one.

Your Reputation with Other Businesses

If you're a B2B company (that is, you supply products or services to *other* businesses), there's a third reputational risk of doing nothing: You may lose their business.

Giants like Apple and Walmart are pressuring their suppliers to embrace the mother ship's emissions goals—and even cutting ties if they don't comply.

Soon, says Anant Sundaram, professor of business and climate change at Dartmouth's Tuck School of Business, "You will not get a seat at the table for anything. If you're Corning, if you don't have climate-friendly policies, there's no way Apple is buying your glass—end of story." (The good news is, he adds: "Corning does.")

The Carbon Tax

Every major company is already doing contingency planning for a carbon-pricing system, considered by many experts to be the quickest and most efficient means to drop the nation's carbon emissions (see page 548). The risk for businesses is that overnight, fossil-fuel operations would become massively more expensive to operate.

Nobody knows when such a tax will arrive, or how much it will cost. This much, however, is true: The more you reduce your company's emissions now, the less you'll have to worry about when carbon pricing comes to pass.

The Cost of Preparing

The usual argument *against* operating a company sustainably is cost. You know: "Coal costs less than renewables." "Gas cars cost less than electrics." "Plastic packaging costs less than sustainable materials."

Some of those assertions may be true in some situations—but only in the short term. Over and over again, case studies have discovered that becoming greener ultimately *improves* profitability and growth. For example:

- **Apple** has cut its total corporate emissions 64% since 2011—while more than doubling its revenues.
- **Walmart** has cut its emissions 30% since 2006—while its revenues have grown 150%.
- **CVS** has managed to cut its carbon emissions 4% since 2018—while simultaneously expanding its stores' retail space by 1%.
- **IBM** managed to reduce its energy consumption by 6% a year between 2013 and 2017—and wound up *saving* $139 million a year.
- **Boeing's** factories have cut emissions 28% since 2009—while simultaneously increasing the number of planes it builds by 66%.
- **The United Kingdom's** carbon emissions have dropped 42% since 1990—during which time its economy has grown 72%.

A 2019 survey of the world's 500 biggest companies estimates that the changing climate offers $2.1 trillion in financial *upside*—opportunity, not cost—from selling new, low-emissions products and services, and gaining a competitive advantage by being green where their rivals are not.

Maybe they're onto something.

Getting Going

Most Fortune 500 corporations now employ an executive they didn't have 30 years ago: a chief sustainability officer, whose job it is to oversee the company's carbon-reduction goals, energy use, and environmental progress. Many of them received their training from another modern creation: MBA programs in sustainability.

But what if you didn't get a master's in sustainability? What if you're not in a Fortune 500 company? Preparing for the new climate is still incredibly important for a smaller company—*more* important, in fact.

FM Global's Lou Gritzo acknowledges that the *uncertainty* of climate change has been a longstanding challenge for businesses. Nobody knows what the exact temperature will be in 2050, or the exact sea level in 2100. That's one reason we wind up reading about multiple future emissions scenarios, like the "business-as-usual scenario," the "worst-case scenario," and so on.

The problem, he says, is that sometimes "businesses get caught up in that uncertainty. And they do the worst possible thing, which is nothing."

Ramboll's Sue Kemball-Cook has heard of companies delaying action because "Well, that's going to happen 30 years down the line." But in fact, a lot of the impacts are happening now. "Flood maps are changing; insurance rates are rising," she points out.

At the very least, Gritzo says, prepare your company for what you believe to be the risks of weather events *now*. Look at the drought rates, the wildfire rates, the intensity of the rainstorms, the number of "once-in-500-years" superstorms and floods in the last couple of years. You can safely assume that it won't get any *better* than it is right now.

That exercise usually gets companies over the "We're not sure exactly how much we need to do, so we're won't do anything" hump.

In a big company, a risk manager may be charged with performing the disaster-prep analysis; in your company, maybe it's you. But *somebody* has to ask: What could do us harm? What could put us out of business tomorrow?

"If they don't ask the questions," Lou Grizo says, "then odds are, they'll experience those hazards playing out at some point."

Chapter 17

Where to Find Hope

IN 2019, THE OCEANS REACHED THEIR HIGHEST TEMPERATURE ever recorded. July 2019 was the hottest month ever measured on land. In the United States alone, the 2019 weather broke more than 120,000 heat records, rain records, flooding records, and even snowfall records.

We're talking global weirding on a major scale. It was 70° in Chicago—in February. On the Fourth of July, it was hotter in Anchorage, Alaska, than in Key West, Florida. A snowstorm knocked down traffic lights and power lines—in Hawaii.

The parade of freakishness pushed on. February 2020 was the driest month ever recorded in California—at the peak of the "rainy season." The twelve months ending June 2020 tied for the hottest year in measured history. In June, wildfires *in the Arctic* released more carbon dioxide than in any month since record-keeping began. And don't forget that business of the Siberian heat wave in June 2020, where the Arctic temperature hit 100.4°.

As 2020 dawned, 58% of U.S. citizens described themselves as "concerned" or even "alarmed" by climate change—triple the number five years earlier.

So why isn't anybody *doing* anything? How can everybody ignore so vast a problem? Doesn't anybody care?

Actually, here's a dirty little clean secret: An overwhelming number of organizations—companies, schools, cities, states—do care, a lot. "You're seeing all kinds of extraordinary things happening, all around the world," says former UN Environment communications director Nick Nuttall. "We should be terrified by the signs, but we should feel unbe-

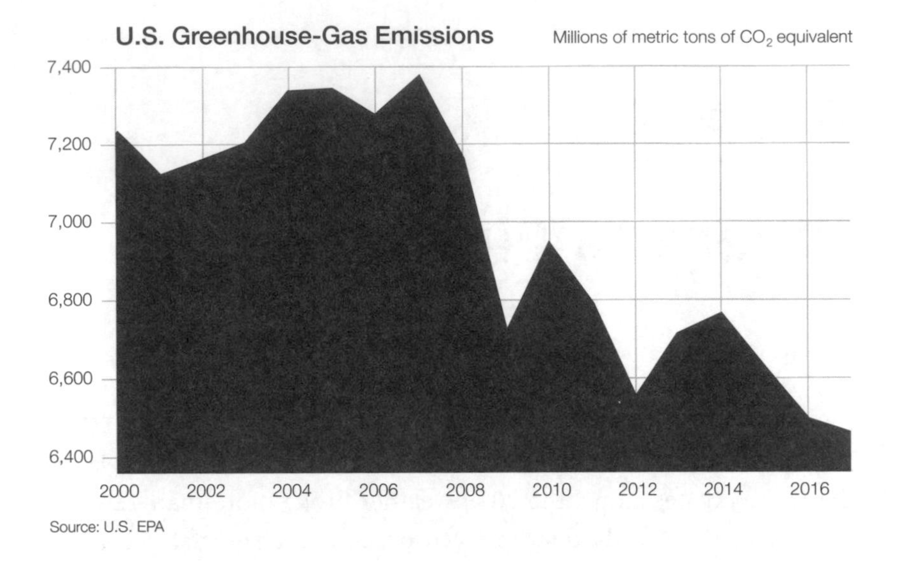

Figure 17-1. U.S. carbon emissions have dropped by 11% since their all-time peak in 2007. Some of that comes from the switch from coal to natural gas and improved car-emission laws.

lievably good about the things that have been achieved. We need to be excited *and* terrified."

These are *big* organizations, big enough that their actions actually amount to something. They're cutting their carbon emissions according to the Paris Agreement's goals or even more, completely ignoring the U.S. government's do-nothing policies. It's a quiet mutiny, in the name of doing the right thing.

Added together, these efforts are making huge strides in reducing the American carbon footprint. Not enough to stave off climate change, not enough to meet the Paris goals—but maybe enough to earn you a glimmer of hope.

Cities and States

Unhappy that President Trump intended to withdraw from the historic global Paris climate agreement, the mayors of hundreds of American mayors simply blew past him. They committed their *cities* to abiding by the Paris agreement, even if the country itself did not.

The group is called Climate Mayors, and it includes the mayors of 438 U.S. cities. They include the obvious suspects—coastal cities like New York City, Boston, Baltimore, Seattle, Portland (both of them), San Francisco, Los Angeles, Sacramento, San Diego, Honolulu, and even Washington, DC.

But they also include big heartland towns like Chicago, Denver, Philadelphia, Columbus, Cleveland, Detroit, Milwaukee, Saint Paul, Salt Lake City, Fort Wayne, South Bend, Kansas City—and even major cities in traditional red states, including Houston, Dallas, Austin, San Antonio, Phoenix, Nashville, Memphis, Louisville, Atlanta, New Orleans, Birmingham, Miami, Tampa, Charlotte, Raleigh, and Boise.

Some of these cities have already exceeded the Paris Agreement targets. Aspen, Burlington, and four other cities already run on 100% clean energy, and 94 more have made commitments to run on 100% by 2035 or sooner.

States are leading the charge, too. California, Hawaii, New Mexico, Washington State, and Rhode Island, and the territory of Puerto Rico have all passed laws requiring that, by 2050 or sooner, all of their electricity will come from renewable or clean sources. Nevada and Colorado have set that as a goal, not a requirement. Nine other states are considering similar laws. North Carolina and Colorado have committed to reducing CO_2 emissions 40% and 26%, respectively, by 2025.

Countries

The U.S. government may be the planet's biggest climate foot-dragger. But the rest of the world, including the 195 countries that signed the Paris Agreement, are charging right ahead in taking action. For example:

- **The United Kingdom** will "end its contribution to global warming by 2050." That is, it will become net-zero carbon (removing as much carbon as it produces) by 2050. Already, the UK countries have cut CO_2 emissions 44% since 1990.

- **China, India, France and the UK** intend to ban gas- and diesel-powered cars and trucks by 2040.
- **The European Union** has set goals for 2030: to cut emissions 40% and to increase use of renewable energy to 32%. Its governing body, the European Commission, has even proposed a path to be carbon neutral by 2050.

  And, more surprisingly, a quarter of the biggest companies in Europe link their CEOs' bonuses to the corporation's climate progress.
- **India** is one of the few countries on track to meet their Paris Agreement goals. (India's goal: by 2030, to get 40% of its power from solar.) The government's original plan was to install 20 gigawatts' worth of solar arrays—enough to power 14 million homes—by 2022. It reached that goal four years early.
- **Norway** intends to cut emissions 40% by 2030 and become net zero by 2050. Half of all cars sold there are electric or plug-in hybrids.
- **48 vulnerable countries** have agreed to get all of their power from renewable sources by 2050. What's remarkable about this coalition is that these aren't rich Western countries. They're low-lying nations in Asia, Africa, Central America, and South America, and island nations all over the world, who are especially vulnerable to sea-level rise, superstorms, and other climate-crisis effects: Afghanistan, Haiti, Philippines, Bangladesh, Kenya, Viet Nam, Colombia, and so on.

Corporations

More than 8,500 companies have disclosed their environmental progress to CDP, the nonprofit reporting group (original name: the Carbon Disclosure Project). In 2019, they reported eliminating 563 million tons of carbon dioxide just from their supply chains, which generally pollutes four times as much as a company's direct operations. And in the process, they saved a cumulative $20 billion.

100% Renewable Energy

By the end of 2019, nearly 300 multinational corporations—with total revenue totaling more than $5.5 trillion—had committed to deadlines for using *zero* fossil fuels for electricity, therefore producing no CO_2. They include Bank of America (by 2020), Capital One (now), Citigroup (now), Coca-Cola (2020) Ford (2035), General Motors (2050), IKEA (2025), Johnson & Johnson (2050), Lyft (now), Morgan Stanley (2022), Nike (2025), Procter & Gamble (2030), Sony (2040), Target (2030), Unilever (3020), Walmart stores (now), and Xcel Energy (2050).

All of these massive, global companies, powering their way into the future without burning *any* coal, oil, or gas? What was once unthinkable is now commonplace.

Carbon Neutrality

The pledges many companies are making to become *carbon neutral, net-zero carbon,* or even *carbon negative* all sound great, but they mean different things:

- **Carbon neutral.** A carbon-neutral company may spew plenty of CO_2 into the atmosphere—but they buy enough carbon offsets to balance the emissions. That is, they pay an offset company to plant trees somewhere—or to *not* cut down existing trees.
- **Net zero carbon** is better. It means that the company itself removes as much carbon from the air as it pumps out—by planting trees, for example, or using carbon-removal technologies.
- **Zero carbon** is better still. It means that no carbon is ever produced in the first place. All the power was produced by solar, for example. There's nothing to "offset."
- **Carbon negative** (also called, confusingly, carbon *positive*) means that the company is removing *more* carbon every year that it produces. That's the best of all.

The Three Scopes

When companies talk about reducing their emissions, you, as a skeptical consumer, should immediately ask: *What* emissions?

Every company is part of a chain—the fabric of commercial life. There are upstream links, meaning the companies that supply parts and make the raw materials. And there are downstream links: whoever buys and consumes your products. Every action a company takes produces ripples that affect other institutions.

So the question is: Do we count the ripples? And how far out?

To clear up this kind of confusion, scientists and environmentalists have created three categories of emissions that a company might track and reduce. Suppose, for example, that you work for a company called CO2Much Corp.

Scope 1 is pollution from CO2Much Corp's own activities, like what comes out of its factories' smokestacks and the tailpipes of its trucks.

Scope 2 pollution is what's generated by power companies for the electricity that CO2Much uses.

Scope 3 is even more indirect. It's *all* the emissions, both upstream and downstream, produced by CO2Much Corp's very existence: All the power used by the part makers and subcontractors. The travel of its workers, and the flights they take, for work. Creation of the materials used in the buildings. All the power used by CO2Much products during their lifetimes in customers' hands.

Scope 3 emissions, obviously, can be vast. Here's a typical example: As of 2020, Microsoft produces 100,000 tons of Scope 1 emissions; 4 million tons of Scope 2 emissions; and *12 million* tons in Scope 3.

"When they say 'carbon neutral,' the good companies mean Scope 1; the great companies mean also Scope 2," says Tuck Business School's Anant Sundaram.

Apple, for example, says that it will be entirely carbon neutral—Scopes 1, 2, and 3—by 2030.

It's difficult to go net-zero or carbon neutral, but it's clearly possible without losing your profitability. Monsanto (2021), Nestlé (2050), Qantas Airlines (2050), Unilever, Verizon (2035), Volkswagen (2050), and Duke Energy, the United States' largest power utility (2050) have all vowed to go net zero or carbon-neutral.

The Tech Companies

The big American tech companies aren't exactly media darlings these days, thanks to their handling of our data—but they're certainly become corporate leaders in measuring and reducing their carbon footprints:

- **Alphabet.** Alphabet (Google and its sister companies) is the largest consumer of renewable energy on earth. It's been carbon neutral since 2007, and has run on 100% renewable energy since 2017. The company has also committed $2.5 billion to building new, large-scale renewable-energy projects.

- **Amazon** aims to use 80% renewable energy by 2024, and 100% by 2030. Its goal is to go carbon-neutral by 2040.

  Amazon also intends to switch to electric delivery vans (100,000 by 2030); has committed $100 million to reforestation projects and $10 billion to fund climate science and advocacy; is building 15 massive wind and solar farms; and is pushing other corporations to sign a pledge that promises decarbonization strategies on the Paris Agreement's time line and becoming net-zero carbon 2040.

- **Apple.** The company has cut its carbon footprint by 35% since 2016. By 2018, Apple's operations worldwide ran on 100% renewable energy, two-thirds of which comes from projects, like solar and wind farms, that Apple itself built. Apple's *products* now consume 70% less power than they did in 2008, too. The company is building massive solar and wind plants in China.

  Apple is also working on improving the record of its factories and parts suppliers, which produce two-thirds of the company's emissions. Apple has pledged that by 2030, its *entire* business—manufacturing, product life cycle, *and* supply chain—will be fully carbon neutral.

- **Microsoft** has been carbon neutral since 2012. By 2025, it intends to use 100% renewable energy. By 2030, it intends to be carbon negative (*removing* more CO_2 than it produces) for *all three scopes* (see page 534). And by 2050, Microsoft intends to have removed all the carbon it has ever produced—since its founding in 1979.

  The company has also invested $1 billion in carbon-removal technologies, which it will certainly need if it intends to meet its 2050 carbon-removal goal.

- **Tesla's** a technology company, too, at its heart—and the only one whose *product* is a carbon-reducing machine. By the end of 2019, all the Teslas sold and driven had kept the equivalent of 500,000 gas cars' pollution out of the atmosphere. The company's goal is to use 100% renewable energy for all operations and manufacturing, although it hasn't specified a year.

The Fine Print

Plenty of big companies haven't announced any carbon-reduction goals at all, including Costco, Comcast, Berkshire Hathaway, and GE. (GE, in fact, is building *more* coal plants overseas.) Keep in mind too, that these are goals, and not everybody meets them.

Still, these huge public commitments are an encouraging sign. They mean that now, somebody important is observing and judging these companies' progress: the public.

Coalitions

Huge teams of organizations, meanwhile, have banded together in various efforts to announce that they will meet or exceed the Paris Agreement goals. Together, these coalitions represent 65% of the U.S. population, 68% of U.S. gross domestic product, and 51% of U.S. emissions. If they were a country, these coalitions would constitute the world's second largest economy.

As a group, their efforts are on track to reducing U.S. carbon emissions 25% by 2030 (from 2005 levels). That's not quite the Paris Agreement goal, but it's an excellent start. These meta-organizations include:

- **We Are Still In** (wearestillin.com) is the biggest climate coalition ever assembled. It's 3,850 entities from all 50 states—cities, states, companies, universities, faith groups, and tribes, totaling $9.46 trillion in market value—who pledge to meet the Paris goals. The efforts of all of these groups are summarized and quantified in an annual report called America's Pledge, led by former California governor Jerry Brown and Michael Bloomberg.

- **United for the Paris Agreement** (unitedforparisagreement.com) is a group of 80 companies that have pledged to meet or exceed the Paris Agreement goals. They include gigantic corporations like Apple, Microsoft, Google, Verizon, Tesla, Coca-Cola, Disney, Unilever, Salesforce, Citigroup, and Uber.

- **Climate Mayors** (climatemayors.org) is the network of 438 American city leaders who intend to lead the country in climate action, described earlier.

- **Global Covenant of Mayors for Climate & Energy** (globalcovenantofmayors.org) is like an international version of Climate Mayors. Its members include more than *10,000* cities from 138 countries that have committed to the Paris Agreement goals. They share expertise, data, and ideas in an effort to help each other, and the world, decarbonize.

- **U.S. Climate Alliance** (usclimatealliance.org) is a network of 25 American *governors* whose states have similar climate-leadership policies. These states represent 55% of the U.S. population and G.D.P. Since 2005, Alliance states have reduced their emissions by 14%—yet their combined economic output grew more than the other states'.

- **We Mean Business** (wemeanbusinesscoalition.org) is composed of 1,176 companies around the world, with a total market cap of $24.8 trillion. Their goal is to inspire fellow business leaders—and governments—to commit to reducing emissions.

- **The RE 100** (there100.org) is companies who have committed to running on 100% renewable energy (RE 100, get it?). RE 100 is part of We Mean Business.
- **Coalition for Climate Resilient Investment** is 34 companies, with assets totaling over $5 trillion. Their goal is to bring together experts in finance and investment to develop practical climate-adaptation techniques.

In your travels through environmental land, you may also encounter references to the CDP, formerly (and more clearly) known as the Carbon Disclosure Project (cdp.net). It's a nonprofit organization, with offices in 50 countries, dedicated to encouraging companies and cities to disclose their climate goals and progress. Each year, CDP gives a letter grade to each company or region for its climate-action excellence.

So far, 8,500 companies, 800 cities, and 120 states and regions have disclosed their climate information. These details are especially juicy for investors, who often want to see how dedicated a certain company is to meeting carbon goals. But you, too, are welcome to skim the results.

The Car Problem

Every gallon of gas your car burns pumps out 19 pounds of CO_2—and that's not counting the 6 additional pounds of CO_2 produced in extracting, refining, and transporting that gallon to you. Altogether, our personal cars and trucks account for about 20% of all U.S. emissions.

The joyous news is that we have an alternative. Electric cars are fantastic—clean, fast, safe, quiet, smooth.

Every carmaker has begun making or developing electric cars, and many will stop developing new gas and diesel cars *entirely* within the next 30 years. They include Volkswagen (by 2026), Mercedes (2026), Volvo (2024), Toyota (2040), BMW (2050), Honda (2022, in Europe), and GM (no date set). Ford hasn't announced plans to abandon gas, but it's spending $11 billion to develop 40 new electric models.

That's fortunate, because a growing list of countries will soon *ban*

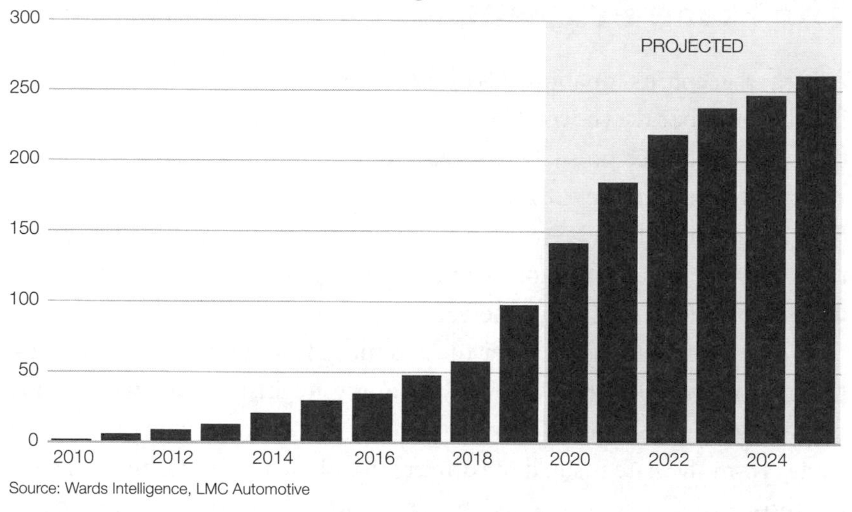

Figure 17-2. The number of electric-car models is about to explode.

the sale of gas-powered cars, on various timelines and with various degrees of fine print. They include Canada (2040), Austria (2020), Scotland (2023), The Netherlands (2025), Norway (2025), India (2030), Germany (2030), Denmark (2030), Iceland (2030), Ireland (2030), Israel (2030), Sweden (2030), Slovenia (2030), Sri Lanka (2040), France (2040), Britain (2035), Costa Rica (2050), Taiwan (2040), Singapore (2040), and even China (no date yet set).

On one hand, EVs (electric vehicles) made up barely 2% of 2019 new-car sales worldwide. On the other hand, the number has quadrupled in three years, and analysts expect that 80% of car sales will be electric by 2030.

Meanwhile, even gas cars are getting better. California, the fifth-largest economy on earth, struck a deal—incredibly, counterintuitively—with Ford, Honda, Volvo, BMW, and Volkswagen. These companies agreed to raise their cars' fuel efficiency *voluntarily*, to an average of 51 mpg by 2026, despite the Trump administration's *loosening* of existing regulations.

Eventually, 22 other states joined California in enforcing the stricter rules, and more car companies are expected to be part of the agreement.

The Plane Problem

Air travel accounts for about 2% of the world's CO_2 emissions, but bringing that number down won't be easy. You can replace gas cars with electric cars, but there's no such thing as a full-size electric jet.

And yet, in October 2013, airline representatives from 193 countries (the International Civil Aviation Organization, part of the United Nations) agreed to cap emissions from international aviation at 2020 levels, with carbon-neutral growth thereafter.

How? By developing cleaner fuels; removing weight from the planes; taxiing around airports with only one engine instead of two; towing planes with trucks instead of engine power; powering parked planes with cables from the gate instead of running the planes' power unit; and adding those little flipped-up winglets to the ends of the wings (improves lift and requires less fuel).

The Federal Aviation Administration is helping, too. After 13 years of planning, a new, GPS-based air-traffic system went into effect in 2020. Among other benefits, it lets planes descend to the runway by coasting along a smooth, efficient diagonal path instead of the traditional stairstep pattern.

The aviation problem will be one of the last to get fixed—but the world is working on it.

The Beef Problem

If you're a typical American, you eat 222 pounds of meat a year. That's 2.5 quarter-pounders a day—about twice the protein the USDA recommends.

Beef is a *huge* problem. Every four pounds you eat releases as much greenhouse gas as your share of a flight from New York to London. If cattle were a country, they'd be the third biggest greenhouse-gas polluter on the planet, right behind China and the United States.

How is that possible?

- **Methane.** Cows and sheep are *ruminant* mammals, which means that microorganisms in their stomachs digest their food, breaking it down into the nutrients their body needs, and the nasty greenhouse gases they don't. (Buffalo, deer, elk, giraffes, and camels are also ruminants, but McDonald's has yet to offer a McCamel sandwich.)

Those gases come out. A lot of people say that cattle farts are the problem, which is hilarious but not accurate. It's mostly burps: as much as *12 gallons of gas an hour per cow.* If you burped that much, you'd be asked to leave the restaurant—or the relationship.

Half of the unburped greenhouse gases is *methane,* which is awful, awful stuff, many times worse than CO_2. The belching of our red-meat creatures makes up 27% of all methane produced by human activity—more methane than the entire fossil-fuel industry.

- **Land.** Cows need a lot of land for grazing—and they get it. About 30% of the *entire ice-free land area of the planet* is dedicated to livestock grazing.

  To make more land for cows, we generally burn or cut down forests, which releases millions of tons of CO_2 that they had been storing.

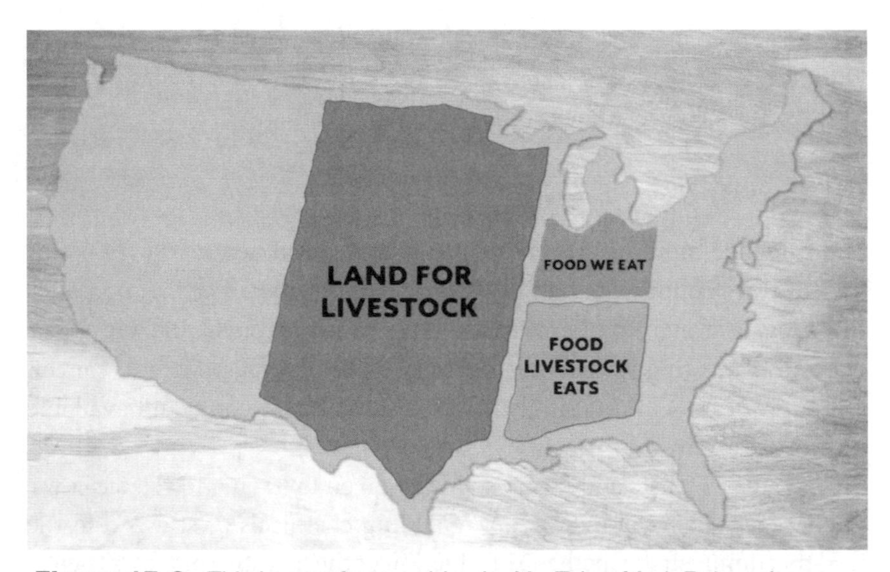

Figure 17-3. This image, from a video by YouTuber Mark Rober, shows the proportion of U.S. land dedicated to raising meat animals (left)—and the land dedicated to growing feed for them (lower right). By comparison, the land dedicated to growing plants that we eat (top right) is very small.

That's why Brazilian farmers set the 80,000 fires that devastated the rainforest in 2019—to make more grazing land for cattle. (In 2020, they picked up the burning pace by another 20%.) They replaced forests, which had been *subtracting* greenhouse gases from the air, with cattle, which *multiply* greenhouse gases.

On a per-calorie basis, we need *160 times* more land to raise cows than to grow potatoes, wheat, and rice.

- **Water.** The stream of stunning statistics is just getting started. Get this: Almost half of all water in the U.S. is used to raise food animals—especially cows. The agricultural system consumes 2,400 gallons of water to bring you one pound of beef. That's 4 times as much water as you'd need for pork, 5 times as much for chicken, and about 100 times as much as you'd need for wheat.
- **Feed.** To give you one pound of beef, a cow eats *eight* pounds of grain—food that people could have eaten directly. "It's a very inefficient way to make protein, compared to just eating plants right out of the ground," says plant-based nutritionist Sharon Palmer.

The bottom line: A finished steak or burger required 28 times as much land to raise as pigs or chickens, used 11 times more water, and pumped 5 times as much greenhouse gas into the atmosphere.

Cutting Back

You might be thinking: This is supposed to be *good* news?

But there is hope in all of this. First, Americans are now eating one-third less beef than in the 1970s, and five times as much chicken. We've learned that eating red meat is strongly linked to heart disease, cancer, and early death. Chicken is also cheaper, quicker to cook, and available in more ready-to-eat forms.

Second, scientists and farmers haven't given up trying to reduce cows' impact on the environment. They're adding chemicals or seaweed to the feed that diminish the methane production of the digestive process. They're developing new breeds that do their stomach conversion more efficiently. They're introducing methane-reducing ingredients to the cows' diet.

Then there's the rise of plant-based beef. The burgers from companies like Beyond Meat and Impossible Foods are not "veggie burgers," which don't fool anyone. These new patties look, cook, feel, smell, and taste astonishingly close to ground beef. Until you cook it, the Impossible Burger even *bleeds* the way raw beef does. In a bun, with condiments, it's virtually impossible to distinguish them from the best dead-cow burgers you've ever had.

And yet they're meatless. Their protein comes from peas, soybeans, beans, and rice. Environmentally, these burgers are home runs. An Impossible Burger's creation, for example, is responsible for only 13% of the water required to make cow burgers, 11% of the greenhouse-gas emissions, 9% of water contamination from runoff, and 4% of the land.

They give you about the same amount of protein and iron as beef, and no cholesterol. The fat and calories are in the same ballpark, though, and they contain much more sodium than beef.

Most people become instant fans. They've driven up Beyond Meat's stock to stratospheric levels, and they buy Impossible Burgers faster than the company can make—and that was before the coronavirus pandemic struck. That's when virus outbreaks tore through the meat industry, sickening and killing workers, closing meatpacking plants, raising prices—and driving up plant-based meat sales an incredible 264%.

Nestlé, Tyson, ConAgra, Kellogg, Hormel, and Kroger have all introduced their own meatless burgers, or will shortly.

Meanwhile, Impossible Foods now offers plant-based sausage and pork, and is working on fish (no mercury, no microplastics). Beyond Meat is focusing on bacon, sausage, chicken, pork, and steak.

The meat industry won't dry up overnight. But the rise of animal-free meat that tastes delicious and has no obvious downsides does suggest that a *reduction* in animal raising is likely. Today, we raise, kill, and eat 300 million cows a year—along with 66 billion chickens, 1.5 billion pigs, and half a billion sheep. Eliminating any of that would be a huge gift to our future.

The Asia Problem

China is the biggest greenhouse-gas polluter on earth, by a huge margin; the United States is number 2, and India is number 3. Asia now pumps out four times the greenhouse gases of Europe, and 2.5 times what the United States puts out. And we don't see *them* scolding each other for eating red meat or flying in planes.

The ground beneath the feet of Asian citizens contains very little oil and gas. What it has a lot of, though, is coal. And so, as the Asian countries' economies grew, "the fastest, safest, and least risky way to grow was digging up coal and burning it," says Vivek Tanneeru. He manages Matthews Asia, a U.S. fund that invests only in Asian companies.

Today, China burns more coal than all other countries *combined.* The result is an environmental disaster, and it's just getting started. Most of China is still relatively poor—India is *very* poor—but prosperity is rising. By 2030, two billion Asian citizens are expected to join the middle class, heating homes, buying cars, and taking flights.

Oh, and eating beef. In China and India, eating beef is a sign of upward mobility, wealth, and success. Beef used to be called "millionaire's meat"; these days, Asians' appetite for cow meat is exploding. In the early 1980s, the average Chinese diner ate 30 pounds of meat a year; these days, they eat almost five times as much—twice as much as Americans do.

If nothing changes, all of these factors could make the Asia emissions problem ten times bigger in the next 20 years.

But burning coal doesn't just produce carbon emissions; it also pumps out soot and toxic gases. And who cares about things like polluted air, water, and food? That growing middle class. "Once these guys don't need to worry about where the next meal is going to come from, they are thinking about the next level of needs," says Tanneeru—like the cleanliness of their air, water, and food.

So yes, China may have the biggest greenhouse-gas problem on Earth. But perhaps paradoxically, it's also become the most aggressive greenhouse-gas *fighter* on Earth, because of its citizens' increasingly vocal demands.

- In 2020, China introduced the biggest national carbon-tax program in the world. At the outset, only power plants are required to participate, but cement, steel, manufacturing, and industrial plants are next in line.

- China is developing more renewable energy than any other country. It owns the world's biggest wind-turbine manufacturer, as well as five of the six largest solar-module manufacturing firms on earth. From 2016 to 2020, it invested $361 billion in renewables. Its goal is to quintuple the number of solar-power plants in China—about 1,000 new plants. By 2030, renewables will reach 20% of all energy consumption, and the country's CO_2 emissions will begin to drop.

- China intends to phase out or limit greenhouse gases like methane, nitrous oxide, and hydrofluorocarbons (HFCs).

- The government offers subsidies and tax benefits to electric car buyers, and is raising miles-per-gallon requirements for gas cars.

- President Xi signed the Paris Agreement, recommitted in 2019, and then, in late 2020, announced a national goal "to achieve carbon neutrality before 2060."

All of these steps forward are part of a weird dance. China has just *lifted* a ban on building more coal plants and is *reducing* subsidies that make it cheaper to build solar- and wind-power plants. And the country's greenhouse emissions are still climbing.

Ah, well. The takeaway is this: China may remain the biggest polluter on earth, but its government isn't ignoring the problem.

The Lightning Round

And now, a tidy bulleted list of other evidence that humanity is not doing *nothing* to fight back.

- 2019 was a huge year for renewable energy. In the United Kingdom and Germany, renewables produced more electricity than fossil fuels did for the first time. In the United States, 2019 was the year when renewables generated more power than coal did for the first time.

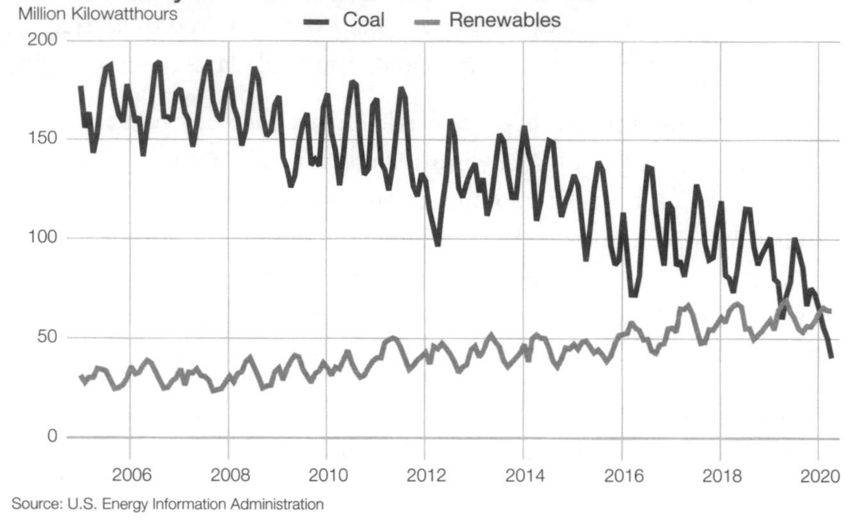

Figure 17-4. In 2019, for the first time in history, the United States got more power from renewable sources than from coal.

- The number of solar-panel installations is growing 50% a year, and the price of solar power has crashed 90% since 2010. It's the fastest-growing power source in the U.S. It's expected to power 48% of the country by 2050.

- Wind power in the United States has doubled since 2010. Wind now generates 7.3% of all U.S. electricity—and in six states, it's over 20%. The price of wind power keeps dropping; it's now the cheapest form of power there is.

  In 2019, renewable energy in the United States hit an all-time high: 18% of all electricity, more than twice the proportion ten years ago. We're getting there.

- More than 1,000 colleges, churches, charities, governments, and other institutions no longer invest in coal, oil, or gas companies. That's $11 trillion no longer invested in fossil fuels—an increase of 22,000% since 2014.

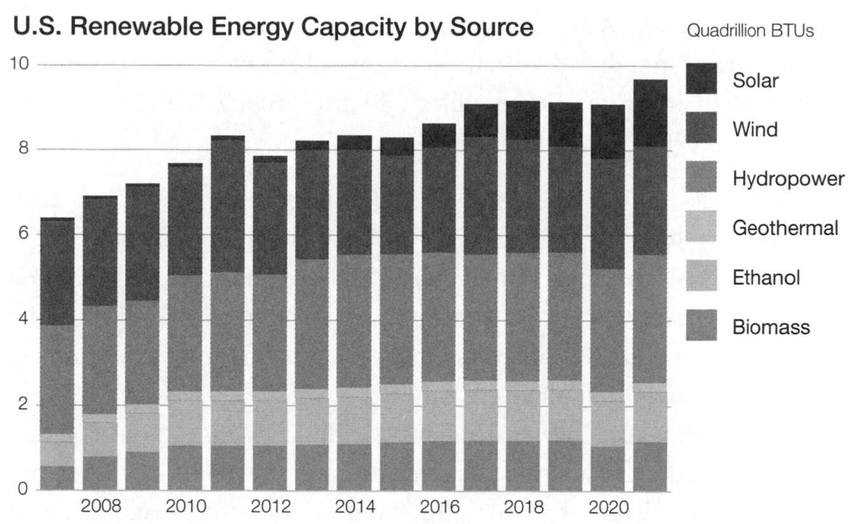

Figure 17-5. In the United States, renewable energy has been steadily replacing fossil fuels for power generation.

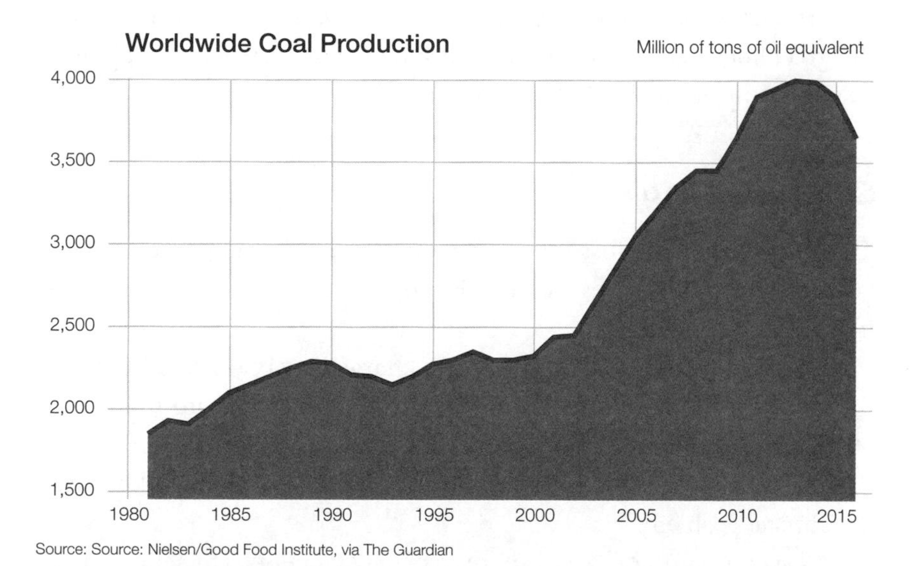

Figure 17-6. Worldwide coal production peaked in 2013, and has taken a sharp drop since.

- ♦ Since 2010, 632 American coal plants have shut down, and no new ones are planned. "We're retiring a coal plant every month," Rocky Mountain Institute's Bruce Nilles told *USA Today*. "Coal will all be gone by 2030."
- ♦ BlackRock, the world's biggest asset manager, announced in 2020 that it would begin pulling out of investments that contribute to the climate crisis, like coal companies. CEO Larry Fink vowed that he'd vote to replace CEOs who aren't making sustainability progress. The expectation is that BlackRock's move will put pressure on rival money managers to do the same.

  Meanwhile, the investment bank Goldman Sachs announced that it would no longer provide financing for new or expanding coal mines and coal-fired power plants, nor for new Arctic oil exploration. It joins 100 other banks around the world who are divesting from coal.

The Big-Idea Solutions

Decarbonizing our power and our cars are important ways to bring down the greenhouse-gas levels in the atmosphere, but they're not happening fast or deeply enough. A lot of people are thinking bigger.

Carbon Pricing

The big problem with fossil fuels is that they're cheap energy. You'll never get the masses to burn less of them just by saying, "Oh, come on, people, it's bad for the planet!"

But suppose that the United States charged a tax on fossil fuels: as they're extracted from the ground, as they're imported into the United States, or as they're burned. An army of experts, academics, and scientists firmly believe that this is the only way to seriously and quickly tackle the climate problem.

If gas, oil, and coal were more expensive, energy efficiency would *mean* something. People would have a gigantic financial incentive to reduce fossil-fuel consumption; innovation would soar. Solar, wind, and even nuclear would be universally cheaper than fossil fuels; carbon emissions would plummet. People would bike or use public transport instead of driving. Businesses would install upgrades to become more energy efficient. Homeowners would patch leaks and install smart thermostats.

Everybody would know exactly how much they'd be saving; they could build those numbers into their forecasts and planning.

"Putting a price on carbon is such a no-brainer," says Anant Sundaram. "Nothing is certain, but by 2030, I'd be shocked if we didn't get there."

And what where would all that new tax money go? It could pay down the deficit, it could fund environmental projects—or it could go right back into Americans' pockets, bypassing the government entirely.

"All proceeds from a nation's carbon fee would be divided equally among its citizens, and returned to all adults through a quarterly dividend check automatically deposited in their bank accounts. Polls reveal that Americans favor this plan by a 2.5–1 margin, including a 3–1 margin among Republicans."

That's the suggestion of the Climate Leadership Council, a bipartisan, nonprofit institute founded by a gang of well-known people (Jim Baker, George Shultz, Ben Bernanke, Steven Chu, Steven Hawking, Lawrence Summers, etc.) and companies, many of whom you might not expect to embrace a carbon tax (ConocoPhillips, ExxonMobil, Ford, GM, etc.). The big oil companies like this plan because they'd be shielded from the plague of lawsuits they're now facing.

Their idea is to send, say, $2,000 to $5,000 a year to every American household, based on its carbon diet. Under this arrangement, people whose lifestyles use less energy will actually make money. They'll be subsidized by people who don't need the money, or who just don't care.

Already, 40 countries and regions have carbon taxes, ranging from under $1 per ton of CO_2 (Poland) to $139 per ton (Sweden).

Does carbon pricing work? When it's high enough, yes. In England, where a carbon tax on the power industry is now $25 per ton of CO_2, greenhouse gas emissions have dropped to their lowest levels since 1890. In British Columbia, Canada, emissions have dropped 15% since the tax was imposed in 2008 (currently $30 per ton)—and the economy is doing *better* than before.

California has a carbon tax that covers power plants, manufacturers, refineries. In the Northeast, nine states have banded together to form the Regional Greenhouse Gas Initiative, which offers carbon exchanges among power plants. The Department of the Treasury calculates that if the United States had a national tax at $49 a ton, by 2028 we'd reduce emissions by 21%.

Safer Nuclear Power

Nuclear power is clean and infinite—what if we could also make it safe and cheap? Here are a couple of ways to do that:

- **Traveling-wave nuclear.** Bill Gates, the U.S. Department of Energy, and the Los Alamos National Laboratory are backing a new nuclear power design that promises far greater safety, lower cost, and less nuclear waste than current systems.

  So get this: 99% of mined uranium is a variety called the U-238 isotope, which our current nuclear plants can't use. We just throw it away. (What our current plants require instead is the far more scarce U-235.)

  That discarded U-238 is exactly what TerraPower's traveling-wave reactor uses as fuel. In fact, this kind of plant could also use the 700,000 tons of *used* fuel rods from today's reactors. They still contain 90% of their energy, and at the moment, they're sitting in waste depositories, unused.

  The other key advantage: Safety. This reactor can't melt down and doesn't involve the kind of uranium that can be used to make bombs.

- **Thorium nuclear.** Most nuclear power plants today use plutonium or enriched uranium as fuel. But thorium could work, too.

  The element thorium is 500 times as abundant in the earth as uranium, and much less expensive. In a thorium reactor, there's much less nuclear waste. The system is meltdown-proof. There's no need for massive cooling systems, either, so you can build the plants anywhere. And here's the capper: The used nuclear material is useless for making weapons.

  "It's a remarkably safe approach to nuclear," says Anant Sundaram. "But the last demonstration plant was built something like 60, 70 years ago."

Why haven't we been using thorium all along? Precisely *because* it can't be weaponized. When nuclear power was just getting started, the United States was interested in simultaneously developing nuclear weapons. Uranium could be used for both.

No thorium plant is yet operating, and there are technical problems to solve. But scientists in India, China, Russia, France, the Netherlands, and the United States are exploring them with varying degrees of investment.

If someone perfects either of these new reactor types, we'd have ourselves a nearly limitless source of power for the planet—carbon-free.

Carbon Removal Technology

We can send a man to the moon (well, we used to)—so can't we invent some technology that sucks carbon dioxide out of the atmosphere?

We already have it. In fact, sucking CO_2 out of the air made it *possible* to send men to the moon; otherwise, they would have suffocated in that little capsule. Same thing in submarines. Can't we build some massive plant that does the same thing on a huge scale?

This "negative-emissions technology" is called direct air capture (DAC, if you're in a hurry). Huge fans push air across chemicals that bind with the carbon dioxide and pull it out of the air. It's not the same thing as factory-smokestack CO_2 capture; these new technologies extract CO_2 from the *atmosphere,* which is a much tougher task.

If we could then bury the extracted CO_2 deep beneath the surface of the earth forever, well, great; it's no longer trapping heat. But more often, the direct-air capture companies intend to turn the CO_2 into fuel, enriched air for greenhouses, plastic, building materials (cement, concrete, asphalt), fertilizer, or carbonation for soda.

Perfecting DAC technology is *really* important. The United Nations' Intergovernmental Panel on Climate Change (IPCC) calculates that we have no chance of meeting the Paris Agreement goals *without* these machines sucking hundreds of millions of tons of CO_2 from the atmosphere.

Already, 11 carbon-removal plants are operating. Companies like these have scored a lot of headlines:

- **Carbon Engineering** pulls CO_2 out of the air and turns it into synthetic fuel for trucks, ships, and planes. Burning that fuel still releases carbon dioxide—but only about one-tenth the emissions of gasoline.

  Bill Gates is an investor, and so are Occidental and Chevron. Carbon Engineering's demo plant in Canada is pulling a ton of CO_2 out of the air every day.

- **Climeworks** has pilot plants in Switzerland and Iceland, which it says is pulling 900 tons of CO_2 annually from the air. The company then sells the gas to local greenhouses and carbonated-beverage companies. Climeworks also has a pilot program to inject the recovered CO_2 into rock formations underground.

- **Global Thermostat,** funded by Goldman Sachs and ExxonMobil, extracts CO_2 from the air and turns it into fertilizer, plastic, and building materials. A pilot plant is under construction in Alabama.

Many more companies—Skytree, Soletair, Oak Ridge National Laboratory, Center for Negative Carbon Emissions, National Energy Technology Laboratory, Infinitree, and MIT—have developed various ways to extract CO_2 from the air, and are at various stages of testing and scaling.

Figure 17-7. A Climeworks direct-air-capture setup.

As you might guess, the big oil companies *love* the idea of sucking carbon out of the atmosphere—because hey, that would mean that they can keep right on selling and burning fossil fuels!

Carbon-from-the-air sure sounds like a silver bullet. We could go on eating our beef and flying our planes, consequence-free. Our fossil-fuel machines fuels would pump out CO_2, and our DAC machines would suck it right back down again.

There are just a *few* obstacles in the way. First, at the moment, DAC costs $600 to remove one ton of CO_2 from the air. To make a dent in global warming, we'd need to remove hundreds of thousands of tons a year. Nobody's going to pay that much.

The DAC companies say they'll eventually get the price down to $100 a ton. At that point, *maybe* the economics would work out.

The second big problem is that most DAC systems require huge amounts of water and power to operate, which presents an environmental problem all its own.

So far, solving those "if onlys" is a long way off. Even DAC's fans (the human ones) believe that we *also* need to decarbonize our buildings, transportation, farming systems, and power plants. We need *both* strategies.

Geoengineering

Direct air capture of carbon dioxide is at least simple, safe, and inexpensive when you compare it to some of the other technological solutions that sometimes make the rounds.

- **Stratospheric aerosol injection.** Suppose we could spray trillions of shiny particles into the atmosphere—limestone dust, sulfur dioxide, artificial clouds of water vapor—so that they reflect some of the sun's heat back into space. That'd cool the earth just like ash from volcanoes. It might slow down the melting of land ice and restore traditional rainfall patterns.

  On the other hand, it wouldn't do anything about the CO_2 that's already in the atmosphere, acidifying the oceans. Blocking the sun would interfere with our solar-power systems. And astronomers would *hate* it.

- **Marine cloud brightening.** Here's another nutty idea: We could send ships out around the oceans, spraying saltwater into the sky with immense cannons. The salt particles would trigger water condensation, forming clouds or making existing clouds brighter and whiter—to reflect sunlight. And you'd be able to put these clouds only where you wanted them.

- **Fertilizing the oceans.** The idea is to dump tons of iron filings into the ocean, hoping that they'd spur the growth of microscopic plankton. More plankton, more photosynthesis—a chemical reaction that consumes carbon dioxide.

- **Bubble the oceans.** We'd send out fleets of ships to churn up the water everywhere they went. The tiny microbubbles would be more reflective than the dark ocean water.

- **Spray-painting the deserts white.** Since we're losing so much reflective ice from the planet's poles, why not replace it with reflective white regions elsewhere?

- **Fly giant mirrors into space**, again, with the intent to reflect the sun away from us.

Papers have been written, computer models run, laboratory experiments set up. China has spent $3 million to set up an exploratory institute. But little of it has been tested in the real world—and maybe we should be thankful for that.

These ideas would be massively expensive, would take decades to get ready, and could have disastrous unintended consequences. What if they backfired, and sent us into an ice age? What if they really messed up the weather, and we couldn't grow stuff? What if they cooled one part of the earth, but caused droughts in other spots?

There's a geopolitical dilemma, too. Whose hand would control the global thermostat? Could leaders tweak the weather in a way that helped their own country, but hurt their enemy's?

Maybe the biggest terror of all, though, is that our government would view these weather hacks as a quick fix to the climate-change problem, and would back off from the one sure, known solution: Cutting carbon emissions in the first place. If we put big bets on geoengineering and kept on pumping out greenhouse emissions in the meantime, and then the geoengineering *failed,* we'd be in deep, deep trouble.

Fixing Food

By 2050, the earth's population will hit 9.7 billion. And if we keep eating what we eat now, we simply won't have enough land to feed them—at least not without cutting down more of the planet's forests (hi there, Brazil!). Especially because it's getting harder to *grow* food in the new climate.

There's no one solution. There are, however, a range of options that, taken together, might get us there.

- **Genetic modification.** The food technology with the greatest potential by far is genetic modification of crops. By manually introducing changes into the DNA of a plant, you can make it much more resistant to disease, drought, or bugs, or you can make it more nutritious. We've been genetically modifying crops for thousands of years, using crude techniques like selective breeding and radiation. In the United States, where almost all corn and soybeans are genetically modified species, we're getting 20 to 30% more food from each plant.

  And yet all over the world, GM foods are currently banned, limited, and feared—an attitude that drives scientists crazy. The U.S. Food and Drug Administration, the American Medical Association, the National Academy of Sciences, and the World Health Organization have all determined that GM foods are safe.

  The resistance may weaken in the coming years; it will have to. And in the United States, that acceptance is accelerating. "The Impossible Burger is engineered already, and that seems to be taking off," points out Diego Rose, professor of nutrition and public health at Tulane University.

- **Bugs.** Yes, yes, we know: You'd never eat bugs. They're gross. Yet 2 billion people in Africa and Southeast Asia never developed our Western cultural distaste for them; they routinely eat more than 1,900 kinds of bugs, like beetles, caterpillars, bees, wasps, ants, grasshoppers, locusts, and crickets.

Pound for pound, insects blow cows right off the barnyard. They have more protein, produce 1% the greenhouse gases, require one-tenth the land area, and consume *one-twelfth* as much feed.

Furthermore, insects are so distant from humans genetically that we're not susceptible to their microbes. They can't give us the diseases that we catch from animals or their feces, like coronavirus, swine flu, anthrax, mad cow, monkeypox, ringworm, and salmonella.

Finally, raising bugs can be very low-tech, within the reach of very poor people.

None of this is likely to convince Westerners to start crunching crickets like candy. But bugs may begin to enter our diets through side doors. Maybe we'd be willing to eat food that contains cricket flour, which is packed with protein and vitamins and doesn't have much of a taste. In fitness and athletic circles, cricket flour is already a popular booster powder for smoothies, shakes, salad dressings, veggie burgers, cookies, and so on.

Or maybe we'd tolerate eating beef from *cows* that were raised, in part, on insects. Companies like AgriProtein, Protix, Ynsect, and Enviro-Flight are building huge facilities to raise bugs as *animal feed*. The idea is to take some pressure off the soy and fishmeal sources that chickens, pigs, and fish eat now, and even feed it to cattle as a protein supplement.

In the United States, bugs-as-feed is permitted for chicken and fish already, and is expected to be approved for other animals soon.

- **Policy changes.** Scientists and governments have been preaching the virtues of plant-based diets for decades, but ad campaigns alone aren't enough. It's hard to change people's food behavior without changes in availability, price, or convenience. And at the moment, climate damage is not priced into our foods.

  For example, as Tulane nutrition professor Diego Rose points out, we've grown accustomed to being able to buy any fruit or vegetable at any time of year, even if they're out of season where we live. "We're air-freighting blueberries to the United States that are grown in greenhouses in Chile!" he says.

The solution is to build the price of the emissions into the price of foods. "We need price changes that the consumer can see, so the consumer can orient away from those high-impact, unhealthy, and, hopefully, high-priced goods, and consume them only every once in a while," says Marco Springmann, a senior researcher on environmental sustainability and public health at the University of Oxford.

Of course, that move would infuriate the public. It would have to be accompanied by an ad campaign that explains the reasons and the benefits—and a *visible* use of the revenues generated by the higher prices.

That strategy worked in Mexico, where more than 70% of the population is obese or overweight. Mexico has among the highest rates of sugary-soda consumption on earth, a significant factor in weight gain. So, in 2013, the Mexican government instituted a 10% tax on sugary drinks.

Amazingly, the public accepted the tax without much fuss. Within two years, soda consumption dropped 15%.

That, says Springmann, is because of a clever governmental move: The money collected from those taxes was put into free, clean drinking fountains in schools and public places. Consumers could *see* what they were getting for the tax.

That could work here, too. "If it becomes more expensive, we'll stop eating cherries at Christmas," says Rose. "We'll start eating more seasonally."

♦ **Restaurant policy changes.** "Lots of people really don't think about what they order, right? They get what the person in front of the queue gets, or the first thing on the menu. That's where businesses come in," says Springmann.

Menus of Change is a joint venture of The Culinary Institute of America and Harvard's school of public health. The target audience is chefs, restaurants, cafeterias, and other food-service decision-makers, and the mission is to shift the world's menus to reduce the reliance on red meat.

A central tenet of the campaign is the Protein Flip: Moving meat, especially red meat, into a less central role. The organization encourages restaurants and chefs to serve smaller meat portions and promote better animals (shrimp instead of poultry; anything instead of beef).

There's a lot of psychology at work here. In one study, researchers eliminated the "Vegetarian" section from a menu and, instead, listed the vegetarian dishes *among* the meat entrées without any special labeling. Once plant-rich dishes were normalized this way, people ordered them more than twice as often.

What's Going to Happen?

On the great spectrum of humanity, you can find all kinds of expectations about the future of our planet's climate. On one end, the optimists: "We'll solve this! We always do!" In the middle, the minimizers: "We'll sacrifice Miami, shift civilization northward a bit, and develop more resilient crops. We'll be fine."

And at the pessimistic end, you find people like Jem Bendell, professor of sustainability leadership at the University of Cumbria, whose 2018 paper "Deep Adaptation" predicts the imminent collapse of society.

"When I say starvation, destruction, migration, disease and war, I mean in your own life. With the power down, soon you wouldn't have water coming out of your tap," he wrote. "You will become malnourished. You won't know whether to stay or go. You will fear being violently killed before starving to death."

Most scientists, however, don't consider our fate quite so sealed.

Humanity has, after all, pulled back from the brink of environmental disaster before. Remember these?

- **Leaded gasoline.** In 1923, General Motors and Standard Oil began adding a lead compound to gasoline to improve its burnability and eliminate the knocking sounds that car buyers found alarming. Unfortunately, lead, as we now know, is a powerful neurotoxin. People who worked with it developed memory loss, coordination loss, sudden rage, convulsions, and death.

  As word spread to the public, Standard Oil fought back by questioning the science. (That response may sound familiar.) It denied that lead had anything to do with the employees' problems.

In 1986, the United States finally banned leaded gas—after 68 million children had absorbed toxic levels of lead, 5,000 adults a year had died, and our IQs had dropped measurably.

- **The hole in the ozone layer.** In the late '70s, Sherwood Rowland, a University of California professor, and Mario Molina, his post-doctoral fellow, discovered something horrifying about CFCs (chlorofluorocarbons), the primary chemical in spray cans, fridges, and air conditioners. These molecules were floating up into the atmosphere and destroying the ozone layer.

  Who would care? We might, because ozone absorbs UV-C. That's the deadly part of the sun's ultraviolet radiation, which, if it reached the ground, would "sterilize the Earth's surface," according to NASA. That would probably not be good.

  The chemical industry, wary of losing $500 million a year in CFC sales, fought back, attacking Rowland's reputation and questioning the science.

  When scientists discovered that, in fact, the ozone layer had already shrunk by 40%, governments finally got the message and took action. The Montreal Protocol, signed by 56 countries, eventually called for a worldwide ban on CFCs.

  Rowland and Molina shared the Nobel Prize in 1995—and the ozone hole began to close up. It's expected to heal completely by 2050.

- **Acid rain.** When we burn fossil fuels, nitrogen-oxide gases waft up into the atmosphere and form acid compounds that rain back down on us. In the 1980s, acid rain was another environmental disaster, making lakes and streams toxic to fish, killing trees, and threatening crops.

  Eventually, Congress made amendments to the Clean Air Act, requiring smokestack scrubbers and low-sulfur coal. Catalytic converters helped with the car problem, and a cap-and-trade system of sulfur dioxide and nitrogen oxides led to even further reductions of those nasty gases.

  It worked: From 1980 to 2008, sulfur dioxide emissions fell 56%, and nitrogen oxides dropped by a third. Lakes and streams slowly began to recover.

Acid rain hasn't gone away entirely, especially in China. But the episode illustrates that the U.S. government is capable of getting its act together to prevent disaster.

The Optimism Balance

You have now read, and will continue to read, daily servings of horrifying news about the climate. Change of *any* kind is stressful and upsetting—changing jobs, moving to a new state, breaking up—but *permanent* change, *global* change, *everlasting* change like climate change can be debilitating.

And yet, as you now know, it's false to say that nobody's doing anything about the problem. *So many* companies, organizations, cities, states, and countries are making changes, passing laws, committing to targets! Decarbonizing our species is going to be expensive and difficult, especially since we face massive resistance by oil companies and all the politicians who benefit from them. But it's becoming obvious that decarbonizing is possible and, in fact, well underway.

So which should you feel? Depressed or hopeful? The informed answer is: "A little of both."

Most of the experts who contributed to this book believe that the Great Decarbonization will continue to completion.

"We are technologically capable of making the transition fast enough to avoid the most extreme outcomes," says science writer Brian Clegg. "It will be bumpy along the way, but we can get and will get there."

The problem, of course, is that we're getting started far too late.

The notion that greenhouse gases might cause the earth to heat up was first posed to scientists in 1861. It began reaching the public in a 1956 *New York Times* story. And it reached the ears of Congress on June 24, 1988, when NASA climate scientist James Hansen testified about his findings from temperature readings around the world. "It is time to stop waffling," he said (in 1988). "The greenhouse effect is here."

As we all know now, we didn't stop waffling in 1988. We spent the next 30 years disregarding, debunking, and denying. And now, it's too late. Even if we stopped burning fossil fuels today, it would take many generations to *stop* heating up the planet, simply because 93% of all the added heat is now in the oceans.

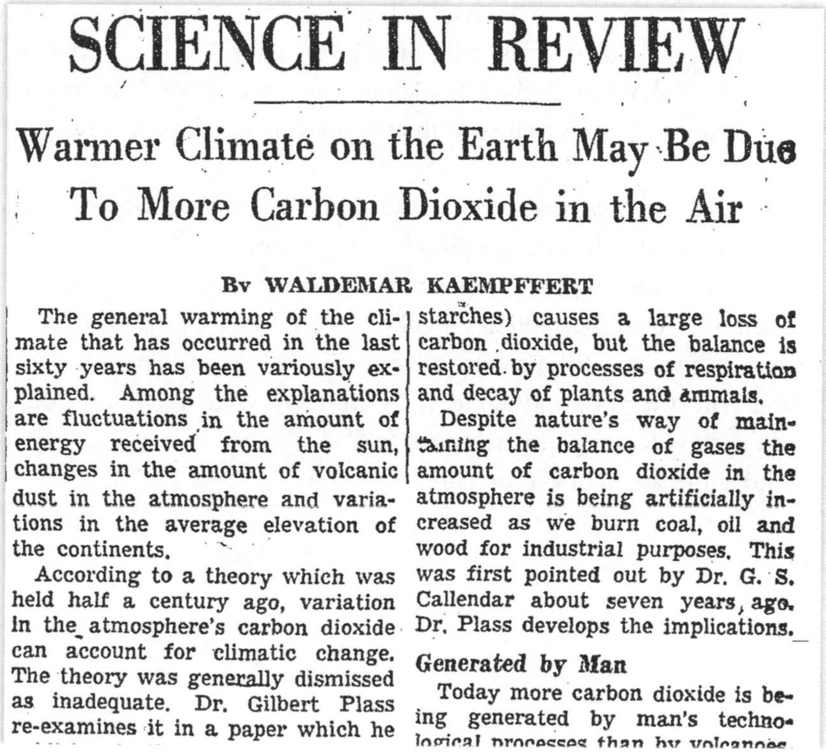

SCIENCE IN REVIEW

Warmer Climate on the Earth May Be Due To More Carbon Dioxide in the Air

By WALDEMAR KAEMPFFERT

The general warming of the climate that has occurred in the last sixty years has been variously explained. Among the explanations are fluctuations in the amount of energy received from the sun, changes in the amount of volcanic dust in the atmosphere and variations in the average elevation of the continents.

According to a theory which was held half a century ago, variation in the atmosphere's carbon dioxide can account for climatic change. The theory was generally dismissed as inadequate. Dr. Gilbert Plass re-examines it in a paper which he

starches) causes a large loss of carbon dioxide, but the balance is restored by processes of respiration and decay of plants and animals.

Despite nature's way of maintaining the balance of gases the amount of carbon dioxide in the atmosphere is being artificially increased as we burn coal, oil and wood for industrial purposes. This was first pointed out by Dr. G. S. Callendar about seven years ago. Dr. Plass develops the implications.

Generated by Man

Today more carbon dioxide is being generated by man's technological processes than by volcanoes

Figure 17-8. The *New York Times* was warning about the link between fossil fuels and the climate in 1956.

We *won't* stop burning fossil fuels today, either. Decarbonizing will probably take 50 to 80 years to complete.

So yes, it's too late to stop what we've started. For the rest of our lifetimes, our children's, and their children's, the great ice sheets will keep melting, the sea levels will keep rising, the hurricanes will grow more deadly, water will become more precious, more wildfires will burn out of control, and the weather overall will get hotter and more violent. Thanks to ticks and mosquitoes, disease will become a more familiar part of our lives; thanks to the shifting seasonal patterns, we can expect more animals to go extinct. As the global population grows, there won't be enough food, water, and livable land to go around; the result will be conflict and suffering.

Even so, the planet will survive. We'll survive, too. Our lives may be different, but our species will muddle through. A lot of things will cost more: food, water, shelter, medicine, insurance, travel. The "more is better" mantra of perpetual consumer growth may take a hit. But we'll be here.

You have the advantage of knowing all of this before it happens. The question is, what will you do with this knowledge?

Will you build a go bag to keep in the front closet? Will you pore through your insurance policies? Will you reconsider where you live, and where you'll retire? Will you change your lifestyle, your diet, your travel habits? Will you guide your children? Will throw your weight around by the way you vote, shop, invest, and speak?

And above all: Will you prepare?

Acknowledgments

I desperately wish I could take credit for this book's concept. But in fact, it was Priscilla Painton's idea, and the moment she proposed the title *How to Prepare for Climate Change,* I knew it was pure gold. She's an executive editor at Simon & Schuster—and, better yet, she was *my* editor. Editor, shaper, facilitator, beta reader, supporter, guard rail, and shepherd.

This project was a massive exercise in discovering and distilling data. Much of this work fell to two amazing researchers:

Christine Brittle, PhD, an expert in qualitative and quantitative research design and analysis. She's the winner of a National Science Foundation dissertation enhancement award, a U.S. Environmental Protection Agency customer service award and bronze medal, and several academic fellowships.

Olivia Noble, a writer, recent graduate of Yale University, and Fulbright grantee in the Czech Republic.

Olivia was so efficient, thorough, and smart in her research assistance that I asked her to serve as the book's photo editor and endnote creator, too.

Information designer David Foster (davidfostergraphics.com) created the graphs and charts for this book. The task involved a lot of sleuthing to hunt down the data we needed, and I'm grateful that he persisted.

Milica Stamenković and Dejan Čuturilov converted ugly data sets into the beautiful risk maps and city-by-city data graphs in Chapter 2. Leo Ramos made the book's line-drawing diagrams. Tim Hipp (transcribeyourpod.com) and Rev.com transcribed the all of the expert interviews.

A long list of wonderful people did favors for the project at just the right times, offering contacts, expertise, or data. They include Christopher Allan Smith, Ed Maibach, Kellen Klein, Victoria Odinotska, Peter Girard, Gavin Foster, Jennie Allison, Stacie Reece, Jon Posen, Christine Schuldheisz, Sean

Sublette, Steve Schaefer, Matthew Tinder, Steve Zenofsky, Ernie Asprelli, Joe DeAngelis, Tammie DeVooght Blaney, Logan Wu, Theo Gray, Jerry Bell, and Julie Van Keuren.

My agent, Jim Levine, has been a cheerful coconspirator on this whole thing. He didn't just define the loop—he stayed in it.

Publisher Jon Karp made me feel so welcome at Simon & Schuster that I wound up wondering why my career took so long to wind up there. And editorial assistant Hana Park made the whole thing hum.

I'm also grateful to the members of U.S. government agencies who lent their insight into federal programs, but who asked not to be named. Keep fighting the good fight!

Above all, I offer love, admiration, and thanks to my children, Kell, Tia, and Jeffrey, for their patience and interest during this time-sink of a project—and to my brilliant bride, Nicki. She's been a stadfast, sunny source of counsel, common sense, humor, and love during the creation of this book—and of our lives together.

The Expert Panel

When I began working on this book, I wasn't an expert on disaster prep, insurance, investment, gardening, ticks, demographics, or even climate change. *Now,* I can speak for hours about the finer points of federal flood insurance. My wife says I'm a blast at parties.

The point is, I needed expert help—and I learned from the best. Each of the researchers, scientists, thinkers, and educators listed here granted me interviews on their areas of specialty. Not a single one turned me down or set a time limit. Each lent their wisdom, humor, and enthusiasm to this project, all in the name of sharing their work, spreading good science, and helping people prepare.

Without the help of these experts, you'd be reading *How to Prepare for Climate Change: The Brochure.*

Amir AghaKouchak is a professor of civil and environmental engineering and earth system science at the University of California, Irvine. He studies natural hazards and climate extremes, and crosses the boundaries between hydrology, climatology, and remote sensing. He has received a number of honors and awards, including the American Geophysical Union's James B. Macelwane Medal. Amir is the editor in chief of *Earth's Future,* has a passion for nature and landscape photography, and uses his photos to create educational materials.

Amy Bach is a nationally recognized consumer advocate and a drum-playing attorney who doesn't think insurance is boring. She co-founded United Policyholders, the national nonprofit that "has your back when insurance matters." Bach, author of *The Disaster Recovery Handbook,* leads workshops for disaster victims on how to collect all available insurance funds as painlessly and quickly as possible. Honored by *Money* magazine as a "Money Hero," and an advisor to public officials and agencies, Bach's expertise can be found on UP's website and in print in *WISE UP: The Savvy Consumer's Guide to Buying Insurance.*

Merle Bombardieri, MSW, is a psychotherapist and coach in Lexington, Massachusetts, and the author of *The Baby Decision: How to Make the Most Important Choice of Your Life*. She's the former clinical director of RESOLVE, the national infertility organization, and has been quoted in the *New York Times*, the *Washington Post*, *Time*, and *Newsweek*. Her writing has appeared in *Our Bodies, Ourselves*, the *Boston Globe Magazine*, *Glamour*, and *Self*. She has lectured at Harvard Medical School, Wellesley, and MIT. She is writing a novel about a surrogate pregnancy in which the wrong sister bonds with the baby.

Alex Cary is the market development manager at the Insurance Institute for Business and Home Safety, providing direct support to communities along the Gulf Coast and in the Great Plains states. Alex began her career with Habitat for Humanity and has worked in the home-building industry for more than 20 years since. She also serves on the board of the nonprofit Smart Home America, which helps communities across the country become stronger and more resilient. She's a licensed home builder in Alabama, where she built her own Fortified Gold home. In her spare time, she enjoys cooking and the outdoors.

Donna Childs is the founder and CEO of Prisere LLC, which develops resilience technology for climate-change risks. The U.S. Department of Energy selected Prisere for a competitive Small Business Innovation Research award for its innovations in climate and disaster resilience of the built environment. Prisere is a participant in the Innovation Corps program of the National Science Foundation, in partnership with MIT. Donna is the author of *Prepare for the Worst, Plan for the Best: Disaster Preparedness and Recovery* and holds degrees from Yale, Columbia, and Harvard.

Brian Clegg (www.brianclegg.net) has a degree in natural sciences from Cambridge University and a masters in operational research. He worked for 17 years at British Airways before setting up a creativity consultancy; for the last 20 years, he has been a full-time science writer. His recent books include *What Do You Think You Are*—the science of what makes you *you*—and *Conundrum*, a book of mind-expanding puzzles. Brian has been called a "green heretic" because he thinks it's important to follow the science, not just do what feels good. He lives in Wiltshire, England, and is an enthusiastic singer of Elizabethan church music.

Steve "Woody" Culleton was a Paradise, California, town council member from 2004 to 2016, and served as the town's mayor three times. He arrived in Paradise in 1981 via Greyhound bus—sober, employable, and responsible after 30 years of drug and alcohol addiction. He met and married his wife; the two them together adopted their granddaughter, and bought their first home in Paradise. During the 2008 Paradise wildfires, he worked at the emergency operation center, and helped to craft new town wildfire plans. Woody plans to run for town council again in November 2020 to help rebuild his community.

Leslie Davenport brings psychology to climate-change solutions within academia and organizations, and she takes it to the streets. She is the author of three books, including *Emotional Resiliency in the Era of Climate Change* (2017), and was a reviewer for American Psychological Association's "Mental Health and Our Changing Climate." She's provided disaster mental-health relief on the front lines of climate events and lobbied for green energy policies. She's on the faculty at the California Institute of Integral Studies and is completing a new book for kids on coping with climate change.

Ann Davidman, marriage and family therapist, parenthood clarity mentor, and author, has worked with women and men for 30 years to help them gain clarity in choosing parenthood or a childfree life. She coauthored *Motherhood—Is It For Me? Your Step-by-Step Guide to Clarity* with Denise L. Carlini. Ann's expertise in choice and mental clarity led her to her next project, *The Art of Decision Making*, which conveys that there is no such thing as a bad decision-maker. When Ann isn't working, you can find her playing tennis or pickleball near her home in Oakland, California.

Thomas Doherty, Psy.D., is a Portland, Oregon, psychologist whose work addresses climate-change concerns and coping. His publications include the groundbreaking paper "The Psychological Impacts of Global Climate Change," which has been cited more than 400 times. Dr. Doherty was a member of the American Psychological Association's Task Force on Global Climate Change, and he founded one of the first environmentally focused certificate programs for mental health counselors in the United States. He provides coaching and online groups to help people grapple with the ways climate change, pandemics, and social justice concerns affect themselves, their families, and their communities.

Patrick T. Drum, CFA, CFP, is a senior investment analyst and portfolio manager of the Saturna Sustainable Bond Fund at Saturna Capital. Before joining Saturna, he led environmental, social, and governance (ESG) research at UBS Institutional Consulting Services, where he was director of fixed-income portfolio management, specializing in investments for global conservation and national wildlife-park endowments. He is a former chair of the United Nations' Principles for Investment (UNPRI) Fixed Income Outreach Subcommittee, and a current member of the UNPRI's Bondholder Engagement Working Group (BEWG). He holds a BA in economics from Western Washington University and an MBA from Seattle University.

John Englander (johnenglander.net) is an oceanographer, consultant, and leading expert on sea-level rise. He's the author of the bestselling *High Tide on Main Street: Rising Sea Level and the Coming Coastal Crisis*, and *Moving to Higher Ground: Rising Sea Level and the Path Forward.* Englander helps businesses, governmental agencies, and communities understand the risks of increased flooding from rising seas, extreme tides, and severe storms. He advocates for "intelligent adaptation" while there is still time to prepare for the higher sea level and changing coastlines. He is the president and founder of the nonprofit Rising Seas Institute.

Dr. Jim Fredericks is a board-certified entomologist. He received his BS in biology education from Millersville University of Pennsylvania, and an MS and PhD in entomology at the University of Delaware. His PhD research involved subterranean termite behavior. Prior to his role with the National Pest Management Association (NPMA), Jim spent more than 11 years working with a large regional pest-control firm as technical director. A Philadelphia native, Jim lives in Fairfax, Virginia, with his wife and two children.

Victor Gensini is an assistant professor in the Department of Geographic and Atmospheric Sciences at Northern Illinois University. He received his PhD from the University of Georgia in 2014 and now teaches various meteorology courses at NIU. His research has focused on the relationship between severe convective storms and climate change. Currently, a majority of his research is examining weather and climate dynamics that explain variability in extreme weather frequency (e.g., hail, tornadoes, heavy rain, heat waves) and analyzing ways to forecast these events at sub-seasonal to seasonal time scales.

Sara Grineski is a professor at the University of Utah in the Sociology Department and Environmental and Sustainability Studies Program. She codirects the Center for Natural and Technological Hazards, where she mentors undergraduate and graduate students conducting research on social vulnerability, environmental injustice, and hazards/disasters. She teaches undergraduate courses in environmental sociology and environmental health disparities. Dr. Grineski is currently researching the effects of air pollution on children's well-being.

Dr. Louis Gritzo is vice president and manager of research with FM Global, one of the world's largest commercial property insurers. He oversees FM Global's team of scientists, who have expertise in fire, explosions, windstorms, floods, earthquakes, risk and reliability, and cyber hazards. Gritzo also oversees FM Global's $250 million, 1,600-acre Research Campus in Rhode Island, the world's largest center for property-loss prevention research. Before joining FM Global in 2006, Gritzo was manager of fire science and technology at Sandia National Laboratories in Albuquerque. He has a PhD in mechanical engineering, with a minor in applied mathematics, from Texas Tech University.

Don Hankins is a professor of geography and planning at California State University, Chico, and field director for the CSU Big Chico Creek Ecological Reserves. His areas of expertise include pyrogeography, water resources, conservation, and Indigenous land and water stewardship. He studies wildland fire with an emphasis on landscape scale, cultural and prescribed fires, and the implications of these fires to biodiversity conservation and environmental change. Don's work includes various aspects of land management, policy, and conservation with a variety of organizations and agencies, including Indigenous entities in North America and Australia.

Christine Harada is the president of i(x) investments, an impact investing platform. She was the U.S. Chief Sustainability Officer under President Obama, responsible for promoting environmental and energy sustainability across all Federal government operations. Ms. Harada started her career as a satellite systems engineer, and later spent a decade as a management consultant at Boston Consulting Group and Booz Allen Hamilton, managing organizational strategies for corporate, nonprofit, and government entities. She holds degrees from the Wharton School, Lauder Institute at the University of Pennsylvania, MIT, and Stanford, and lives in Los Angeles with her husband and two sons.

Mathew Hauer is an assistant professor of sociology at Florida State University who studies the impacts of climate change on society. His recent research has focused on how migration induced by sea level rise could reshape the U.S. population distribution. The *New York Times, National Geographic, Time Magazine, Popular Science, USA Today*, and others have featured his research. Before coming to Florida State, Dr. Hauer spent eight years directing the Applied Demography Program at the University of Georgia, where he provided demographic research to local, state, and federal governments.

Christopher W. Heidrick, CPCU, ANFI, CFP, is the founder of Heidrick & Company Insurance in Sanibel, Florida, and Trusted Flood, a wholesale insurance brokerage. He is the chair of the Flood Insurance Task Force for the Independent Insurance Agents and Brokers of America (IIABA), past chair of the Flood Insurance Producers National Committee (FIPNC), and a past board member for the Florida Association of Insurance Agents (FAIA). Locally, Chris has served as chairman of the Sanibel and Captiva Islands Chamber of Commerce, vice chairman of the City of Sanibel Planning Commission, and on several nonprofit boards. He holds a BS in Economics from Pennsylvania State University.

Richard Heinberg is senior fellow at the Post Carbon Institute, and author of several books on energy and the environment, including *Afterburn: Society Beyond Fossil Fuels*, and, with David Fridley, *Our Renewable Future*. He's won an award for excellence in energy education and has been published in *Nature* and the *Wall Street Journal*. Heinberg's work is cited as one of the inspirations for the international Transition Towns movement, which seeks to build community resilience ahead of climate change and society's possibly wrenching shift away from its dependence on fossil fuels. He and his wife, Janet Barocco, live in Santa Rosa, California.

Caroline Hickman is a psychotherapist who teaches at the University of Bath. She's a member of the Climate Psychology Alliance (CPA), for whom she is creating "Climate Crisis Conversations: Catastrophe or Transformation," a podcast about climate psychology, eco-anxiety, and grief. She is currently researching young people's feelings about the climate and biodiversity crisis. Caroline is thankful for her co-researcher Murphy, who needs daily walks in the woods, and whenever she has the chance, she spends time underwater hanging out with fish, preferably sharks. She is also slightly obsessed with seagrass and its CO_2-absorbing properties.

Jed Horne shared Pulitzer Prizes with others on the staff of *The Times-Picayune* for coverage of Hurricane Katrina. *Breach of Faith*, his account of the disaster and recovery, was declared "the best of the Katrina books" by NPR. An earlier book, *Desire Street*, about a Louisiana death row case, was a finalist for the American Bar Association's Silver Gavel award and for the Edgar, awarded by the Mystery Writers of America. In semiretirement, Horne has been writing fiction under a pen name and enjoying life in a Mexican village 7,000 feet above rising seas.

Dr. Lucy Jones is founder and chief scientist of the Dr. Lucy Jones Center for Science and Society, a research associate at the Seismological Laboratory of Caltech, and author of *The Big Ones: How Natural Disasters Have Shaped Us* (Doubleday, 2018). With a BA in Chinese Language and Literature from Brown, and a PhD in seismology from MIT, Dr. Jones furthers resilience to natural hazards through scientific research and collaborations with policy makers, including 33 years with the U.S. Geological Survey, where she created the first Great ShakeOut. She plays the viola da gamba and composes for gamba ensembles, including *In Nomine Terra Calens (In the Name of a Warming Earth)*.

Jesse M. Keenan is a social scientist and associate professor of Real Estate at the School of Architecture at Tulane University. His principal research focus is climate-change adaptation and the built environment. Dr. Keenan holds concurrent appointments to the Intergovernmental Panel on Climate Change (IPCC) and as a Visiting Scholar at the Perry World House at the University of Pennsylvania. Keenan's books include *Blue Dunes: Climate Change by Design*, *Climate Change Adaptation in North America: Experiences, Case Studies and Best Practices*, and *Climate Adaptation Finance and Investment in California*. He lives in Philadelphia, Pennsylvania, and was the first person ever to exhibit an Excel spreadsheet at the Museum of Modern Art, New York.

Dr. Susan R. Kemball-Cook is a principal in Ramboll's Novato, California, office. Her expertise includes climate-change impact assessment and climate modeling. Sue has performed climate-change assessments as part of due diligence, climate risk disclosure, environmental impact assessments, and flood resilience measure design. She received her BA in physics from Yale University and her PhD in atmospheric science from the University of California, Davis. She likes to surf and lives in Pacifica, California, so sea-level rise hits home for her.

Tom Knutson is a climate scientist with NOAA's Geophysical Fluid Dynamics Laboratory in Princeton, New Jersey, where he leads the Weather and Climate Dynamics Division. He is a fellow of the American Meteorological Society, and has served as the chair of the World Meteorological Organization's Expert Team on Tropical Cyclones and Climate Change. He led a recent assessment on tropical cyclones and climate change, which was published in the *Bulletin of the American Meteorological Society* in 2019. He was the lead author of the "Detection and Attribution of Climate Change" chapter in the *U.S. Climate Science Special Report.* His recent research has been on detecting human influence on regional precipitation and atmospheric circulation trends.

Erik Kobayashi-Solomon, founder of IOI Capital, is an entrepreneur and investor who in 2018—after 20-plus years in the investment banking, hedge fund, and independent research worlds—resolved to apply his investing talent to helping humanity face the existential threat of global warming. Erik believes that capitalism represents the economic manifestation of our greatest strength—adaptability—and sees venture capital as an important tool in our species' self-preservation. His greatest joy is figuring out how to connect investment dollars to innovative entrepreneurs working on solutions to mitigate and adapt to the effects of this century's increasingly perilous climate reality.

Mike Kreidler is Washington State's Insurance Commissioner, reelected to his sixth term in 2020. He's the longest-serving insurance commissioner in the United States. Since 2007, Commissioner Kreidler has chaired the National Association of Insurance Commissioners' Climate Risk and Resilience Work Group. In 2015, he joined the Paris Pledge for Action, and his office became a supporting institution for the UNEP FI Principles for Sustainable Insurance Initiative, the largest collaboration between the United Nations and the insurance industry. In 2016, Commissioner Kreidler joined the UN Environment's Sustainable Insurance Forum, a network of insurance regulators working to strengthen their response to sustainability challenges.

Linda Masterson wrote *Surviving Wildfire: Get Prepared, Stay Alive, Rebuild Your Life* for homeowners because it had to be written. As an agency creative director, business writer, researcher, and author of the groundbreaking handbook *Living with Bears*, she was uniquely qualified to share all the wildfire lessons she learned the hard way. So instead of taking her full-settlement insurance check and getting on with her life, she went to work educating, motivating, and scaring people into action. Today, she lives on Florida's Gulf Coast, where summer and fall are hurricane-prep season. She still loves the outdoors; check out her latest people-persuasion mission at www.bearwise.org.

Tom Mather, PhD (aka the TickGuy) is professor of public-health entomology at the University of Rhode Island. He's the director of URI's Center for Vector-borne Disease and its TickEncounter Resource Center. His research interests include blood-sucking arthropods and tick-transmitted diseases, but his current effort is developing an anti-tick vaccine that prevents infection with any tick-borne disease germ. His nationally prominent website TickEncounter.org includes a variety of features aimed at increasing literacy about ticks and tick-borne diseases. When not crawling around the woods purposely collecting ticks, he enjoys training for and running Spartan obstacle course races with his daughters.

Rachel Minnery, FAIA, is an architect and senior director of Resilience, Adaptation, and Disaster Assistance at The American Institute of Architects (AIA). She develops policy, design guides, and education for the organization's 95,000 members and the building industry. Her previous experience includes design and management in the nonprofit and private sectors, focusing on environmentally and socially responsible design. She is the cofounder of Architects Without Borders Seattle, and has led groups of building-safety volunteers to disaster-stricken places. In her spare time, she escapes the city to spend time in the great outdoors, running and hiking with her husband and daughter.

Nan Fischer is a gardener and garden writer based in Taos, New Mexico. She's the author of the upcoming *Adapting to Climate Change in the Home Garden: Creating Climate Resilience for the Future*. She writes about gardening for her local paper, the *Taos News*, and regularly contributes to *Mother Earth Gardener* magazine, among others. Nan owns a small organic nursery called nannie plants, and she also founded the Taos Seed Exchange. She's always up to the challenge of starting her plants in spring and seeing them through to putting up produce for winter. She views each growing season as a new experiment in today's climate chaos.

Tony Nester has been a survival instructor for 28 years, and is the author of numerous books and DVDs. He has provided survival training for the military special operations community, U.S. Marshals, the FAA, the Travel Channel, the *New York Times, Backpacker* magazine, and the film *Into the Wild.* He has a BA in anthropology, and has also spent many years living among native cultures. When not on the trail or sleeping in a cave with his students, Tony lives in Colorado Springs and writes postapocalyptic thrillers under his pen name, JT Sawyer.

Nick Nuttall has more than 30 years of experience writing and communicating environmental issues, first as a journalist on *The Times* of London and then as director of communications at UN Environment HQ in Kenya and UN Climate Change in Germany. Nick was the spokesperson for the United Nations Paris Agreement of 2015. Today, he splits his time doing freelance strategic communications for a variety of environmental organizations, including the annual Earth Day event We Don't Have Time; trying to see more of his kids; supporting innovative eco-arts projects; a "Ladette" backing vocalist for Berlin-based singer/song writer Bernadette La Hengst; and trying to stay calm avidly supporting English soccer team Burnley.

Sharon Palmer, MSFS, RDN, writes and speaks about plant-based nutrition and sustainability. She has authored more than 1,000 articles in publications like *Better Homes and Gardens, O, The Oprah Magazine*, and *LA Times*, and blogs at sharonpalmer.com. She's the author of *The Plant-Powered Diet: The Lifelong Eating Plan for Achieving Optimal Health, Beginning Today* (2012); *Plant-Powered for Life: Eat Your Way to Lasting Health with 52 Simple Steps & 125 Delicious Recipes* (2014), and two new books arriving in 2021. Sharon lives in the Ojai Valley of California, where she enjoys tending her organic garden, shopping at the farmers' market, and cooking for friends and family.

Mary Beth Pfeiffer is a fire-in-the-belly Hudson Valley journalist whose investigative reports on Lyme disease in 2012 went viral. In countries worldwide, she found legions of sick patients who had been failed by tests and treatments. Medicine has underestimated the power of ticks, she contends, and humankind set them loose. She is author of *Lyme: The First Epidemic of Climate Change,* with the emphasis on *Epidemic*, though she says *Climate Change* may get us all first. She warns her grandsons of what lurks in the grass, and has installed 20 solar panels to fuel a plug-in car.

Diego Rose is a professor and the director of nutrition at Tulane University's School of Public Health and Tropical Medicine. He teaches graduate courses in nutrition assessment and food and nutrition policy. His current research examines the environmental and health consequences of diets in the United States and the effects of simulated changes on these outcomes. Before joining the faculty at Tulane, he worked for the USDA's Economic Research Service on U.S. food assistance policy, and in Mozambique and South Africa on food security and nutrition. Diego has degrees in nutritional science (BS), public health (MPH), and agricultural economics (PhD), all from the University of California, Berkeley.

Vivek Shandas, PhD, hails from the "sweat belt" of south India, and now serves as a professor of climate adaptation at Portland State University. He examines assumptions about building our cities in an era of unprecedented climate dysfunction. He also serves on several local and national advisory boards, and as principal at CAPA Strategies, LLC, a global consulting group that helps communities prepare for climate-induced disruptions. During his spare time and with his family, he revels in the mountains and waters of the Pacific Northwest, and pines for late-night vegan hot dogs anywhere he travels.

Kelly Helm Smith works at the National Drought Mitigation Center, which is at the University of Nebraska. She studies how we understand and measure the effects of drought. She has official degrees in history, journalism, community and regional planning, and natural resource sciences, and describes her career path as actually more of a spiral. Advantages of that approach are that you can't see too far ahead, and that sometimes things come full circle. She lives in Lincoln, Nebraska, and as a fair-weather cyclist, she manages to take advantage of the city's excellent bike trails at least several times a year.

Laurence C. Smith is the John Atwater and Diana Nelson University Professor of Environmental Studies at Brown University. He is a fellow of the American Geophysical Union and the John S. Guggenheim Foundation. His climate-change research has been featured in the *New York Times*, the *Wall Street Journal*, *The Economist*, the *Los Angeles Times*, and the *Washington Post*, among others. His book *The World in 2050* won the Walter P. Kistler Book Award and was a Nature Editor's Pick of 2012. His next book, *Rivers of Power*, about the future of water, was published in 2020.

Marco Springmann is a senior researcher on environmental sustainability and public health at the University of Oxford in the UK. He is interested in the health, environmental, and economic dimensions of the global food systems, and often uses systems models to provide quantitative estimates on food-related questions. Most recently, he was part of the EAT-Lancet Commission on Healthy Diets form Sustainable Food Systems, which developed recommendations for a planetary health diet that is both healthy and compatible with the environmental limits of our food system. He lives in London and can occasionally be seen dancing at warehouse raves.

Dr. Benjamin Strauss, president, CEO, and chief scientist at Climate Central, is a sea-level rise expert and architect of Climate Central's maps, tools, and flood-risk visualizations. Strauss has testified before the U.S. Senate, his research has been cited by the White House and two UN Secretaries-General, and in 2019, he coauthored the top-ranked climate research paper in the world for online and media attention. It earned coverage in more than 120 countries and features in the *New York Times*, the *Washington Post*, and Reuters. Strauss has appeared on CNN, MSNBC, BBC News, and NPR. He holds a PhD in ecology and evolutionary biology from Princeton University, an MS from the University of Washington, and a BA from Yale University.

Anant Sundaram is a professor at the Tuck School of Business at Dartmouth. His expertise is in business valuation and in understanding how climate change impacts companies. He created the first MBA course on business and climate change in a U.S. business school. He is widely published in academic journals, and is a coeditor of the *Handbook of Business and Climate Change* (2021). Anant created the Fossil Fuel Beta to measure the stock price impact of companies' exposure to fossil fuel prices, for *CFO* magazine. He is developer of *uValue*, a free iOS business valuation app. Anant is a fusion jazz fanatic, and lives in bucolic Hanover, New Hampshire.

Vivek Tanneeru is a portfolio manager at Matthews Asia, an independent investment specialist firm. Since 2015, he has managed the firm's Asia ESG Strategy, which invests in companies that meet one or more of its environmental, social, and governance standards. He also championed the firm's joining of the United Nations Principles for Responsible Investing (UNPRI) initiative. Before joining Matthews Asia, Vivek held positions at Pictet Asset Management in London, The World Bank, and Arthur Andersen Business Consulting. He holds an MBA from the London Business School and a master's in finance from the Birla Institute of Technology & Science in India.

Jason Thistlethwaite is an associate professor in the School of Environment, Enterprise and Development (SEED) at the University of Waterloo; a senior fellow at the Centre for International Governance Innovation (CIGI); and associate director for Partners for Action (P4A). He is the co-lead of the Climate Risk Research Group, which collaborates with cities, residents, businesses, and NGOs to develop strategies for community resiliency. Jason lives with his family in Waterloo, Canada, where you can find him talking local weather, enjoying craft beer, and chasing his kids.

Dr. Sue Varma (*@doctorsuevarma*) is a board-certified psychiatrist in private practice in Manhattan. She is a clinical assistant professor of psychiatry at New York University (NYU) Langone Health and a distinguished fellow of the American Psychiatric Association. She is the recipient of the inaugural Sharecare Emmy Award, and is now nominated for her role in *CBS This Morning*'s "Stop the Stigma" campaign. She is the former medical director of the 9/11 mental health program at NYU. Sue is a self-proclaimed foodie who also enjoys swimming in oceans around the world (and sometimes with sharks!).

Dr. Sara Via is a professor and climate extension specialist at the University of Maryland. After 35 years of research on the evolutionary genetics of insect pests, her growing concern about climate change led her to leave research behind in 2015. Now Sara teaches people across Maryland about climate impacts and solutions in agriculture, gardening, biodiversity, and health. She also works on ways to increase carbon sequestration in agriculture, a top natural climate solution. Most recently, Sara is experimenting with strategies for climate-friendly suburban landscaping that will slash emissions, sequester carbon, boost water quality, and reduce flooding.

Leslie B. Vosshall is the Robin Chemers Neustein Professor, head of the Laboratory of Neurogenetics and Behavior, and director of the Kavli Neural Systems Institute at Rockefeller University. She is a molecular neurobiologist known for her work on the genetic basis of innate behaviors in both insects and humans, with special expertise in chemosensory receptors. Her laboratory is investigating how mosquitoes find and bite people and what can be done to stop them. Dr. Vosshall is a Howard Hughes Medical Institute investigator, an elected AAAS fellow, a member of the National Academy of Sciences, and David Pogue's sister-in-law.

Lucas White is a portfolio manager and partner at GMO. He focuses primarily on the energy, metals, agriculture, and water sectors, and runs a climate-change strategy, a resources strategy, and a resources long/short strategy. His research involves pragmatic analysis of the nature and scale of the clean-energy transition. Prior to joining GMO in 2006, he worked at Standish Mellon Asset Management and MFS. Mr. White earned his BA in economics and psychology from Duke University and is a CFA charter holder. He currently resides in the Boston area.

Dr. Jalonne L. White-Newsome is a teacher, equity advocate, philanthropist, researcher, speaker, passionate environmentalist, and, most important, a proud mom. As a senior program officer in environmental philanthropy at the Kresge Foundation, she manages grantmaking related to climate change, flooding, and health, while teaching master's-level public health students at George Washington University—from home—in West Bloomfield, Michigan. Her expertise ranges from chemical engineering to the impacts of climate change on low-income communities and communities of color. She is working on two books for young people—and in her spare time manages her people-centered consulting firm, Empowering a Green Environment and Economy, LLC.

Alex Wilson is the founder and president of the Resilient Design Institute in Brattleboro, Vermont, having first gotten involved in resilience work following Hurricane Katrina in 2005. He has worked on resilience initiatives for New York City, Boston, and most recently, the Climate Ready DC Program in Washington, DC. Wilson also founded BuildingGreen, Inc., a B Corp–certified consulting and publishing company in Brattleboro. When he's not sitting behind a computer, he might be found walking the trails on his Vermont farm or watching birds from his canoe. He's the coauthor of four Appalachian Mountain Club books in the Quiet Water Canoe and Kayak Guide series.

Erol Yayboke is deputy director and senior fellow with the Project on Prosperity and Development at the Center for Strategic and International Studies (CSIS). His research interests include U.S. foreign assistance, good governance, climate change, migration, forced displacement, and innovation-led economic growth. Previously, he served on the Hillary Clinton presidential campaign and was a research manager at the Center for International Development at Harvard University's Kennedy School of Government. His field experience includes trips to more than 60 countries on all seven continents. He teaches a graduate-level course on state fragility at the Syracuse Maxwell School of Citizenship and Public Affairs. He lives with his wife and two daughters in Washington, DC.

Notes

The endnotes for this book—the bibliography of articles, books, and journals that fueled its research—are over 100 pages long.

To save your back muscles (and a few trees), we've posted the endnotes online, at **www.simonandschuster.com/p/how-to-prepare-for-climate-change-bonus-files.** As a handy extra bonus of that approach, all of the links in those online endnotes are clickable.

Illustration Credits

Figure I-1	Photo courtesy of BIG Architects.
Figure I-2	Reuse of all content from Rijkswaterstaat is free and permitted. https://www.rijkswaterstaat.nl /english/disclaimer/index.aspx.
Figure 1-1	Courtesy of votesmart.org.
Figure 1-2	Courtesy of meetup.com.
Figure 2-8	Courtesy of David Pogue.
Figure 2-11	Courtesy of David Pogue.
Figure 2-12	Photo by tspdave (David Mark).
Figure 2-13	Photo courtesy of Justin Russell.
Figure 2-14	Courtesy of Boulder CVB.
Figure 2-15	"Cityscape of Boise" by Alden Skeie, courtesy of Good Free Photos. https://www.goodfreephotos.com/united-states/idaho/boise/cityscape-of-boise-lighted-up-in-boise-idaho.jpg.php.
Figure 2-16	"Portland and Mt Hood" by Amateria1121. Creative Commons license. https://cs.m.wikipedia.org/wiki/Soubor:Portland_and_Mt_Hood.jpg.
Figure 2-18	"Buffalo Skyline" by Peter Stergion. Creative Commons license. https://commons.wikimedia.org/wiki/File:Buffalo_skyline_2014.jpg.
Figure 3-3	Courtesy of Generac Power Systems.
Figure 3-5	Photo courtesy of Constant Water.
Figure 3-6	"Painted Barrels! (#0646)" by regan76. Creative Commons license. https://www.flickr.com/photos/j_regan/9005225009/in/photolist-eHL8Ea-21bjJr5-g4NGUu-g4NBwX-8mpRBU-xCmf1U-ef2JgQ-2gdwnrz-6MJAJ3-4YL4qv-89sn5R-77whab-iVzVGe-26cCXMb-2f4rps9-bVZHfV-fgE4s3-8mzEeb-7owNJp-x4JVuU-6webE4-gW8oDp-8dTy83-agp3pf-6eFzD9-o8kweZ-atNP87-542e5y-48GzMN-7kEnfe-77whcN-73yKUq-77wh7y-og1joi-ddHZZM-9KK4Gt-vYSbf7-ahACzM-o1k8As-xe2qq6-7a6jSH-nRcEd9-2ikWHEK-2gbUZ28-5Msngx-oefUxu-der1pN-ddJ5RK-xhknDm-amBd57.
Figure 3-7	FEMA Technical Bulletin 1, "Openings in Foundation Walls and Walls of Enclosures." https://www.fema.gov/media-library-data/20130726-1502-20490-9949/fema_tb_1__1_.pdf
Figure 3-8	Photo courtesy of Nancy Leonetti, LBI House Raising & Contractors.

Figure 3-9 Home-raising in Beach Haven, NJ: Photo courtesy of LBI House Raising & Contractors, Inc.

Figure 3-12 "Quebec City—Metal Doors & Roofs" by David Ohmer. Creative Commons license. https://www.flickr.com/photos/the-o/2241977920/in/photolist-4saYj5-4s75Mz-4saYjf-4saYj9-4q7HVJ-4s75MH-4saUxU-4saUxN-4q3mWX-hZHxyo.

Figure 3-14 Photo courtesy of Insurance Institute for Business & Home Safety.

Figure 3-15 Courtesy of the California Department of Forestry and Fire Protection. https://www.readyforwildfire.org/prepare-for-wildfire/get-ready/defensible-space.

Figure 3-18 "Bluff Neighborhood, Historic Homes of Beaufort, South Carolina" by Ken Lund. Creative Commons license. http://j.mp/387R96K.

Figure 4-1 "Container garden on the patio," Thomas Kriese. Creative Commons license. https://www.flickr.com/photos/thomaspix/2432153264/in/photostream/.

Figure 4-2 Courtesy of USDA.

Figure 4-3 Photo courtesy of Audreen Williams, Master Gardener Foundation of Clallam County.

Figure 4-5 "Permeable Pavers" by Center for Neighborhood Technology. Creative Commons license.

Figure 4-6 Photo of courtesy of Eartheasy.

Figure 5-2 "Verticrop" by Valcenteu. Creative Commons license. https://commons.wikimedia.org/wiki/File:VertiCrop.jpg.

Figure 7-2 "Greta Thunberg" by Anders Hellberg, Creative Commons license. https://commons.wikimedia.org/wiki/File:Greta_Thunberg_4.jpg.

Figure 8-1 Source: JP Morgan, via Wall Street Journal.

Figure 8-7 "Urban Harvest Tour—Rain Water Barrels" (CC BY-SA 2.0) by jbolles. https://www.flickr.com/photos/jbolles/4718021158. Photo courtesy of WaterPrepared.com.

Figure 8-8 Courtesy of ThePlumbingInfo.com.

Figure 8-9 Courtesy of AquaTabs.

Figure 8-10 "kitchen pantry contents" by Mr Thinktank. Creative Commons license. https://www.flickr.com/photos/tahini/4048839268.

Figure 8-11 Courtesy of Biolite.

Figure 8-12 Courtesy of ASPCA.

Figure 8-13 Air Force photo by Joel Martinez (public domain). https://www.jbsa.mil/News/News/Article/463234/wellness-packages-offered-in-on-base-treatment-for-pets/.

Figure 8-17 Photo by Senior Airman Brigitte N. Brantley. Courtesy of the U.S. Air Force.

Figure 9-2 Top photo: Glacier National Park Archives. Photographer unknown. Public domain. www.usgs.gov/media/images/grinnell-glacier-overlook-1940-0. Bottom photo: Lisa McKeon, USGS. Public domain.

https://prd-wret.s3-us-west-2.amazonaws.com/assets/palladium/production/s3fs-public/thumbnails/image/USG_0376cr_2013_L_0.jpg.

Figure 9-4 Source: Pew Trusts.

Figure 9-5 Source: US Global Change Research Program.

Figure 9-6 "Flooding in Waterloo, Iowa"—Photography: Don Becker, USGS. Public domain.

Figure 9-9 "Flooding in Cedar Rapids, Iowa" by Don Becker, USGS. Public domain. https://www.flickr.com/photos/usgeologicalsurvey/2593475733/.

Figure 9-10 https://www.youtube.com/watch?v=BwBcLz61tg4 (Posted by subasurf).

Figure 9-11 "Flood Damage to House," posted to Flickr by ICMA Photos under a Creative Commons license. https://www.flickr.com/photos/icma/3608228084.

Figure 9-12 "Storm Downed power lines and trees from Storm Sandy." Posted by Arlington County under Creative Commons license. https://www.flickr.com/photos/arlingtonva/8138919297.

Figure 9-13 Courtesy of Liz Roll Photography.

Figure 9-14 "WestendMoldyLivingRoom." Posted by Infrogmation under Creative Commons license. https://commons.wikimedia.org/wiki/File:WestendMoldyLivingroom.jpg.

Figure 9-15 Photo courtesy of ServiceMaster Restore.

Figure 10-1 Courtesy of Scripps Media, Inc.

Figure 10-2 Source: CDC.

Figure 10-3 Source: National Climate Assessment.

Figure 10-4 Source: RCC-ASIS via Climate Central.

Figure 10-5 Source: CDC.

Figure 10-10 Courtesy of Tesla.

Figure 11-1 "20130109-USACE-UNK-0011," U.S. Department of Agriculture. Creative Commons license. https://www.flickr.com/photos/usdagov/8365634300/in/photolist-DvQPtj-eeipGe-dQFN31-cPcwR7-dK9A7z-dK9AyP-dK9A44-dKf4C9-dKf5bj-dK9A1c-dKf571-dK9AdZ-dK9zWR-h9RPt9-h9RJHr-dK9AiZ-h9RJpa-h9RMKN-dGRGQf-h9RJwK-dGRGJs-dBiEsV-dK9AgM-dBiMYZ.

Figure 11-2 Source: National Drought Mitigation Center.

Figure 11-4 Source: CircleofBlue.org.

Figure 12-1 Source: USGS, National Atlas.

Figure 12-2 "Hurricane Sandy Aftermath—Howard Beach—10/30/2012," Pamela Andrade. Creative Commons license. https://www.flickr.com/photos/skdecember/8143872869.

Figure 12-3 The Secure Door Brace, a product of Storm Supply Depot (securedoorbraces.com), Pompano Beach, FL. Photo courtesy of Mike Wittlin

Figure 12-4 Gina Kelly / Alamy Stock Photo

Figure 12-5 "nss10210," NOAA Photo Library. Creative Commons license. http://j.mp/2si9YF7.

Figure 13-1 Photo by John McColgan, courtesy of the U.S. Department of Agriculture. https://commons.wikimedia.org/wiki/File:Deerfire_high_res_edit.jpg.

Figure 13-2 "180110-Z-F3881-015.jpg," by Staff Sgt. Cristian Meyers of the California National Guard. Creative Commons license. http://j.mp/2RriASC.

Figure 13-3 Source: National Interagency Fire Center.

Figure 13-4 Source: NOAA.

Figure 13-5 "Satellite image of Thomas Fire.jpg." NASA Earth Observatory images by Joshua Stevens, using MODIS data. Public domain.

Figure 13-9 Courtesy of Christopher Allan Smith.

Figure 14-1 Source: CDC Vital Signs, May 2018.

Figure 14-2 Source: CDC.

Figure 14-5 Photo by James Gathany courtesy of the CDC; public domain. https://phil.cdc.gov/details.aspx?pid=9875.

Figure 14-6 Source: CDC.

Figure 14-7 Source: EPA.

Figure 15-1 "Protesters attempt to avoid tear gas on Mohammed Mahmoud Street," ©2012 Alisdaire Hickson, Creative Commons license. www.flickr.com/photos/alisdare/6821225189/in/album-72157628036164125/.

Figure 16-1 Courtesy Flood Control International.

Figure 16-2 "Floodwaters inundated Rojana Industrial Park in Ayutthaya Province, Thailand." U.S. Marine Corps photo by Cpl. Robert J. Maurer.

Figure 17-1 Source: EPA.

Figure 17-2 Source: Wards Intelligence, LMC Automotive.

Figure 17-3 Courtesy of Mark Rober.

Figure 17-7 Courtesy of Climeworks.

Figure 17-8 Courtesy of the *New York Times*

Index

Page numbers in *italics* refer to illustrations, maps, and graphs

About the Author

David Pogue was the *New York Times* weekly tech columnist from 2000 to 2013. He's a five-time Emmy winner for his stories on *CBS News Sunday Morning*, a *New York Times* bestselling author, a five-time TED speaker, and host of 20 *NOVA* science specials on PBS.

He's written or cowritten more than 120 books, including dozens in the Missing Manual tech series, which he created in 1999; six books in the *For Dummies* line (including *Macs, Magic, Opera,* and *Classical Music*); two novels (one for middle-schoolers); his three bestselling Pogue's Basics book series of tips and shortcuts (on *Tech*, *Money*, and *Life*); and his new how-to guides, *iPhone Unlocked* and *Mac Unlocked*, published by Simon & Schuster.

After graduating summa cum laude from Yale in 1985 with distinction in music, Pogue spent ten years conducting and arranging Broadway musicals in New York. He has won a Loeb Award for journalism, two Webby awards, and an honorary doctorate in music. He lives with his wife Nicki and their blended brood of five spectacular children in Connecticut and San Francisco.

For a complete list of Pogue's columns and videos, and to sign up to get them by email, visit https://authory.com/davidpogue. On Twitter, he's @pogue; on the web, he's at www.davidpogue.com. He welcomes civil email exchanges at david@pogueman.com.

FAMILY EMERGENCY INFO

FAMILY CONTACTS

Home address: ______________________________

Name ______________________
Cell: ______________________
Email: ______________________
Chat app: ______________________
Notes: ______________________
____

Name ______________________
Cell: ______________________
Email: ______________________
Chat app: ______________________
Notes: ______________________
____

Name ______________________
Cell: ______________________
Email: ______________________
Chat app: ______________________
Notes: ______________________
____

Name ______________________
Cell: ______________________
Email: ______________________
Chat app: ______________________
Notes: ______________________
____

WORKPLACES + SCHOOLS

Name: ______________________
Website: ______________________
Address: ______________________
Best phone #: ______________________
Emergency pickup plan: ______________________
____

Name: ______________________
Website: ______________________
Address: ______________________
Best phone #: ______________________
Emergency pickup plan: ______________________
____

School Name: ______________________
Website: ______________________
Address: ______________________
Best phone #: ______________________
Emergency pickup plan: ______________________
____

School Name: ______________________
Website: ______________________
Address: ______________________
Best phone #: ______________________
Emergency pickup plan: ______________________
____

EMERGENCY MEETING PLACES

In case we can't reach each other

In the house *(away from windows)*

Where: ______________________________

Details: ______________________________

In the neighborhood

Where: ______________________________

Details: ______________________________

In town

Where: ______________________________

Details: ______________________________

Out of town

Where: ______________________________

Details: ______________________________

EMERGENCY CONTACTS

Relative/Friend/Neighbor ______________________________

Out-of-Town Relative/Friend ______________________________

Police: 911, or ______________________________

Fire: 911, or ______________________________

Doctor: ______________________________

Doctor: ______________________________

Pediatrician: ______________________________

Health Insurance: ______________________________ Policy: ______________

Hospital/Clinic: ______________________________

Pharmacy: ______________________________

Homeowners Insurance: ______________________________ Policy: ______________

Flood Insurance: ______________________________ Policy: ______________

Veterinarian: ______________________________

Kennel: ______________________________

Electric Company: ______________________________

Gas Company: ______________________________

Water Company: ______________________________

House of worship: ______________________________

Other: ______________________________

EMERGENCY NOTES: ______________________________

____

____